Path Dependence
and Creation

■ ■ ■

Arthur Brief and James Walsh, Series Editors

Ashforth • *Role Transitions in Organizational Life: An Identity-Based Perspective*

Beach • *Image Theory: Theoretical and Empirical Foundations*

Garud/Karnøe • *Path Dependence and Creation*

Lant/Shapira • *Organizational Cognition: Computation and Interpretation*

Thompson/Levine/Messick • *Shared Cognition in Organizations: The Management of Knowledge*

Path Dependence and Creation

Edited by

Raghu Garud
New York University

Peter Karnøe
Copenhagen Business School

Psychology Press
Taylor & Francis Group

New York   Hove

This edition published 2012 by Psychology Press
27 Church Road, Hove, East Sussex, BN32FA
Simultaneously published in the USA and Canada by Psychology Press
711 Third Avenue, New York, NY 10017

Lawrence Erlbaum Associates, Inc., Publishers
10 Industrial Avenue
Mahwah, NJ 07430

Cover design by Kathryn Houghtaling Lacey

Library of Congress Cataloging-in-Publication Data

Path Dependence and Creation / Edited by Raghu Garud,
    Peter Karnøe.
            p. cm.—(LEA's organization and management series)
            Includes bibliographical references and index.

1. Technological innovations. 2. Technological innovations—
    Management. I. Garud, Raghu. II. Karnøe, Peter. III. Series.

T173.8.P37 2000
658.5'77—dc21   00-044256
                                                         CIP

Psychology Press is an imprint of the Taylor & Francis Group, an informa business

First issued in paperback 2012

ISBN13: 978-0-415-65071-7 (pbk)
ISBN13: 978-0-805-83272-3 (hbk)

10 9 8 7 6 5 4 3 2

To our path creating children,
Amalie, Karoline, Keshav, and Nandita

—RG
—PJK

Contents

Foreword

Arthur P. Brief
Tulane University

James P. Walsh
University of Michigan

Imprinting, QWERTYnomics, increasing returns, vicious circles, virtuous cycles, and path dependence—the field is captivated by the role of history and experience as they inform contemporary action. Appreciative of these insights, yet mindful of the power of human agency, Raghu Garud and Peter Karnøe have made a wonderful contribution to our understanding of technological innovation and change. Well aware of the recent work in evolutionary theory and the science of chaos and complexity, they challenge the sometimes deterministic flavor of much of this work. They are interested in uncovering the place of agency in these theories that take history so seriously. In the end, they are as interested in path creation and destruction as they are in path dependence.

Garud and Karnøe have gathered a first-rate collection of scholars from around the world to consider these issues. We are treated to a wonderful array of theoretical and empirical writing. Our new insights about agency are developed in a host of settings. We will now see such relatively well-known industries as the automobile, biotechnology, and semi-conductor in a new light. Also, we are invited to learn much more about medical practices, wind power, lasers, and synthesizers. Ultimately, this is a book that is theoretically important and downright interesting. We are very pleased to publish it in Lawrence Erlbaum Associates' Organization and Management Series.

Preface

Like many human initiatives, this edited volume too is an outcome of purposive action and is illustrative of a central theme that runs through all its chapters—path creation. The volume was conceptualized to address an observation that human agency is neither unbounded nor is it absent. Although we may be creatures caught in webs of significance of our own making (Geertz, 1973–1974), we do possess the capacity to untangle these webs to create new paths.

Implicit in this imagery is the role that history plays in shaping human behavior, a facet that is explicitly dealt with in the literature on path dependence (cf. Arthur, 1994; David, 1985). According to this perspective, we can only know in hindsight which of the many events that have transpired in the past imprints the flow of current and future events. That is, not only are we prisons of the past, but we may even have contributed to its creation unknowingly.

The role of history in shaping human choice, explicit in the literature on path dependence, is implicit in every theoretical framework that we employ in the social sciences. It is important to explicate this role as theoretical structures can potentially trap us into certain modes of thinking and behavior. It is to explore this role and our ability to shape history in the making that we organized a workshop in Denmark with generous help from the Center for Interdisciplinary Studies in Technology Management (CISTEMA). We decided to begin with path dependence as a common point of departure as this would offer invited participants a way to better articulate their own perspectives and positions.

If "interpretive flexibility" (Pinch & Bijker, 1987) is the hallmark of success, we can say with great confidence that the workshop exceeded our expectations. No facet of human existence was left unexamined. We discussed with great passion a range of topics from the origins of the human tie to the genesis of institutional arrangements such as modern intellectual property rights. We explored these phenomena not just from different disciplinary perspectives but from multiple levels of analyses as well. At the end, a central question that brought us together was, How should we conceptualize the nature and scope of human agency given that we are creatures caught in complex webs of our and others' making?

The chapters in this book represent each author's epistemological and ontological position on this issue. Indeed, the variety of positions taken by the contributors made our jobs as editors all the more difficult. We found that, depending upon the vantage points of the persons describing the phenomenon, the same phenomenon could be described in terms that were consistent with literature on path dependence or could be described in ways that recognized real time active agency. Specifically, "outsiders" were more likely to view outcomes of processes as being serendipitous. In contrast, "insiders," were more likely to describe the active roles that they played in the genesis of novelty.

If it were just differences in the vantage points that determined whether or not a particular phenomenon would be viewed as being path dependent or as an outcome of purposive action, there is not much to celebrate regarding the role human agency plays in the genesis of novelty. However, there is a deeper issue involved. Insiders are able to endogenize "time." "relevance structures," and "objects." By doing so, they are able to develop the generative impulse required to commit themselves to an initiative and, in the process, enact their environments (Weick, 1979; March, 1998). Consequently, those who embrace an insider's perspective are able to increases the likelihood of the success of their initiatives as compared to those who embrace an outsider's perspective.

But aren't these insiders "optimistic martyrs" in-the-making, one could ask (Dosi & Lovalo, 1997). And the answer is that there is every danger that these insiders could over-commit themselves to their initiatives oblivious of the larger structures within which they are embedded and from which they attempt to depart. Such over-commitment lies in contrast to under-commitment that one is likely to witness from those who adopt an outsider's perspective; one that does not conceptualize entrepreneurship as a process that endogenizes time, relevance structures and objects.

Indeed, realizing that path creation has to reconcile the enabling and constraining facets of the structures that we create in the course of our actions, many contributors to this volume have implicitly adopted Giddens' (1979) notion of structuration. That is, structures are both medium and outcome of human action. From this perspective, we are all embedded in structures that we have created and from which we may attempt to depart. However, such departure is mindful of the ways in which existing structures may respond. Too much change could provoke the system to generate unnecessary resistance that can thwart an initiative. Too little change may not be able to generate the momentum required to overcome the inertia and resistance to set new paths in motion.

In summary, entrepreneurs can gain greater strategic choice by acknowledging that they are circumscribed by the very structures that they and others' have created and by recognizing that these structures are negotiable and malleable. This notion of embedded agency is an important facet of path creation that this volume offers. It requires a realization that we are embedded in structures from which we attempt to mindfully depart.

As several papers in this volume point out, embedded agency and the duality implicit in structuration suggests that successful entrepreneurs are neither insiders nor outsiders, but, boundary spanners instead. It is not just at the boundaries of communities and their relevance structures, but at the boundaries of time and objects as well. An ability to be at the boundaries of communities, objects and time provides entrepreneurs with greater degrees of freedom to strike at the right time, at the right place with the right initiatives.

All this, then, brings us back to the composition of this book. It consists of many arguments at multiple levels of analyses and attempts to connect with different audiences across disciplinary boundaries. Our motivation in putting this volume together is to foster a debate on how humans can strategically manipulate and create history rather than be its prisoner. In attempting to accomplish this task, we have many to thank besides those who have contributed to this book. A complete list of names would be too long to offer here. Those who have helped shape this volume will recognize their contributions and we are indebted to them. And, we are indebted to our families for their unyielding support for this project. It is to them we dedicate this book.

—*Raghu Garud*
—*Peter Karnøe*

REFERENCES

Arthur, B. (1994). *Increasing Returns and Path Dependence in the Economy.* The University of Michigan Press.

David, P. (1985). Clio and the economics of QWERTY. *Economic History, 75,* 227–332.

Dosi, G., & Lovallo, D. (1997). Rational entrepreneurs or optimistic martyrs? Some considerations on technological regimes, corporate entries, and the evolutionary role of decision biases. In R. Garud, P. Nayyar & Z. Shapira (Eds.), *Technological innovation: oversights and foresights* (pp. 41–70). Cambridge, England: Cambridge University Press.

Geertz, C. (1973). *The interpretation of cultures.* New York: Basic Books.

Giddens, A. (1979). *Central problems in social theory.* Los Angeles: University of California Press.

March, J. G. (1998, September 20–22). Research on organizations: Hopes for the past and lessons from the future. Samples for the Future. Lecture Sunday, Cubberley Auditorium at the SCANCOR Conference.

Pinch, T., & Bijker, W. E. (1987). The social construction of facts and artifacts: Or how the sociology of science and the sociology of technology might benefit each other. In W. E. Bijker, T. P. Hughes & T. J. Pinch (Eds.), *The social construction of technological systems: New directions in the sociology and history of technology.* Cambridge, MA: MIT Press.

Weick, K. E. (1979). *The social psychology of organizing* (2nd ed.). Reading, MA: Addison-Wesley.

About the Contributors

Andrea Bassanini is an Economist at the Economic Department of the OECD in Paris. In 1998, he completed a PhD in Economics at La Sapienza University in Rome, after completing graduate studies in Oxford, Stanford, and Vienna. His main field of research is economic growth.

Joel A. C. Baum is Canadian National Chair in Strategic Management at the Rotman School of Management (with a cross-appointment to the Dept. of Sociology)University of Toronto, where he teaches corporate strategy, competitive dynamics, and organization theory. Studying economic phenomena from the point of view of a sociologist, Joel is concerned with how institutions interorganizational relations and managers' understandings shape the dynamics of competition and cooperation among organizations. Recent publications on multiunit organizations, multimarket strategy, and alliance networks have appeared in the *Academy of Management Journal*, *Strategic Management Journal*, *Administrative Science Quarterly*, *Management Science*, *Social Forces*, and *Social Science Research*. He recently co-edited three books, Multiunit Organization and Multimarket Strategy (*Advances in Strategic Management*, Vol. 18, JAI Press) with Henrich Greve, Economics Meets Sociology in Strategic Management (*Advances in Strategic Management*, Vol. 17, JAI Press) with Frank Dobbin, and *Variations in Organization Science: In Honor of Donald T. Campbel* (Sage) with Bill McKelvey. Joel is a member of the editorial boards of *Administrative Science Quarterly* and *Academy of Management Journal,* and is editor-in-chief of *Advances in Strategic Management* (JAI Press). He is the 2001 program chair for the Academy of Management's Organization and Management Theory Division, and recently served as chair of the INFORMS College on Organization Science.

Koenraad Debackere holds MSc and PhD degrees in Electrical Engineering and Management. He is a full professor in Technology and Innovation Management at the University of Leuven. He has received several international awards and nominations for his research activities in the area of technology and innovation management. He obtained Best Research Paper Awards from the American Academy of Management and the Decisions Sciences Institute. He has authored more than 70 articles and book chapters in this field. He has been involved in projects for the European Commission as well as Belgian and Dutch government and multinationals.

Niels Dechow is a PhD student at the Department of Operations Management, Copenhagen Business School. His project is regarding the development and implementation of ERP systems, but he is also interested in performance management and in technology management. Previously, he was a consultant with Deloitte & Touche and was research associate on several research projects. His work has been published in Danish journals.

Giovanni Dosi is currently Professor of Economics at Sant'Anna School of Advanced Studies of Pisa in Italy. He is the editor of Industrial and Corporate Change as well as Honorary Research Professor at the University of Sussex, England. Major works include *Technical Change and Economics Theory* (edited with C. Freeman, R. Nelson, G. Silveberg & L. Soete) 1988; *The Economics of Technological Change and International Trade* (with K. Pavitt & L. L. Soete) 1990; and *Innovation, Organization and Economic Dynamics, Selected Essays*, 2000.

Raghu Garud is interested in exploring the nexus between technology, strategy and organization. His work explores path creation, metamorphic change, new organizational forms, technological path creation, economies of substitution, researcher persistence, and technology embeddedness. He has co-authored and edited several books about technology: *Technological Innovation: Oversights and Foresights* (co-edited with Zur Shapira & Praveen Nayyar) Cambridge University Press; *Cognition, Knowledge, and Organizations*, JAI Press (co-edited with Joseph Porac), and *The Innovation Journey* (with Andrew H. Van de Ven, Douglas Polley, & Suresh Venkataraman) Oxford University Press. He is program chair for the Technology and Innovation Management Division for the 2001 Academy of Management meetings in Washington, D.C.

James J. Gillespie is an advanced degree candidate at Northwestern University's Kellogg Graduate School of Management. He also holds a JD from Harvard University. His research focuses broadly on organization theory, strategic management, and behavioral decision-making.

Paul M. Hirsch is the James Allen Professor of Strategy and Organization at Northwestern University's Kellogg Graduate School of Management. He is also a member of Northwestern's Departments of Sociology and Communications Studies. Prof. Hirsch's articles have appeared in a wide variety of journals, including *Administrative Science Quarterly, American Journal of Sociology, Accounting Organization and Society,* and *Theory and Society.* He is co-executive editor of the *Journal of Management Inquiry.* In 1998, he was selected Distinguished Scholar by the Organization and Management Theory Division of the American Academy of Management.

Peter Karnøe has research interests in exploring constructive views on technology and organizing, especially understanding the processes by which patterned collectives of humans and non-humans become interwoven, evolve and temporarily stabilize. His publications explore the emergence of modern wind power, generative processes associated with path creation, the institutional embeddedness of technological innovation, the business systems framework and economic sociology, field research methodology. He has co-authored or edited several books: *Mobilizing Resources and Generating Competencies* (co-edited with Peer H. Kristensen & Poul H. Anderson) Copenhagen Business School Press; and *On the Art of Doing Field Studies* (co-edited with Finn Borum, Peer H. Kristensen & I. B. Andersen) Copenhagen Business School Press.

Rene Kemp is a senior research fellow at the Maastricht Economic Research Institute on Innovation and Technology (MERIT) working on topics of environmental policy and technical change, technological transitions to environmental sustainability, green innovation policy, and evolutionary theories of technical change. He is the author of *Environmental Policy and Technical Change,* that is regarded as a major contribution on the topic. He is currently engaged in a study about transitions and transition management for the fourth Dutch environmental policy plan (NMP4), and is writing a book about niche management for sustainability with Remco Hoogma, Johan Schot, and Bernhard Truffer.

Martin Kenney graduated from Cornell University in 1984. He is a Professor of Human and Community Development at the University of California, Davis, and is Senior Project Director at the Berkeley Roundtable for the International Economy. His interests include the high technology industry. He is the author of three books and has edited *Understanding Silicon Valley.* He is completing a book on the U.S. venture capital industry. He has been a visiting scholar at Cambridge University, Copenhagen Business School,

Hitotsubashi University, Osaka City University, Kobe University, and the University of Tokyo. He has authored more than 90 articles.

Joseph Lampel is a Professor of Strategic Management at Nottingham University Business School. In 1990, he received his PhD from McGill University. He received an MS from the University of Montreal and a BS from McGill University. Lampel's research focuses on strategic learning and technological change. His current research is on knowledge-intensive project-based firms. His work has appeared in *Strategic Management Journal*, *Journal of Management*, *Sloan Management Review*, *R&D Management*, and *Advances in Strategic Management*. He is co-author (with Henry Mintzberg & Bruch Ahlstrand) of *The Strategy Safari: A Guided Tour Through the Wilds of Strategic Management* (Prentice-Hall, 1998).

Richard N. Langlois is Professor of Economics and (by courtesy) Professor of Management at the University of Connecticut, Storrs. He attended Williams, Yale, and Stanford. Before coming to Connecticut in 1983, he was affiliated with the Center for Science and Technology Policy and the C.V. Starr Center for Applied Economics at New York University. Professor Langlois's research has ranged widely, with one focus on the theory of the firm. He has attempted to bring together the transaction–cost approach to organization with an evolutionary account of economic behavior. A milestone on this research track was the publication (with Paul Robertson) of *Firms, Markets, and Economic Change: A Dynamic Theory of Business Institutions* (London: Routledge, 1995), which articulates the theory of dynamic transaction costs and the theory of modular systems. A focus of Langlois's work has been the economic history of technology. He has written on such industries as semiconductors, semiconductor equipment, and software. His history of the microcomputer industry won the Newcomer Award as the best article in *Business History Review*, 1992.

Jan Mourtisen is a Professor of Management Control, Department of Operations Management, Copenhagen Business School. His research interests include the management control of technology development, intellectual capital and knowledge management, immaterial assets and the capital market, and performance management in the new economy. His work has been published in books and journals including *Accounting, Organizations, and Society; Organization; Scandinavian Journal of Management; International Journal of Technology Management;* and *Management Accounting Research*.

Joseph Porac is Professor of Organization and Management at Emory University's Goizueta Business School. His current research focuses on the micro underpinnings of macro organizational behavior, particularly knowledge and cognition in organizations, market dynamics, and executive compensation. Joe's papers have appeared in most of the management journals, and he has served on the editorial boards of *Administrative Science Quarterly*, *Academy of Management Review* and the *Journal of Management Inquiry* and has been a co-editor for special issues of *Organization Science* and *Journal of Management Studies*. Currently, he is Associate Editor for *Administrative Science Quarterly*. Joe is a member of several divisions of the Academy of Management, having participated in a varied sessions and consortia over the past 15 years at annual conferences. He was the OMT program chair for Chicago, 1999, and is currently OMT's division chair elect.

Hayagreeva Rao is Professor of Organization and Management, Goizeuta Business School, and an Adjunct Professor in the Department of Sociology, Emory University. He completed his PhD in Organizational Behavior at Case Western Reserve University and his research analyzes the social foundations of economic outcomes. His research has been published in the *American Journal of Sociology*, *Administrative Science Quarterly*, *Academy of Management Journal*, *Organization Science*, *Strategic Management Journal*, and the *Journal of Marketing*. Recent publications include Embeddedness, Social Identity and Mobility: Why Firms Leave NASDAQ and Join NYSE, *Administrative Science Quarterly*, June, 2000 (co-authored with Gerald Davis & Andrew Ward), and *Power Plays: How Social Movements and Collective Action Create New Organizational Forms* (co-authored with Mayer Zald & Cal Morrill), *Research in Organizational Behavior*. He is a Consulting Editor of the *American Journal of Sociology*, Senior Editor at *Organization Science*, and serves on the editorial boards of *Administrative Science Quarterly* and the *Academy of Management Journal's* Special Research Forum, Extending the Frontiers of Organizational Ecology.

Arie Rip was educated as a chemist philosopher. After a period of research in physical chemistry at the University of Leiden, he developed a program of research and teaching in chemistry and society at Leiden University, and became one of the key figures in Science, Technology, and Society teaching and research, both nationally and internationally. He was Professor of Science Dynamics at the University of Amsterdam (1984–1987) and is now Professor of Philosophy of Science and Technology at the University of Twente. He has been active in international scholarly societies, including serving as Secretary of the European Association for the Study of Science and Technology (1981–1986) and as President of the Society for Social

Studies of Science (1988–1989). He was and is a member of science policy committees of various kinds, and has advised on science policy in England, France, Australia, and South Africa. He was a member and then chair of the Panel of Economic and Social Sciences in the HCM and TMR programs of the European Union. He is active in Research and Development Evaluation, Technology Assessment, and Science and Technology Foresight. He was member of the board of the Rathenau Institute (formerly the Netherlands Organization for Technology Assessment; NOTA). His own research has focused on science and technology dynamics and science and technology policy analysis. In recently published works two clusters of themes are distinguished: ongoing changes in modes of knowledge, production in national research systems, policy instruments in context, and processes of technology development and embedment in society and the ways to influence, or modulate, them for the better. This is background analysis for more practically oriented Constructive Technology Assessment.

José Antonio Rosa is an Assistant Professor of Marketing at the Weatherhead School of Management, Case Western Reserve University. Rosa completed his PhD at the University of Michigan, and holds degrees from Dartmouth College and General Motors Institute. His research has been published in the *Journal of Marketing, Industrial Marketing Management*, and *Advances in Strategic Management*.

Vernon W. Ruttan is a Regent's Professor in the Department of Economics and Applied Economics and an Adjunct Professor in the Hubert H. Humphrey Institute of Public Affairs at the University of Minnesota. He attended Yale University (BA, 1948) and the University of Chicago (MA, 1952; PhD, 1954). He has served as a staff member of the President's Council of Economic Advisors (1961–1963) and as President of the Agricultural Development Council (1973–1978). Ruttan's research has been in the field of agricultural development, resource, economics, and research policy. He authored *Agricultural Research Policy* (University of Minnesota Press, 1982)with Yujiro Hayami), *Agricultural Development: An International Perspective* (Johns Hopkins Press, 1996). His latest book is *Technology, Growth and Development* (Oxford University Press, 2001). He is currently working on a book, *Social Science Knowledge and Economic Development*, scheduled for publication by the University of Michigan Press in 2002. Ruttan was elected a fellow of the American Academy of Arts and Sciences (1976), the American Association for the Advancement of Science (1986), and to membership in the National Academy of Sciences (1990).

Deborah A. Savage received her BS in economics from George Mason University (1982) and her MA (1984) and PhD (1993) degrees in economics from the University of Connecticut. She was a Robert Wood Johnson Foundation Health Policy Scholar at Yale University, 1995–1997, and remains a visiting fellow in the Yale Health Policy Program. Savage has taught at Babson College, Connecticut College, Trinity College, and elsewhere, and is currently Associate Professor of Economics at Southern Connecticut State University, where she was recently named most outstanding teacher in the School of Business Administration. Savage's research interests lie at the intersection of health economics and the economics of organization. Her PhD dissertation draws on the dynamic-capabilities approach to the economics of organization to propose a novel theory of the professions. Her current work focuses on the economics of quality monitoring in healthcare organizations and on the theory of strategic networks.

Michael Scott Saxon is a Research Director at Jupiter Research. He has worked for AC Nielsen and BBDO-Chicago in research capacities. Saxon has an MS in Business Administration from the University of Illinois at Urbana Champaign and other degrees from Cornell University. His current interests include identifying key drivers of Internet usage among consumers, assessing the impact of the Internet on product research and purchase channel preference, and quantifying the impact of Internet advertising on consumer purchasing behavior. His research has been published in the *Journal of Marketing* and *Administrative Science Quarterly*, and has been presented at American Marketing Association and Marketing Science Institute conferences.

Johan Schot is a professor in history of technology at Eindhoven University of Technology and University of Twente, and directing the Foundation for the History of Technology (FHT). He is program leader of the national research program on the history of technology in the Netherlands in the 20th century. This program results in seven volumes, entitled *Techniek in Nederland, 1890–1970*. Two volumes were published in 1998 and 2000. He is also the co-founder (1991; with Kurt Fischer of Clark University), and 1992–1998 European Coordinator of the (now worldwide-operating) Greening of Industry Network. He was project coordinator of several EU-funded international projects. His research work and publications include history of transport technology, theory of innovation, and diffusion and policy studies.

Brian S. Silverman is an Assistant Professor in the Competition and Strategy area at Harvard Business School. Previously, he served on the faculty of the Rotman School of Management at the University of Toronto. Professor Silverman's research interests involve the management of innovation. Silverman earned his PhD from the Haas School of Business at the University of California, Berkeley.

Jitendra Singh is the Saul P. Steinberg Professor, Department of Management and Vice Dean, International Academic Affairs at the Wharton School, University of Pennsylvania. He has been a faculty member at Wharton since 1987, having moved here from University of Toronto, Canada, where he was an Associate Professor in the (now) Rotman School of Business. He received his PhD from Stanford Business School, 1983. His earliest education was in natural and mathematical sciences and he received his BS from Lucknow University, India, in 1972. He received his MBA from the Indian Institute of Management, Ahmedabad, India, in 1975. His current interests focus on the intersection of strategy, technology, and organization, with a specific focus on high technology organizations. He has published more than 35 papers in leading management journals, and currently serves on the editorial boards of *Asia Pacific Journal Management, Strategic Management Journal,* and *Organization Science.* Previously, he has also served on the editorial boards of *Academy of Management Journal* and *Administrative Science Quarterly.* He has edited *Organizational Evolution: New Directions* (Sage, 1990), and *Evolutionary Dynamics of Organizations,* co-authored with Joel Baum (Oxford University Press, 1994).

Jelena Spanjol is a doctoral candidate in Business Administration at the University of Illinois at Urbana–Champaign, and has earned other degrees there as well. Her research interests include the sustained innovation capabilities of firms involved in new product development, the role of fantasy in consumer products advertising, and product market evolution. Her research has been published in the *Journal of Marketing* and presented at various American Marketing Association and Association for Consumer Research conferences.

Urs von Burg received his PhD in 1999 from the University of St. Gallenin, Switzerland. His academic interests include the data networking industry, the emergence and commercialization of new technologies, standardization, and the role of venture capital in the creation of new industries. He is author of *The Triumph of Ethernet: Technological Communities and the Battle for the LAN Standard* (Stanford University Press, forthcoming), as well as co-author of many scholarly articles. Von Burg was a visiting scholar at the University of California in Davis and Berkeley. Currently, he is a business analyst with Aureus Private Equity AG, a venture capital company in Switzerland.

1

Path Creation as a Process of Mindful Deviation

Raghu Garud
New York University

Peter Karnøe
Copenhagen Business School

Think different
　　　—Slogan for Apple's iMac advertising campaign

Panasonic—just slightly ahead of our time
　　　—Slogan for Panasonic advertising campaign

We live in an era of continual change. We are bombarded by new products and technologies, some of which have the potential to fundamentally change our lives. It is not surprising, therefore, that many people are becoming increasingly interested in the genesis of novelty.

One perspective acknowledges the historical antecedents of novelty. Our present and future choices are conditioned by choices we have made in the past. *Novelty*, from this perspective, is not a negation of the past, but an elaboration and extension in specific directions depending on the particular sequence of unfolding events. Stated differently, the emergence of novelty is a path dependent phenomenon (David, 1985; Arthur, 1988).

The insight that novelty has historical antecedents is a refreshing one. It offers us a way of understanding the emergence of novelty in process terms rather than having to resort to functional explanations. Moreover, it provides us a way of viewing social action as being temporally located and socially embedded.

Despite these strengths, a path dependence perspective has important implications for human agency that are problematic for a theory of entrepreneurship. Path dependence suggests that temporally remote events play a key role in the development of novelty and that these events gain significance only post hoc. Indeed, proponents of a path dependence perspective often celebrate historical accidents to explain the emergence of novelty. They relegate human agency to choosing to go with a flow of events that actors have little power to influence in real time.

Departing from path dependence, we offer a contrasting perspective that we term *path creation*. In our view, entrepreneurs meaningfully navigate a flow of events even as they constitute them. Rather than exist as passive observers within a stream of events, entrepreneurs are knowledgeable agents with a capacity to reflect and act in ways other than those prescribed by existing social rules and taken-for-granted technological artifacts (Schutz 1973; Blumer 1969; Giddens, 1984). We believe entrepreneurs attempt to shape paths, in real time, by setting processes in motion that actively shape emerging social practices and artifacts, only some of which may result in the creation of a new technological field.[1]

Path creation does not mean entrepreneurs can exercise unbounded strategic choice. Rather, entrepreneurs are embedded in structures that they jointly create (Granovetter, 1985) and from which they mindfully depart. *Mindfulness* implies the ability to disembed from existing structures defining relevance and also an ability to mobilize a collective despite resistance and inertia that path creation efforts will likely encounter. Indeed, entrepreneurship is a collective effort where paths are continually and progressively modified as new technological fields emerge.

Facets associated with path creation are implicit in several bodies of work. In economics , for instance, path creation is implicit in notions of dynamic efficiency and dynamic equilibria (cf. Schumpeter, 1942; Hayek, 1948; Kirzner, 1992). Literature at sociopsychological levels offer the concept of enactment—of how humans literally "put things out there" (cf. Weick, 1979). Complementing this perspective are those offered by social constructivists who explore the social and cognitive processes involved in the creation and diffusion of new technologies (cf. Bijker, Hughes, & Pinch,

---

[1]We define a technological field as representing a pattern of relationships between artifacts and humans related to any product–market domain (Karnøe, 1999). Actors in a technological field may have different structures of relevance, and yet, be a part of the same technological field. In our use of the term *technological field*, we depart from a common meaning system implicit in the concept of an organizational field in neoinstitutional theory (Scott 1995, p. 130).

1987). Additionally, forecasting literature offers *scenario thinking* as a process where practitioners work backward to fulfill a projected future state (Porter et al., 1991). Even in population ecology literature, there is an appreciation of "quantum speciation," or in other words, how mutants create new ecological spaces to grow and prosper (cf. Astley, 1985; Rao & Singh, chap. 9, this volume).

By stressing path creation, we draw attention to phenomena in the making—that is, the temporal processes that underlie the constitution of phenomena.[2] Such a perspective assumes reciprocal interactions between economic, technical, and institutional forces that constitute technological artifacts and actors involved. Thus, social orders, institutional rules, and artifacts are both the medium and outcome of human endeavors (Giddens, 1984; Berger & Luckmann, 1966).

We begin this chapter with a brief description of path dependence and why its articulation is problematic for conceptualizing issues around human agency. We then provide an overview of path creation in contrast to path dependence. To develop a deeper understanding and appreciation of path creation, we explore how entrepreneurs are embedded in day-to-day activities that involve the production and consumption of objects that take on specific meanings. Path creation occurs as entrepreneurs disembed out of these activities in ways that mobilize, rather than alienate, constituents of a technological field. After explicating these processes, we explore the implications of path creation for key issues such as learning and commitment.

PATH DEPENDENCE

The origins of the path dependence perspective can be traced to David's (1985) description of the evolution of the letters on a typewriter keyboard. His description suggested that actors of that time chose to address the problem of jamming typewriter keys by employing the QWERTY layout. The problem disappeared over time with the adoption of the ball keyface mechanism and the use of personal computers. Yet, we continue using the QWERTY keyboard.

Path Dependence As NonErgodic Processes

David (1985) ascribed this phenomena to *technology inter-relatedness, economies of scale and quasi-irreversibility* of efforts.[3] These three elements are the

---

[2]As Latour (1992:120) pointed out, "Both economics and sociology arrive on the scene after the decisive moments in the battle.... Since the explanation of an innovation's path cannot be retrospective, it can only spring from the socio-logics of programs and anti-programs."

[3]Technology inter-relatedness means the complementarity and compatible of system parts. Economies of scale alludes the benefits that arise from size. Quasi-irreversibility of efforts means difficulties associated with redeploying an asset for alternative purposes.

basics of what he termed *QWERTYnomics*. Suggesting our use of the QWERTY keyboard can only be explained by employing an historical perspective, David (1985) offered path dependence as a concept:

> A path dependent sequence of economic changes is one of which important influences upon the eventual outcome can be exerted by *temporally remote events*, including happenings dominated by *chance elements* rather than systematic forces. Stochastic processes like that do not converge automatically to a fixed point distribution of outcomes, and are called *non-ergodic*. In such circumstances, *"historical accidents"* can neither be ignored nor reality quarantined for the purpose of economic analysis; the *dynamic process itself takes on an essentially historical character"* (p. 332, italics added).

More generally, path dependence alludes to a sequence of events constituting a self-reinforcing process that unfolds into one of several potential states (see Bassanini & Dosi, chap. 2, and Ruttan, chap. 4, this volume, for descriptions of path dependence and its origins). The specific state that eventually emerges depends on the particular sequence of events that unfold.[4] Those who propose path dependence suggest that phenomena are sensitive to small differences in the underlying sequence of events. Consequently, a steady accumulation of small differences can result in the technological field locking onto a trajectory.

We can gain an intuitive feel for processes underlying path dependence by considering Polya Urn dynamics (Arthur, 1994). The *Polya Urn* contains balls of different colors. The dynamics unfold from a simple replenishment rule—the probability of adding a ball of one color equals its current proportion. With such a rule, a slight imbalance in the proportion of balls can result in the urn eventually containing balls of only one color.

Arthur (1996) suggested that many contemporary phenomena are driven by such increasing returns logic. Driven by network externality effects (Farell & Saloner, 1986; Katz & Shapiro, 1985), phenomena begin exhibiting Polya Urn dynamics. Small accidents are magnified as complex, non-linear interactions between customers, producers, and regulators at the boundaries of an object eventually result in the emergence of a dominant standard. Sunk costs, learning effects, and coordination costs are forces from the past or present that can explain lock-in to a trajectory over time (Arthur, 1988). Future expectations about the performance of a new technological trajectory rarely have the power to unlock.

More broadly, path dependence alludes to the difficulties associated with specific technological trajectories that economic, technical, and institutional forces generate. An appreciation of these forces can be found in many litera-

---

[4]In contrast, a path independent phenomenon is one where the sequence of events have no implications for the eventual outcome (Langlois & Savage's analysis of the American Medical Profession in this volume comes close to describing such a path independent process).

ture streams (Hirsch & Gillespie, chap. 3, this volume). Some have directed our attention to organizational routines that guide behavior (cf. Cyert & March, 1963; Nelson & Winter, 1982). Institutional theorists have explored how characteristics of economic systems depend on their institutional contexts (cf. North, 1990; Whitley, 1992; Karnøe, Kristensen, & Andersen, 1999). Other theorists have explored how activities in economic and social systems are dependent on institutionalized rules (cf. Powell & DiMaggio, 1991). In technology studies, path dependence is apparent in the concept of technological trajectories (cf. Dosi, 1982). In the organizational ecology field, imprinting effects may determine the evolution of organizations (Stinchcomb, 1965; Hannan & Freeman, 1977; Baum & Singh, 1994).

Path dependence has been usefully employed at different levels of analysis. For instance, it has been used to explain the emergence of regions such as silicon valley (cf. Saxanian, 1994; Kenney & Burg, chap. 5, this volume), the self-referential processes associated with the functioning of business systems (cf. Whitley, 1992), the development of technological trajectories as a field gains momentum (cf. Dosi, 1982; Hughes, 1983), and problems and paradoxes in punctuated organizational change (Sastry, 1997). These studies provide excellent accounts of how specific institutional orders emerge and become stabilized.

Controversy Within the Paradigm

As we can see, path dependence is a powerful perspective being used increasingly to explain the emergence of novelty. However, even as we acknowledge the benefits of adopting such a process perspective, an important controversy has surfaced. The roots of this controversy can be traced to the origins of the path dependence perspective. Path dependence was articulated to counter Neoclassical economists' assumption of optimal choice. Specifically, proponents of the path dependence perspective have suggested that historical accidents result in phenomena locking onto choices that perpetuate market inefficiencies. It is this challenge, from the proponents of the path dependence perspective, that Liebowitz and Margolis (1990) question.[5]

To develop a critique of the path dependence perspective and its claims about market inefficiencies, Liebowitz and Margolis made a distinction between weak and strong forms of path dependence. *Weak* forms of path dependence entail durability and false regret (a situation where information gained post hoc may suggest an earlier sub optimal choice). *Strong* forms en-

---

[5]Liebowitz and Margolis challenge the completeness of David's description of the emergence of the QWERTY keyboard.

tail true regret—making suboptimal choices with full information. Although ceding path dependence because of durability and false regret, Liebowitz and Margolis argued that it is true regret that needs to be demonstrated in order to show perpetuation of market inefficiencies. As Liebowitz and Margolis pointed out, it is practically impossible to demonstrate inefficiencies arising out of true regret.

Undoubtedly, there are merits to Liebowitz and Margolis' arguments. However, it is unfortunate that this debate has become mired in this controversy as the path dependence perspective has much to offer in regard to thinking about paths as process. Moreover, the polemics of the debate around market inefficiencies has obscured a more fundamental facet of entrepreneurship. Specifically, in their quest to develop new paths entrepreneurs deviate from existing artifacts and processes intentionally despite the perceived inefficiencies these deviations may create.

PATH CREATION

The need to escape myopic selection pressures of markets by designing technological fields that are inefficient by today's standards was recognized by none other than Schumpeter (1942). As a part of his theorizing on the process of creative destruction, Schumpeter stated that "any system designed to be efficient at a point in time will not be efficient over a point in time." Systems designed to be efficient in the present will be associated with relevance structures (Schutz, 1973) that are likely to discourage experimentation because of associated inefficiencies. Experimentation requires time for new ideas to be refined and grow even as new institutional and market preference structures coevolve (Van de Ven & Garud, 1993). Therefore, time is an important part of Schumpeter's (1942) creative destruction process.

Path Creation as Mindful Deviation

Schumpeter's gales of creative destruction were articulated to offer insights about macroeconomic processes. These insights transfer easily to generate insights about entrepreneurship as well. Specifically, entrepreneurs may intentionally deviate from existing artifacts and relevance structures, fully aware they may be creating inefficiencies in the present, but also aware that such steps are required to create new futures.

Such a process of mindful deviation lies at the heart of path creation. Because deviations can be threatening to existing orders, entrepreneurs exercise judgment regarding the extent that deviations may be tolerated in the present and may also be worthwhile to create new future.[6] Entrepreneurs

---

[6]We have been influenced by Tsoukas (cf. 1996).

recognize that the extent to which they deviate from existing objects, relevance spaces and the present need to be synchronized for path creation to occur. In sum, mindful deviation implies disembedding from the structures that embed entrepreneurs.[7]

Path Dependence or Path Creation?

A juxtaposition of path creation and path dependence provides an intuition for our perspective. In path dependence, the emergence of novelty is serendipitous. Events that set paths in motion can only be known post-hoc. Consequently, the role of agency can be viewed as one of entrepreneurs watching the rearview mirror and driving forward. Stated differently, although path dependence focuses on a sequence of specific microlevel events, it does not have an explicated theory of agency.

Path creation attempts to remedy this. Agency takes on greater importance by bringing into play not only the social and institutional processes inherent in path dependence, but, more importantly, the sociocognitive processes of enactment that are involved in the creation of new states (Weick, 1979; Garud & Rappa, 1994). In sum, an understanding of path creation processes provides a way of understanding how entrepreneurs escape lock-in.

Entrepreneurs set path creation processes in motion in real time. Specifically, they attempt to shape institutional, social, and technical facets of an emerging technological field.[8] But, to the extent that they are unable to generate momentum with their own approaches, path creation requires an ability on the part of entrepreneurs to shift their emphasis to alternative approaches that may have greater promise. This ability to create and exercise options, we believe, is crucial.

Entrepreneurs creating new paths are not necessarily driven by a search for optimality (see also Rosenberg, 1994, p. 53). For those creating paths, errors are red herrings as there are no preexisting universal benchmarks that

---

[7]Deviations in any of the embedding dimensions (viz. objects, relevance structures, and time) may set path creation processes in motion. For instance, any change in institutional regulations may set in motion a sequence of adjustment in objects over time. Similarly, as March (1998) suggested, if actors in a technological field are able to mobilize time as a resource, it may set in motion exploratory acts that, in turn, change institutional arrangements. The symmetry involved in entrepreneurs being able to set path creation processes in motion by being able to manipulate any of the embedding dimensions should not be much of a surprise. After all, entrepreneurship involves managing coevolutionary dynamics that begin when they attempt to disembed from their embedding dimensions. Indeed, this symmetry is advantageous in developing a more complete theory of entrepreneurship where one might conceptualize path creation processes being set in motion by the action of actors in domains of use, production and governance, and one which eventually encompasses a collective with heterogeneous interests. See chapter by Kemp, Rip and Schot, this volume, for how regulators can set path creation processes in motion.
[8]See Hirsch (1975), Rao (1994).

can flag the outcomes of an exploratory act as mistakes. Instead, entrepreneurs creating paths explore the creation of new dimensions of merit that, in time, may set in motion a sequence of events (Garud & Rappa, 1994). Rather than errors and mistakes, advocates of a path creation perspective may use terms such as *experimentation* and *exploration*, wherein any action is a probe into the world even as it is being created (March, 1991a; Weick, 1999). As March (1971) suggested, we may need a "technology of foolishness" in order to make advances with technologies.

In such a conceptualization, what is of value becomes endogenized within an overall process of entrepreneurship. That is, criteria that establish value about facts and artifacts do not lie in a market that is an overall arbiter of what is good and bad, but instead, become endogenized as a pattern of stabilized relationships within an emerging technological field. Thus, the diverse actor-groups involved, including producers, users and regulators, create their own set of practices and relevance structures[9] that coevolve with technological artifacts (Schutz, 1973). From this perspective, the question of whether markets are efficient becomes secondary to a more important one—Where do specific product markets come from? (Kirzner 1992; Koppl & Langlois, 1994; Ventresca & Porac, 2000).

Epistemological and Ontological Differences

Differences between path dependence and path creation perspectives are striking because they represent different epistemologies and ontologies. Path dependence assigns too much weight to history; it inadequately characterizes the fragility of any path as it is produced and reproduced through microlevel practices where social rules and artifacts are enacted (Giddens, 1984). Those who view phenomena as being path dependent are like outsiders looking in at the emergence of novelty. As outsiders, agents may more likely embrace a logic of consequentiality (March, 1994) anchored on present ways of evaluation. Using this benchmark, any deviation from present acceptable social practices are mistakes that may not survive (Christensen, 1997). Those mistakes that survive are, therefore, seen from an outsiders' perspective as

---

[9]Schutz (1973, p. 240) stated that the merits of any object must be understood as "relational notions and have to be defined in terms of the domain of relevance to which they pertain. Only within each of these domains of relevance can degrees of merit and excellence be distinguished. Moreover, that which is comparable in terms of the system of one domain is not comparable in terms of other systems, and, for this reason, the application of yardsticks not pertaining to the same domain of relevance leads to logical or axiological (moral) inconsistencies." This is closely related to the concept of provinces of meaning. Schutz pointed out "We speak of provinces of *meaning* and not of sub-universes because it is the meaning of our experiences and not the ontological structure of the objects which constitutes reality." This concept of relevance structures is similar to Blumer's notions of how meanings are attributed to objects by fellow men.

chance events whose significance can only be known in hindsight. Temporal myopia, then, leads to a perception of intertemporal serendipity.

For entrepreneurs attempting to create paths, the world is constantly in the making.[10] Indeed, entrepreneurs creating new paths are more likely to embrace a logic of mindful deviation. Such logic involves spanning boundaries between structures of relevance. On one hand, entrepreneurs are insiders possessing knowledge of a technological field and an appreciation of what to deviate from and the value of pursuing such a strategy. On the other hand, they are outsiders (Blumer, 1969) evaluating how much they can deviate from existing relevance structures. And because many deviations are perceived as threatening, entrepreneurs have to buy time, with which and within which to protect and nurture new ideas and create new provinces of meanings. From this perspective, ideas are carefully evaluated on an ongoing basis and even those that are abandoned may play a role in shaping ideas that survive over time (Garud & Nayyar, 1994). Temporal elasticity is linked with intertemporal acumen.

In sum, a shift from path dependence to path creation occurs as entrepreneurs endogenize objects, relevance structures, and time. As objects, relevance structures, and time become strategic variables, there is a shift from conceptions of path dependence as ways of describing our past worlds to conceptions such as path creation as ways of shaping our current states to create new futures. Entrepreneurship involves an ability to exercise judgment and choice about time, relevance structures and objects within which entrepreneurs are embedded and from which they must deviate mindfully to create new paths.

ENTREPRENEURS AS EMBEDDED AGENTS

The extent that human actions are embedded in existing structures lies at the heart of an age-old debate on strategic choice. In technology studies literature, there is a widely held view that humans are embedded in a larger technological field that they have helped create. Technological fields repre-

---

[10]Here we see an important overlap with Dewey's pragmatism. Dewey's view is that entities are interdependent and interrelated and that any isolation of entities are mental constructions. Dewey puts a primacy on the entities in interaction: "The materials of our everyday surroundings need to be woven together so that they do not merely accumulate, but rather culminate in a set of habits that provide meaningful ways of interacting with those surroundings." (From Boisvert, 1998, p. 124). Boisvert comments that in Dewey's perspective, we live in, and constitute, a world that is continually in the making, "Affairs are never frozen, finished, or complete. They form a world characterized by genuine contingency and continual process. A world of affairs is a world of actualities open to a variety of possibilities" (Boisvert, 1998, p. 24). Furthermore, Boisvert points out "Since all entities are entities-in-process, they are continually being influenced and altered by the relationships into which they are immersed. The various projects we undertake, relationships into which we enter, and struggles which we undergo, help shape who we are" (Boisvert, 1998, p. 23).

sent ongoing patterns of relationships between heterogeneous entities that include objects and actors (Callon, 1986).[11]

Objects constituting these fields are the physical manifestations of human efforts to tame and shape nature. They include primary and complementary objects that are required to create a useable product (Teece, 1987). Moreover, it is appropriate to include human behaviors and organizational routines required to create and maintain links between these disparate network of objects so that they can seamlessly work together.

Different actors are involved in the creation and maintenance of a technological field. Each actor enacts a frame of reference comprising a set of beliefs, standards of evaluation, and behaviors (Bijker, 1987; Dougherty, 1992; Karnøe & Garud, 1998). Three stylized frames that play a role in technology development are frames on *production*, *use*, and *governance*. For instance, frames on *production* may include beliefs about the future potential of a technological trajectory with respect to its form and function. Frames on *use* may consist of the multiple meanings that can be attributed to a technological artifact when in use. *Governance* frames may include the value of the technology trajectory to multiple stakeholders on one hand, and the effect of specific policy instruments and funding to shape the development of a technology on the other hand.

Different actors in a technological field enact their realities based on their frames. Depending on their vantage point as regulators, users, and producers, agents begin to identify and ascribe specific meanings to the objects constituting the technological field. Eventually, these meanings become deeply internalized within actors.

As they enact their realities, actors interact with one another to negotiate the relevance of objects and behaviors that constitute the technological field. A debate ensues between these actors that results in the creation of institutionalized practices and meanings. These institutionalized practices and meanings, in turn, affect individual actors by shaping their frames and actions. A technological field takes on shape and meaning as an outcome of these intersecting processes.[12]

These processes are reflective of a broader proposition on structuration (Giddens, 1979, 1984). That is, structure is both medium and outcome of action. Rules and resources, drawn on by actors in their interactions, are re-

---

[11]Several scholars have embraced this perspective within a literature stream that is commonly referred to as the "actor network theory" (see for instance, Callon, 1986; Latour, 1987; & Law, 1992).

[12]Blumer presented three premises of symbolic interactionism: "The first premise is that human beings act towards things on the basis of meanings that the things have for them. The second premise is that the meaning of such things are derived from, or arises out of the, social interaction that one has with one's fellows. The third premise is that these meanings are handled in, and modified through, an interpretative process used by the person in dealing with the things he encounters" (Blumer, 1969, p. 2).

constituted through their interactions. An important implication is that objects do not posses any intrinsic meaning in themselves. Objects and their meaning are produced and reproduced in communities of practice (Blumer, 1969; Brown & Duguid, 1991).

Over time, as constituent elements of a technological field begin working with one another, they become aligned and begin to reinforce one another (Callon, 1992; Hughes, 1983; Molina, 1999). Meanings of objects constituting these fields emerge through a process of negotiation and provisionally stabilize (Bijker, Hughes, & Pinch, 1987). Indeed, in our quest to find simplicity in complexity, these meanings and practices become taken for granted (Hughes, 1983). Entrepreneurs then become embedded in self-reinforcing processes of a technological field that they have helped generate.

Entrepreneurial Challenges

This discussion provides a finer appreciation of the many challenges of entrepreneurship. For instance, an entrepreneur may become so deeply embedded in these technological fields that a vision of the future that is different from the present is difficult to muster. Embedded actors continue reproducing existing practices because they may avoid new tests (Weick, 1979, p. 149). Or, the impulse to exploit what has already been created is so great that the impulse to explore and create new structures may reduce or disappear (March, 1991a). For these reasons, an actor may not be able to develop the generative impulse required to set path creation processes in motion.

Notwithstanding the difficulties associated with recognizing and creating new opportunities, they are just the first of a number of challenges. Deviations may disturb the status quo, thereby setting in motion a coevolutionary reaction from interdependent actors with heterogeneous preferences and frames (Callon, 1986; Law, 1992; Latour, 1987). *Coevolution* occurs as two or more parts of a field evolve together, not perfectly, but with slippages across time and space. In doing so, the coevolving parts may both enable and constrain each other through feedback that can be negative or positive. Moreover, feedback can be non-linear when a response is not directly proportional to the stimulus. Non-linear feedback to deviations create interactively complex systems where deviations can either de-amplify and dissipate or amplify and spin out of control (Masuch, 1985).

For instance, unfavorable responses from powerful threatened actors can generate negative feedback. Even without negative feedback, generating momentum within a network of cospecialized objects is difficult enough. Indeed, the very competencies within a technological field can become entrepreneurial traps (Levitt & March, 1988; Leonard-Barton, 1992). To complicate matters, these changes are often attempted within a short time frame, during which entrepreneurs are unable to develop their insights or explain them in appropriate ways to significant stakeholders (Dierickx & Cool, 1989). Moreover, negative dynamics are generated and stoked by the very behaviors of entrepreneurs (Weick, 1979). Specifically, those who can muster the enthusiasm and mindset to depart from existing embedding structures may be so enthused by their act of insight that they begin pursuing it with a single-minded purpose. In doing so, they are likely to disregard feedback others may provide, and, thereby, miss out on an opportunity to mobilize others.

Even if entrepreneurs are able to generate momentum around their ideas, the process may spin out of control. A process may spin out of control when an interactive, complex system generates unmanageable processes that drive the system to unanticipated, unacceptable end states (Masuch, 1985).[13] Perrow (1984) characterized the negative outcomes of such unanticipated, unmanageable processes as representing normal accidents.

In sum, the embeddedness of action generates several challenges for entrepreneurs. Not only do they have to disembed from embedding structures, they have to also overcome the resistance they may generate in the process. Moreover, they have to mobilize elements of the network in which they are embedded to further their efforts while preventing the process from spinning out of control. It is no surprise that path creation processes are fraught with failure!

To address these challenges we probe deeper into path creation processes. Paraphrasing Pettigrew (1992), we offer an understanding of entrepreneurship in a way that (a) acknowledges the embeddedness of actions, (b) explores temporal interconnections between processes, (c) provides a role in explanation for context and action, (d) is holistic rather than linear, and (e) links process analysis to the location and explanation of outcomes.

PATH CREATION PROCESSES

How might entrepreneurs overcome the constraining effects of the dimensions that potentially imprison them?[14] An answer, we suggest, lies in an ability to endogenize objects, relevance structures and time. Such an ability generates agency for entrepreneurs in their being able to disembed from existing technological fields as they shape emerging ones.

---

[13]It is here that one can see how we intend to use some of the apparatus that the path dependence perspective offers. Path dependence is also built around non-linear dynamics. Whereas proponents of the path dependence perspective are interested in describing phenomena shaped by non-linear dynamics, we are interested in the implications of these dynamics for action within the system.

[14]Changes in objects and rules that constitute a technological system can also set in motion a chain reaction that brings about change in the technological field (Callon, 1992, p. 141). In this chapter, we are interested in understanding the role of human agency in navigating and shaping such coevolutionary processes.

To develop this proposition and motivate our discussions, we will use a well-known story of path creation—the development of Post-it® Notes—for illustrative purposes. Most accounts of its development suggest that it was an "accident." In this sense, these are outsiders' accounts consistent with a path dependence perspective. However, an interview with an insider, Spence Silver (a 3M scientist who discovered the weak glue that is used on Post-it® Notes), offers a glimpse of how such accidents are consistently cultivated and nurtured to create something of value.[15]

Mobilizing Molecules

Reflecting on his experiences with the development of Post-it® Notes, Silver vehemently denied that his discovery was a mistake that worked. Rather than a random act of discovery, Silver described his discovery as a cultivated breakthrough that occurred because he chose to deliberately deviate from existing ways of mixing molecules:

> In the course of the exploration I tried an experiment with one of the monomers in which I wanted *to see what would happen if I put a lot of it into the reaction mixture.* Before we had used amounts that would correspond to conventional wisdom. The key to the Post-it® adhesive was doing this experiment. *If I had really seriously cracked the books and gone through the literature, I would have stopped. The literature was full of examples that said you can't do this.*
>
> People like myself get excited about looking for new properties in materials. I find that very satisfying, to *perturb the structure slightly and just see what happens.* I have a hard time talking people into doing that—people who are more highly trained. Its been my experience that people are reluctant just to *try, to experiment*—just to see what will happen!" (from Nayak & Ketteringham, 1986, pp. 57–58; italics added)

His experimentation paid off as he created a substance that he thought looked beautiful under a microscope. This finding aroused his intellectual curiosity. This curiosity quickly led to an intuitive appreciation of the potential value of what he had stumbled on. Silver had created "a solution looking for a problem" (from Lindhal, 1988, p. 14).

Silver's act of insight is reminiscent of Pasteur's famous adage—"Fortune favors the prepared mind." It was because of Silver's prior professional knowledge in monomers that he could carry out a systematic experiment. And, when he stumbled on something different, he could appreciate its potential value.

As this description suggests, insights emerge by building on past experiences, not by negating them (Schutz, 1973; Bijker, 1987). Indeed, in offering

---

[15]Sitkin and Brown make a similar point in their presentation of the Xerox case at the 1999 Academy of Management meetings in Chicago.

their perspectives on entrepreneurship, many have noted how continuity and change are somehow paradoxically associated. For instance, Schumpeter (1934) considered entrepreneurship as acts reconstituting existing resources to create new ones. Similarly, his contemporary and colleague, Usher (1954), argued that innovation is a cumulative synthesis of evolutionary ideas that lead to revolutionary outcomes.

Acknowledging the importance of continuity in the entrepreneurship process, and indeed recognizing its constraining effects, some have sug-gested the need to deframe (Dunbar, Garud, & Raghuram, 1996). *Deframing* implies appreciating cognitive embeddedness to depart from existing webs of significance (Geertz, 1973) in mindful ways. In a similar vein, others have suggested discrediting (Weick, 1979) and unlearning (Hedberg, 1981; Starbuck, 1996). *Discrediting* implies purposely reversing or breaking causal structures of associations—as Weick suggested, "when you believe, you must disbelieve" (see Grove [1996] for an example of how he and those at Intel discredit). *Unlearning* implies a break from the past and consequently an ability to break away from the iron cage of history.

Silver's ability to simultaneously employ and disembed from his profes-sional knowledge base was impressive enough. To appreciate the true signifi-cance of Silver's story, however, we must realize the corporate context in which he was embedded. Silver was working at 3M Corporation, a firm that celebrated glues that stick. As Nayak & Ketteringham (1986, p. 61) suggested in their write-up of the origins of Post-it® Notes, "In this atmosphere, imagin-ing a piece of paper that eliminates the need for tapes is an almost unthinkable leap into the void." For many, a natural impulse in this firm would have been to look for glues that stuck and ignore, or actively reject, glues that did not. The fact that Silver could perceive and create an opportunity inherent in an object that would have been alien to most at 3M suggests a remarkable ability to disembed from localized contexts of meaning.

Mobilizing minds

Such disembedding is only the first of many challenges associated with en-trepreneurship. Most deviations are met with apathy at best and resistance at worst. Indeed, Silver and his colleagues encountered these impulses in equal measure despite 3M's institutionalized appreciation for innovation. Most at 3M said, "what can you do with a glue that does not glue?" Those in manufacturing showed more active resistance as is evident in the following description:

> *What added to the difficulty was the natural resistance of people.* The engineers in 3M's commercial tape division were accustomed to tape—which is sticky all over on one side and then gets packaged into rolls. To apply glue selectively to one side of the paper

and to move the product from rolls to sheet, the engineers would have to invent at least two entirely unique machines" (From Nayak & Ketteringham, 1986, p. 66; italics added)

Silver encountered similar resistance and indifference from those outside 3M. He and Walt Kern, an attorney at 3M, had to convince patent attorneys that 3M had actually discovered something new and valuable. At that time, Post-it® Notes did not exist. Preferences had yet to evolve, institutionalized ways of using Post-it® Notes had yet to congeal, and 3M's capabilities to produce Post-it® Notes were not even on the radar screen. In short, there was nothing that was real (Pinch, in this volume, describes a similar situation with the development of the synthesizer). 3M's patent application was rejected twice—the second time coming back stamped THIS REJECTION IS FINAL in capital letters. Silver remembers telling a 3M attorney, "I know this is new. I've never seen anything like this before. We're just not convincing this examiner about what's going on" (from Lindhal, p. 16).

This description of resistance and indifference to new ideas is typical of entrepreneurial processes. From our vantage point, Silver's deliberate experimental perturbation of molecular structures in turn resulted in the perturbation of existing relevance structures. Such coperturbations are likely to occur in any entrepreneurial context. Consequently, entrepreneurs often encounter apathy and resistance. How entrepreneurs deal with these forces is what is important to the emergence of a path.

Silver was undaunted by the resistance and indifference that he encountered. Describing himself as a "zealot at times in order to keep interest alive" (Nayak & Ketteringham, 1986, p. 60), Silver attempted to talk with anyone who would listen—technical directors, other scientists, the tech group of which he was a part. He hoped to enlist help and support to develop something of value from glue that did not glue. In short, Silver was trying to mobilize a collective to identify "a problem for his solution."

In writing about Silver's efforts, Nayak & Ketteringham (1986, p. 56) highlighted: "Faced with an irrational commercial challenge, Spence Silver applied an unnatural irrationality to the Post-it® adhesive." Indeed, Silver's efforts highlights a paradoxical quality that entrepreneurs possess. On one hand, Silver was a zealot trying to keep an idea alive. On the other hand, he was ready to share and modify his ideas as he went about seeking problems to complement his solution. Persistence[16] with flexibility is an important part of path creation. It offers another vantage point on mindfulness, one where "A fixed view of the future is in the worst sense ahistorical" (Mitchell, 1940).

Some individuals at 3M described Silver's flexibility with persistence as *tenacity*. Corporate scientist Larry Clemens, Silver's colleague, pointed out,

---

[16]See Amabile (1996) for insights on how intrinsic motivation can give rise to persistence.

"Silver is the definition of tenacity. He got rejected on that adhesive many times, but he stuck to it. He really felt that people were missing an opportunity." Clemens added, "What have I learned from working with Silver? I learned tenacity pays off" (From Lindhal, 1988, p. 17).

Boundary Spanning. Silver actively cultivated this paradoxical property by being a *boundary spanner*. He offered: "I've always enjoyed crossing boundaries. I think it's the most exciting part of the discovery process" (from Lindhal, 1988, p. 16). This excitement was a recognition that any new idea has to be meaningfully translated for, and with, others.

Translation is a key element in literature on actor network theory (cf. Callon, 1992, 1986; Law, 1992). Callon defined a successful process of translation as one that generates a "shared space."[17] This shared space is generated by presenting an idea in ways that are understandable by others. Indeed, entrepreneurs may present the same idea in different ways to different constituencies at appropriate points in time.[18] In doing so, entrepreneurs attempt to enroll others by drawing strategically on others' past experiences and evoking pictures of possible futures (see Van Looy, Debakere, & Bouwen, chap. 12, this volume).

Besides the creation of a shared space, translation also implies the transformation of the idea through interactions. Such transformation is required to overcome resistance and indifference. It also sets the basis for generating buy-in required to mobilize a critical mass around an idea.

In this regard, Silver was not just a skillful entrepreneur in the technical sense of the word as manifest in his ability to mobilize molecules, but was also a skillful social entrepreneur as manifest in his ability to mobilize minds. In highlighting the importance of social skills required of institutional entrepreneurship, Fligstein (1997) suggested:

> Social skill is the ability to relate to the situation of the 'other.' This means that, wherever a given strategic actor has interests, he or she must take other people's interests into account if organizational fields are to come into existence and remain stable. Skilled social action revolves around finding and maintaining a collective identity of a set of social groups and the effort to shape and meet the interests of those groups. (p. 398)

This discussion offers us an opportunity to specify the nature of agency associated with entrepreneurship. Clearly, entrepreneurs cannot do what they choose in pursuing their narrow selfinterests. Rather, entrepreneurship

---

[17]This conceptualization is similar to the notion of organizational field in institutional theory as comprising of a shared set of meanings (Scott, 1995). However, we believe a shared space in the technological field does not necessarily mean an unitary relevance structure. It implies a negotiated, and sometimes precarious, understanding between people with different frames of relevance.

[18]We build on Feldman and March (1981).

is a collective enterprise where a shared space is created and nurtured by members of a community who derive different meanings from their involvement (see similar arguments in Hirsch & Lounsbury, 1997).

Besides an ability to translate, a boundary spanners' role offers other benefits. For instance, a boundary spanning perspective offers entrepreneurs with an opportunity to look at the idea dispassionately even as they remain steadfastly resolute about the overall potential of their ideas. This tenacity provides entrepreneurs an ability to present their ideas to others with conviction even while incorporating feedback generated to modify their ideas.

Indeed, a more accurate description of the process of creation would be to consider the process as a "bisociation of ideas" as boundary spanners connect. This is consistent with Usher (1954) and Koestler's (1964) description of the genesis of novelty as a process of cumulative synthesis. Extolling the virtues of boundary spanning as a catalyst for such cumulative synthesis, Silver offered:

> I think it is the most exciting part of the discovery—when you bring two very different areas together and find something completely new. I worked for a very long time on a project called the quartz crystal microbalance with surface chemist Morgan Tamsky and Bob Oliveira, who was a biochemist and knew a lot about immunology. *This was a real nifty synthesis of a bunch of different disciplines where we crossed a lot of boundaries.* (from Lindhal, 1988, p. 16; italics added)

Generating Momentum. At 3M, path creation processes began gaining momentum when Silver was able to convince the first of many champions, fellow scientist Bob Oliveira, to join his quest. Silver and Oliveira began planning how they might sell the material. At one time, the weak glue was applied to a bulletin board on which small pieces of paper could be stuck. This initiative received a lukewarm reaction from those at 3M.

In 1974, almost 10 years from the discovery of the weak glue, another 3M scientist, Art Fry, became involved. To be sure, the weak glue molecules had evolved further, yet the fundamental problem remained—it was glue that did not glue.

Silver had sought Fry's help to identify a problem for his solution. Fry's insight occurred during a choir rehearsal at his church. Constantly losing his place in his song book, Fry had a flash of insight—Silver's weak glue could be a solution to his problem. Fry thought of applying the weak glue to pieces of paper that could be stuck in the song book as a temporary, permanent bookmark.

For those unfamiliar with the entire history, it might appear that the development of Post-it®Notes was a smooth and straight forward process subsequent to Fry's act of insight. However, Fry's act of insight was just a beginning. Subsequent champions involved in the development of Post-it®

Notes encountered indifference and resistance from people in and outside 3M. The project could have failed at any time. Indeed, the entrepreneurial cycle involving disembedding, translation, and mobilization of minds and molecules was often repeated again and again.

These observations are consistent with Usher's (1954) observations on the genesis of novelty. Based on his study of 100 years of mechanical innovations, Usher suggested that acts of insight occur as entrepreneurs set the stage, but are invariably followed by a process of critical revision. Critical revision is followed by a new cycle as entrepreneurs perceive other problems and opportunities. Indeed, in the process of critical revision, entrepreneurs may come to a realization that the original idea is not feasible and must be modified or abandoned. Consequently, judgments of whether to persist or desist at different stages of the entrepreneurial journey is an integral part of path creation.

Coevolution of Minds and Molecules

In the Post-it® Notes case, attempts at mobilizing minds led to the mobilization of molecules instead. As we mentioned, weak glue molecules had, at first, been applied to bulletin boards on which pieces of paper were stuck. Rather than think of the problem as one of selling sticky bulletin boards, Fry was able to disassociate the glue from the board, and instead apply it directly to paper. What is equally intriguing is that Silver, appreciating the value of Fry's insight, was flexible enough to ignore the bulletin board and apply his solution to a problem Fry had discovered. As one 3M employee commented, "we don't kill ideas, but we deflect them" (from Peters & Waterman, 1982, p. 230).

The coevolution of minds and molecules is a key proposition in the actor network theory literature (Callon, 1986; Latour, 1987; Kreiner & Tryggestad, 1997). Such a coevolution of minds and molecules requires flexible minds and objects. *Flexible minds* implies an ability to change structures of relevance in the process of mobilization and translation. It builds on the value of generating interpretive flexibility (Bijker, Hughes, & Pinch, 1987) where the same set of ideas are evaluated and used in different ways.

Flexibility with objects may be gained by *chunking* them (we use chunking as a complementary term to *tuning* which Baum & Silverman use in chap. 7, this volume). Chunking of objects offer several benefits to those attempting to set path creation processes in motion. For instance, entrepreneurs can partition technologies in meaningful ways (Garud and Kumaraswamy, 1995). Entrepreneurs can exercise judgment as to how much of their deviations can be presented and communicated to key stakeholders in such a way that they are not threatened by them, but are instead galvanized by them. Indeed, chunking provides entrepreneurs with an opportunity to share different chunks with different people at different times, and in the process, shape the emerging preferences of key stakeholders. As specific chunks are presented to different social groups, entrepreneurs generate feedback that can be incorporated to make appropriate adjustments to the objects being shaped. Indeed, as they experiment with different chunks, entrepreneurs can decide what chunks to keep and what chunks to abandon.

In sum, by chunking objects, entrepreneurs are able to perturb the technological field even as it is being created. As a consequence, new landscapes emerge in the very act of trying something. Feedback that is generated from such a probe becomes the basis for making appropriate changes as new possibilities open up or close down.

As this description suggests, entrepreneurial ideas are modified many times, over time. Indeed, many ideas may be abandoned or shelved during the entrepreneurial journey. Entrepreneurs use their judgment regarding how much they should persist and when to pull the plug, all the while learning from their mistakes.

Such a process embraces a real options approach to the navigation of complex dynamic flow of events (Luehrman, 1998). Options value are realized because stepwise investments generate sequential outcomes that serve as a basis for deciding whether to continue, modify, or abandon a course of action. Entrepreneurs generate a set of compound options that are revealed by choices at each stage of a complex journey.

Indeed, these processes begin to address the problem of the field spinning out of control. As mentioned previously, a system may spin out of control as the interactively complex system generates unmanageable processes that drive the system to unanticipated and unacceptable end states. By chunking technologies, an entrepreneur gains greater control over a potentially chaotic process. This happens because their position is at the center of an overall architecture the entrepreneur orchestrates. As others gain access to some chunks, the entrepreneur can begin developing and deploying additional chunks.

Virtuous Cycle. Indeed, Silver continued to orchestrate the process by being at the center of an emerging technological field. Silver's tenacity and an opportunity to associate and bisociate at 3M led to the building up of a momentum as minds and molecules were mobilized. In Silver's words: It was more like a *slow crescendo* of things, typical of the discovery process. Things build up and you *begin to see the options the discovery creates.* (from Lindhal, 1988, p. 14; italics added)

Among the many others who played key roles in the Post-it® Notes saga were Nicholson and Ramey from the marketing division. Nicholson and

Ramey also experienced some of the same struggles as people before them had encountered. To translate the idea, Nicholson and Ramey hit on the idea of offering free samples to others to play with. Eventually, because of Nicholson and Ramey's efforts, Lew Lehr, CEO of the company, was enlisted. In turn, Lehr was successful in enlisting other CEOs. Reflecting on this process of translation where a glue that did not glue was eventually conceptualized as offering a key business opportunity, Fry commented:

> There are so many hoops that a product idea has to jump through. It really takes a bunch of individuals to carry it through the process. It's not just a Spence Silver or an Art Fry. *It's a whole host of people.* It's a classic 3M tale. I couldn't have done what I did without Silver. And without me, his adhesive might have come to nothing. (from Lindhal, 1988, p. 17; italics added)

We can only imagine the number of times that molecules and minds were translated within and outside 3M before they all became coaligned to structure a world where Post-it® Notes have become taken for granted. What is apparent in the attributions implicit in the accounts of those involved in the process is that the innovation rightly belonged to a number of people associated with a process that unfolded over a long period of time. This is invariably the case with most acts of entrepreneurship (Braun & Macdonald, 1978; Latour, 1987). Consequently, it is important to conceptualize human agency as a relational concept, one that recognizes entrepreneurship as a collective enterprise and as an outcome of processes that translate and mobilize heterogeneous elements to generate a technological field. Those who attempt to create new paths have to realize they are part of an emerging collective and that core ideas and objects will modify as they progress from hand to hand and mind to mind. Eventually, what may emerge from these processes may be very different from what was conceptualized initially (see Porac, Rosa, Spanjol, & Saxon, chap. 8, this volume, for the emergence of a consensual system between many constituencies that led to features of the automobile that are used). In this regard, there are accidents, but they are a series of cultivated breakthroughs waiting and planned to occur, each breakthrough setting the stage for another in an overall process of cumulative synthesis. In such a virtuous cycle, "normal" accidents have a positive connotation, as compared to the negative one implicit in the use of the term by Perrow (1984) to describe a vicious cycle.

In sum, path creation, in the case of Post-it® Notes, involved the disembedding of an individual from localized structures of relevance and provinces of meaning, overcoming the inertia and momentum that he encountered, mobilizing others to work on an idea that was transformed over time, all the while being flexibly resolute with a vision of what might be possible. Silver and his colleagues were successful in setting in motion a virtuous, coevolutionary process (Masuch, 1985). Not only were they successful in mobilizing the minds of people, but they were also able to mobilize the molecules that constitute Post-it® Notes.

Mobilizing Time

We could end here with our story of Post-it® Notes as a revelatory case for explicating our perspective on path creation as a process of mindful deviation. To do so, however, would be to miss out on an opportunity to dwell on an important facet of path creation implicit in the development of Post-it® Notes and one that is present in every entrepreneurial initiative—the role of time.

Post-it® Notes did not emerge overnight—it took 12 years from Silver's first discovery before Post-it® Notes were mass manufactured. Some question how such a process could have taken so much time. Notwithstanding this debate, it is apparent that path creation, as a process, must be thought of as unfolding over time.

Elucidating the importance of time as a resource underlying the unfolding of these coevolutionary processes, Silver suggested:

> things don't happen all of a sudden. *It's a process.* You're in the process of doing experiments. You're getting analytical data, sending samples off to different groups. These groups give you analytical feedback and you do some more experiments. *It all takes time.* (from Lindhal, 1988, p. 15; italics added)

Indeed, one theme that appears repeatedly in all accounts of the development of Post-it® Notes is the need for, and an ability to, marshal time as a resource. Here, it is easy to connect with Schumpeter's (1942) views on time as a resource with which, and within which, entrepreneurship flourishes. There are many others who have recognized the importance of time as a key resource for entrepreneurship, including Francis Bacon (1625) who recognized the importance of time when he suggested in his essay titled *On Innovation*, "As the births of all living creatures are, at first, misshapen, so are all innovations...." Extending Bacon's metaphor, it takes time for a caterpillar to become a butterfly, and the transformation process is clearly not straightforward.

It is with this recognition of time as a resource that 3M has wisely chosen to institutionalize the importance of imagining the future in order to create it (Coyne, 1996). In many instances, managers are chided for taking short timeframes. Such a focus on time as a resource is despite or because of 3M's status as a large corporation that produces many products that must meet the needs of the marketplace. Perhaps, employees at 3M are intuitively aware of what March (1998) suggested—long timeframes are key to the exploration of ideas.

Time, Timing, and Temporality.   These discussions have important implications for path creation. Specifically, entrepreneurs anchored in today's business practices are less likely to gain the generative impulse to explore. We can gain additional insights by reversing the relation between time and exploration. Specifically, any exploratory act requires an appropriate timeframe within which, and with which, novelty emerges. Combining these two propositions generates a third—timeframe and degrees of novelty must be matched. If too little time is slotted for the deviation, it will either be half done or may not be perceived as an interesting novelty to anyone involved. In contrast, too much time may result in trivialization of the idea or in a situation where those attempting to create a path are unable to generate the necessary momentum required to get the project through (Hughes, 1983).

There is a connection between segments of time that entrepreneurs might mobilize and their status as boundary spanners (Garud & Ahlstrom, 1997). An outsider is likely to have a short timeframe. As a consequence, they are less likely to explore. The insider, however, is likely to work within long timeframes. Whereas such a perspective provides a generative impulse to explore, it may also result in an escalation of commitment to a failing course of action (Staw, 1976).

Entrepreneurs who employ a boundary spanner's perspective may more likely mobilize an appropriate chunk of time consistent with the scope of their deviation. Moreover, if one were to adopt such a perspective, it is possible to view the overall process as a series of small experiments. The feedback from each experiment serves as the basis for modifying the original idea, even as additional champions are mobilized. Tenacity as a boundary spanner, then, addresses the thin line between persistence and undue persistence.

An ability to mobilize time as a resource offers another key benefit that has to do with timing and temporality (Schutz, 1973; March, 1991b). To mobilize time implies an ability to call on history in strategic ways (see Mouritsen & Dechow, chap. 13, this volume). It also implies an ability to evoke images of the future in strategic ways (March, 1998; see Lampel, chap. 11, this volume). When entrepreneurs mobilize time in this manner, it becomes a friend rather than an enemy. Time (and timing) reduces downside risk and prevents needless deployment of resources. Time becomes a resource that offers entrepreneurs options to strike at the right time and place. A manager at 3M explained:

What does it all mean? Among other things, it means living with a paradox: persistence support for a possible idea, but *not foolishly overspending* because 3M, above all, is a very pragmatic company. It typically works this way: The champion, as his idea moves out of the very conceptual stage and into prototyping, starts to gather a team about him. It grows to say 5 or 6 people. Then, suppose, (as is statistically the likely case) the program hits a snag. 3M will likely cut it back quickly, knock some people off the team. But as the mythology suggests, the champion is—if he is committed—is encouraged to persist, by himself or perhaps with one co-worker, at say a 30% or so level of effort. In most cases, *3M has observed that the history of any product is a decade or more long before the market is really ready.* So the champion survives the ups and downs. Eventually, often the market does become ripe. His team rebuilds. (from Peters & Waterman, 1982, p. 230; italics added)

Coevolution of Minds and Molecules Over Time

Implicit in the conception of path creation is strategy as bricolage (Karnøe & Garud, 1998; Garud & Karnøe, 1999). *Bricolage* embodies loose coupling between actions and structure (Giddens, 1984), wherein actors probe their worlds, even as they create it, through local negotiation processes to spawn global orders. When we allow for practical experimentation coupled with thoughtful modifications (a process of bricolage), we allow the evolution of a technological field in an emergent way (Karnøe, 1996). In this conceptualization, actors navigate a flow of events by being mindful of when to persist and when to desist, when to credit and when to discredit, and when it might be possible to make changes in the boundary conditions, all the while aware they are placing bets, the outcomes of which can be only described in probabilistic terms (see also Van de Ven, Polley, Garud, & Venkataraman, 1999).

Indeed, 3M appears to be a place where bricolage is encouraged, as evidenced by these observations:

Our approach is to make a little, sell a little, make a little bit more ... Big ends from small beginnings ... spend just enough money to get what's needed next to incrementally reduce ignorance ... lots of small tests in a short interval ... development is a series of small excursions ... . (from Peters & Waterman, 1982, p. 231; italics added)

This statement provides an opportunity to clarify what we mean by mindfulness. In using the concept for specifying agency, we do not wish to imply that entrepreneurs' minds are full of details corresponding to an unyielding vision of the future. Instead, by *mindful*, we mean that entrepreneurs are conscious of their embeddedness and are able to depart from, and indeed, employ, embedding structures in meaningful ways. One 3M vice-president lucidly suggested:

We don't constrain ourselves with plans at the beginning when ignorance is highest. Sure, we plan. We put together meticulous sales implementation plans. But that's after we know something. At the very front end, why should we spend time writing a 250 page plan that tries to drive out ignorance before having first done some simple tests on customer premises or in a pilot facility somewhere. (from Peters & Waterman, 1982, p. 232; italics added)

Thus, fully formed plans and visions are not preconditions for entrepreneurial action. Instead, plans and visions emerge as a part of the entrepreneurial process. Appreciating the seemingly irrational sentiment of this position and recognizing its power in the entrepreneurial process, March (1971) aptly suggested that entrepreneurial insights may arise from a technology of foolishness. Indeed, one can hear Schumpeter's voice echoing our use of mindfulness:

> The assumption that business behavior is ideally rational and prompt, and also that in principle it is the same with all firms, works tolerably well only within the precincts of tried experience and familiar motive. It breaks down as soon as we leave those precincts and allow the business community under study to be faced by not simply new situations, which also occur as soon as external factors unexpectedly intrude but by new possibilities of business action which are yet untried and about which the most complete command of routine teaches nothing. (Schumpeter, 1939, Vol. I, pp. 98–99)

Art Fry offered this insight about 3M as an example of a company that encourages bricolage:

> At 3M we've got so many different types of technology operating and so many experts and so much equipment scattered here and there, that *we can piece things together when we're starting off.* We can go to this place and do "Step A" on a product, and we can make the adhesive and some of the raw materials here, and do one part over here, and another part over there, and convert a space there and make a few things that aren't available. (from Nayak & Ketteringham, 1986, pp. 66–67; italics added)

The process of mindful reuse and recombination of resources embedded in technological fields is similar to those offered by other researchers. For instance, Prahalad and Hamel (1990) alluded to the emergence of new competencies from a combination of others. Recognizing the challenges of navigating through complexity, others have offered notions such as the science of muddling through (Lindbloom, 1959) or, logical incrementalism (Quinn, 1978). Mintzberg, Raisinghani, and Theoret (1976) are process proponents who recognized the importance of bricolage in dealing with emergent strategies. In a similar vein, Burgelman's (1983) work offered considerable insights on autonomous approaches in contradistinction to the idea of induced approaches. More recently, Brown and Eisenhardt (1997) offered observations on how product development efforts can unfold in an emergent fashion within minimal structures across product generations. Weick (1999) offers improvization as a way of navigating and shaping emerging processes.

DISCUSSION

Entrepreneurs confront a complex flow of events where outcomes are seldom predetermined. To gain some agency in navigating and shaping the flow of these events, we offered a perspective on path creation processes. In such unfolding processes, agency is gained by endogenizing time, relevance structures and objects. More precisely, entrepreneurship requires an ability to span boundaries of relevance structures, translate objects and mobilize time as a resource. As entrepreneurs endogenize time, relevance structures, and objects, they generate power to strategically manipulate and mobilize these elements. Path creation, then, is the binding of objects, relevance structures, and time into an overall co-evolutionary process.[19]

Our perspective is not a recipe for entrepreneurial success. We offer a perspective based on a process logic of mindful deviation rather than on a variance logic of consequentiality. Ironically, an exclusive focus on outcomes can mute feedback generated during the entrepreneurial journey, thereby reducing the likelihood of obtaining a favorable outcomes (Garud & Karnøe, 1999). Entrepreneurs navigate a complex flow of events in real time, fully aware that success and failure are the two sides of the same entrepreneurial coin (Pinch & Bijker, 1987). Our process perspective on mindful deviations suggests that failures and accidents are powerful learning stimuli if entrepreneurs do not lock themselves into a logic of consequentiality (March, 1997).

Indeed, our perspective acknowledges there are many constraints on human agency associated with entrepreneurship. For instance, unfolding structurational processes suggest that entrepreneurs are creatures caught in webs of significance of their own making (Geertz, 1973). In this context, agency involves the ability to discredit and disembed from the structures that enable and constrain entrepreneurs.

Moreover, attempts at disembedding are likely to generates vicious or virtuous coevolutionary cycles. *Vicious* coevolutionary cycles are generated when negative feedback dampens entrepreneurial initiatives or when positive feedback generates so much momentum that the process spins out of control. Agency, from this perspective, requires the deployment of social skills in ways Fligstein suggested, embodied in a readiness on the part of entrepreneurs to present and modify ideas to create a shared collective space.

In summary, our perspective should be viewed as one that sensitizes those attempting to create paths to the dimensions that embed them, and from

---

[19]Both success and failure must be explained with the same model (Pinch & Bijker, 1987). We believe efforts at creating paths succeed when there is a binding of objects, relevance structures, and time tied into an overall coevolutionary process that results in the emergence of a technological field. Failures are more likely to occur when entrepreneurs are unable to create these linkages. They are not necessarily a result of an intrinsic property of an artifact.

which they need to disembed even as they mobilize. We cannot prescribe to what extent entrepreneurs should deviate from existing objects, nor can we say precisely how they should mobilize time, or which specific boundaries they should span and when. These are the challenges entrepreneurs must grapple with (in a mindful way) as they deviate from existing technological fields.

Implications

Our perspective has several implications for entrepreneurship. Consider learning, for instance. One view is that entrepreneurs should open themselves up to feedback. Another is that entrepreneurs should close themselves to feedback because entrepreneurial acts imply departing from existing embedding structures. Clearly, there is tension between these positions as captured by the tension between commitment and flexibility (Ghemawat, 1991).

Path creation suggests a mid-ground. As entrepreneurs chunk up objects, time, and relevance structures, they create a series of chain linked *deviation steps*. Each deviation step explores a deviation with a matched timeframe and relevant social groups. Having initiated a deviation step, entrepreneurs may shield themselves from negative feedback to make progress, and generate momentum. Once a deviation step is completed, entrepreneurs are more receptive to feedback and may reassess progress and plan modifications of subsequent deviation steps. Indeed, an appreciation of the tension between learning and creation is evident in the following statement by a CEO of Excite: "We don't worry as much about making the right decisions as we worry about making the decisions right."

A key question is—How large should these deviation steps be? One answer is to keep them as small as possible to avoid escalation of commitment yet large enough to gain meaningful feedback. Such a process embraces a real options approach to the navigation of complex dynamic flow of events (Kumaraswamy, 1996).

Our perspective has implications for other facets of entrepreneurship. For instance, entrepreneurship is not a negation of the past nor its simple extrapolation. It is a reconstitution and transformation of the past in such a way that continuity and change are both preserved in the act of path creation. That is, entrepreneurs are always attempting to disembed from structures that they are embedded in while reusing some of the rules and resources.

A corollary to this observation is that entrepreneurship is not a random act of genius but is a disciplined effort involving many. Entrepreneurs have to work with others by coopting them into a collective process. Here, the notion of agency is a relational one where credits belong to many who offered their input over a period of time.

Path creation also underscores the fragility of stability. To appreciate this, we return to path dependence. An important claim in path dependence literature is that technological fields become locked in a trajectory because of increasing returns (Arthur, 1988; David; 1985). Consequently, from a path dependence perspective, there are many insurmountable, first-mover advantages. However, contemporary phenomena suggest that second-movers not only catch up, but race ahead of first-movers. For instance, Microsoft has been able to match and eventually overcome Netscape's first-mover advantages with Internet browsers. Similarly, in the case of cochlear implants, a biomedical prosthetic device, multichannel implants could catch up with single-channel implants despite FDA approval and its lead in the marketplace (Garud & Rappa, 1994).

Our perspective on path creation has implications for how entrepreneurs might design embedding dimensions of technological fields to set in motion self-organizing processes (March, 1991; Nonaka, 1994; Stacey, 1993). For instance, they can manipulate the level and type of resources deployed for exploration, the number and kinds of rules that are in play, the flexibility in the interpretation of rules, rules for changing the rules, and the like. In addition, the type of coupling between activities is another strategic variable (Weick, 1976). Specifically, loose coupling between activities sponsors coevolutionary dynamics where there are slippages in time and space between actions in one arena, and activities and actions in another. Manipulating these dimensions of technological fields can shape entrepreneurial processes such that outcomes are neither random nor determined, but a result of path creation processes.[20]

CONCLUDING COMMENTS

We began the chapter with a process perspective on the genesis of novelty—path dependence. We described how path dependence highlights the role played by history in the genesis of novelty. We also noted how it falls short of conceptualizing the roles of actors in creating history in real time. In fact, in many studies, the role of human agency in the generative process is ignored. It is to address this lacuna that we articulated path creation.

History is important for path creation. However, the place and role of history changes. In path dependence, temporally remote events shape the emergence of novelty. With path creation, attention is focused on the efforts of entrepreneurs who seek ways to shape history in the making. First, they offer strategic interpretations of history. Second, they actively shape

---

[20]See Garud & Jain (1996), Dooley & Van de Ven (1999), and Baum & Silverman (2000) for descriptions of different embedding states.

emerging structures of relevance and objects, and, in the process, leave an imprint on development efforts. Third, they evoke images of the future to make history in a self-fulfilling manner.

Acknowledging entrepreneurship as path creation reminds us that entrepreneurs are well aware of history and know they cannot do whatever they chose. On the contrary, entrepreneurship requires an appreciation that any effort is part of a larger ongoing and evolving process. To shape and influence these processes, entrepreneurs locate themselves at the boundaries of objects, relevance structures, and time. We conceptualize the entrepreneurial role as mindful deviation for it is the entrepreneur who breaks away from the constraints imposed by accepted approaches and articulates, then promotes, new alternatives.

INTRODUCTION TO THE CHAPTERS

> The living language is like a cowpath: it is the creation of the cows themselves, who, having created it, follow it or depart from it, or depart from it according to their whims or their needs. From daily use, the path undergoes change. A cow is under no obligation to stay in the narrow path she helped make, following the contour of the land, but she often profits by staying with it and she would be handicapped if she didn't know where it was or where it led to (White, 1957).

The chapters in this volume approach path dependence and path creation from different disciplinary perspectives including evolutionary economics, institutional theory, complexity theory, technology sociology, and organizational sociology. They address these issues at different levels of analyses ranging from the development of regions such as Silicon Valley to the development of theoretical perspectives themselves. Many provide detailed accounts of unfolding processes to illustrate their points on path dependence and path creation. Individually and collectively, the chapters represent a unique set of articles that discuss and debate issues surrounding path dependence and path creation.

Following are brief overviews of the chapters. Our objective is not to repeat arguments in the chapters but to provide a sense of how each chapter adds to emerging views on path creation. All chapters claim to depart from path dependence, yet, some depart more than others. We organized the book to reflect this progressive shift. Moreover, the arrangement of chapters also served the purpose of introducing the reader to a deeper understanding of path dependence in the economics literature before grappling with other literature streams.

Part I, Path Dependence and Beyond, sets the stage for understanding of path dependence in economics literature, and how and why we must extend our understanding of human agency involved in shaping the emergence of novelty. Part II, From Path Dependence to Path Creation, departs from the epistemological and ontological positions implicit in path dependence. Part III, Path Creation as Coevolution, offers chapters that provide an appreciation of coevolutionary processes that entrepreneurs have to manage in their efforts to create new paths. Finally, Part IV, Path Creation as Mobilization, explores how entrepreneurs might endogenize objects, relevance structures, and time in their efforts to create new paths.

Path Dependence and Beyond

In chapter 2, When and How Chance and Human can Twist the Arms of Clio?, Bassanini and Dosi explicate issues around path dependence as they appear in economics literature. In doing so, the authors highlight the overtones of determinism implicit in the path dependence perspective. Specifically, proponents of the path dependence perspective appear to place too much emphasis on initial conditions in shaping the emergence of novelty; they often neglect the power that chance events and human will can play in unlocking paths. Arguing for a stochastic approach, Bassanini and Dosi offer several forces that might result in unlocking paths. These forces include: (a) new technological paradigms, (b) heterogeneity among agents, (c) coevolutionary nature of socioeconomic adaptation, and (d) invasion of new organizational forms from other contexts.

In chapter 3, Unpacking Path Dependence, Hirsch and Gillespie point out there is a long, rich tradition in the social sciences of examining the role of history in shaping contemporary phenomena. Hirsch and Gillespie suggest that our current emphasis on contextual, historical, and evolutionary perspectives in the social sciences is a return to theories that dominated in the late 1990s that were bypassed for perspectives that celebrated the rational actor (and others) that were built around structural perspectives. The authors suggest scholars of innovation and technology should incorporate and integrate differential valuations accorded history and temporality across social science disciplines, especially from anthropology, economics, history, management, political science, and sociology. An important contribution of chapter 3 is that an awareness of differential weights accorded to history by each discipline can potentially liberate proponents of each discipline from path dependencies. Stated differently, an understanding of the metaframing implicit in each theoretical perspective is an important first step for scholars to generate agency in their abilities to create new theoretical paths. In the end, Hirsch and Gillespie argue for path as process, meaning technology is a forever emergent, nonrecursive product of path creation, path destruction, and path dependence.

In chapter 4, Ruttan addresses the economics discipline to trace path dependence's historical development. Ruttan's work on induced innovation models of change has inspired many to think about, and articulate, positions on path creation (cf. Ruttan, 1979). In Sources of Technical Change: Induced Innovation, Evolutionary Theory, and Path Dependence, Ruttan identifies critical junctures in the development of theories of technological change and locates path dependence in the larger mosaic associated with economic theory development. He illustrates how there has been different uses of history in the economic discipline and suggests how and why we must think of path creation models as we move forward.

From Path Dependence to Path Creation

In chapter 5, Kenney and Burg apply the path dependence perspective to study the evolution of Silicon Valley. Consistent with path dependence, the authors' explanation highlights the sensitivity of unfolding processes to initial conditions. At the same time, Kenney and Burg extend the traditional path dependence model by exploring how institutions such as venture capital, law firms, and marketing firms coevolved. Kenney and Burg suggest that this institutional context of resource formation and firm structuration is a process that is continually evolving with technologies and industrial forms. A key contribution is to illustrate multiple path dependent processes at work, each shaping the other. Path dependence, if constructed as a relatively one-dimensional model, does not do justice to the way Silicon Valley emerged as an ecosystem.

In Standards, Modularity, and Innovation: The Case of Medical Practice, Langlois and Savage (chap. 6), discuss a familiar class of increasing return processes: the setting of standards. In this case, however, the standards are not common technological ones, but rather, are standards that are understood as behavioral routines. Standardized routines are, in fact, a well-known form of social institution that gives rise to increasing returns. The case of setting medical standards involves the coordinative standards that have guided the medical profession and the normative standards that assure quality. Langlois and Savage argue that, because of the overwhelming efficiency advantages of a decentralized professional structure, the medical standards of the early 1900s were rules and routines that guided local practitioners, rather than a top-down monitoring system (e.g., through hospitals or professional associations with strong central authority). The chapter argues that this was a desirable system because it proved open enough to allow rapid learning of skills and the invention of new practices and technology. This instance of standard-setting was fundamentally a case of path dependence because alternatives were imagined and tested. This is not, however, a case much like QWERTY is supposed to have been, in that—in the authors' view—it was not minor changes in initial conditions that tipped the balance, but rather, a clear-cut advantage to the system adopted—advantages of both static and dynamic sorts. It is only now, with the changing architecture of revenue in health care that the decentralized, institution-based system of normative standards may have given way to centralized monitoring in the era of managed care.

Baum and Silverman (chap. 7) apply complex adaptive systems theory to technological evolution. To do so, they shift attention from competitive outcomes (i.e., content) to innovation trajectories (i.e., process). Baum and Silverman describe how innovation trajectories produced by competitive interorganizational systems can be related to concepts from complexity theory, and illustrate the range of possible innovation trajectories (or macrostructures)—ordered, chaotic, and random—that such systems can produce. They highlight the tension between exploitation of knowledge gained (path dependence) and exploration of novel actions (path creation) and describe how innovation processes characterized by chaotic behavior balance these tensions, permitting adaptive functioning of competitive interorganizational systems. Can innovation process dynamics thwart lock-in or provoke de-locking on such technological fitness landscapes? Baum and Silverman suggest that it should be possible to tune the innovation process to avoid the danger of becoming trapped in poor local optima. For example, lock-in may be avoided by partitioning the problem into subtasks, each of which optimizes while ignoring the effects of its actions on the problems facing other subtasks. Subtask boundaries permit constraints from other subtasks to be ignored, helping avoid becoming trapped on poor local optima. Overall performance arises as collective, emergent behavior of the interacting, coevolving subgroups. Such coevolutionary problem solving is not useful for simple problems but increases in value as landscapes become less differentially rugged. This proposal is equivalent to recommending that organizations facing difficult problems divide into departments, profit centers, and other quasiindependent suborganizations to improve performance. In summary, Baum and Silverman believe research on tuning innovation process dynamics to technological problem domains will provide basic new insights on technological evolution.

Path Creation as Coevolution

In chapter 8, America's Family Vehicle: Path Creation in the US Minivan Market, Porac, Rosa, Spanjol, and Saxon, argue that markets are fundamentally sociocognitive in nature. Markets are created when potential buyers and sellers connect around an artifact, and in the process, represent the artifact as a conceptual system that defines its attributes, uses, and value.

Porac and his colleagues suggest that a sociocognitive conceptualization of markets provides a robust frame for answering theoretical questions that have, so far, been intractable. For example, how and when is a market created? In a sociocognitive perspective, such conceptualizations emerge and stabilize through conversations and narratives between producers and consumers; specific artifacts and behaviors become associated with consensually understood market categories. Even as they study the creation of new markets, the authors claim that a weak form of path dependence shaped the automakers choices of car design and use-situations when the minivan market was recreated in the early 1980s. For instance, new product markets were based on a mobilization and reuse of old product categories form the 1940s. Porac and his colleagues suggest processes involved with path creation by addressing questions such as "How do market categories evolve and change and how and why do product categories die?" Porac and his colleagues suggest that change occurs in the knowledge structures to which markets cohere such that new attributes become associated with existing artifacts or new artifacts become assimilated into existing structures. Categories die when a market's underlying knowledge structures no longer cohere in a meaningful, profitable way.

In their study of the construction of new paths in the automobile and biotechnology industries, Rao and Singh (chap. 9) focus on the political–institutional processes whereby goals, authority structures, technological artifacts, and consumers are mobilized to create paths. In their chapter, The Construction of New Path Creation: Institution-Building Activity in the Early Automobile and Bio-Tech Industries, Rao and Singh turn away from random theory to embrace a institutional cultural-frame. Such a perspective is coevolutionary, one where social agents have some power to generate new paths. Their study illustrates an important point about emerging technological fields: that technology, preferences, and social groups do not pre-exist. Indeed, new technological fields are realized through a process of mobilization and testing, and settling controversies among engineers making the hardware function according to evaluation standards, that emerge in a coevolutionary manner. The process of mobilization involves building legitimacy among involved social groups where settling controversies produces temporal closure around a new technology.

In chapter 10, Constructing Transition Paths Through the Management of Niches, Kemp, Rip, and Schot explore how political intervention can create new technological paths. Their key concept, *strategic niche management*, focuses on the role played by state regulators to create and nurture technological paths. To do so, state regulators must shape coevolutionary processes associated with the emergence of artifacts, user groups and institutional rules. Niches are protected spaces where entrepreneurs reuse accumulated knowledge and capabilities where regular market conditions do not apply because of special R&D and market subsidies. Gradually, protection may be erased and real market conditions introduced. Energy and technology policies that shaped the emergence of the wind turbine fields in Denmark and the United States between 1974–1990 is an empirical basis for these recommendations.

Path Creation as Mobilization of Resources

Chapter 11, by Lampel, Show-and-Tell: Product Demonstrations and Path Creation of Technological Change, offers technological dramas as an approach to shape relevance structures to generate momentum for a new technology. Lampel uses historical case studies to examine processes underlying these technological dramas. Technological dramas may trigger collective adaptive expectations around a technological trajectory. Lampel suggests that innovation success depends on bridging the specialized domains of inventors on the one hand and the larger world of investors and consumers on the other. This happens by appealing more to the noncalculative part of the human mind (affect, imagination, or fantasy), so choices are based on a commitment to the future, as much as a proper evaluation of the present. Through such dramas, technology entrepreneurs attempt to initiate a bandwagon to jumpstart generation of a technological trajectory. These dramas generate images of technologies that circulate, through private and public channels of communications, and give shape to the identity of the new technology—even before it has been established as an accepted part of economic reality.

In chapter 12, Innovation as a Community-Spanning Process: Looking for Interaction Strategies to Handle Path Creation, Van Looy, Debackere, and Bouwen examine microlevel process associated with boundary spanning between and across communities of practice. Communities of practice are characterized by shared beliefs, evaluation routines and artifacts. Consequently, they create powerful path dependencies that might inhibit path-breaking innovations. To understand the effect of boundary spanning actions on path dependence and path creation, Van Looy and his colleagues suggest paying closer attention to the antecedents and the consequences of microlevel interaction patterns between communities of practitioners. At this level of analysis, the fragility of stability becomes clearer. Indeed, the authors' empirical study shows how a spectrum of community spanning interaction patterns lie at the origin of path creation processes.

Mouritsen and Dechow's chapter 13, Technologies of Managing and the Mobilization of Paths, illustrates how world class supplier relationships produce new practices and organizational rules in two firms that they study.

Based on Giddens structuration and actor-network theory, they suggest that world class, as a concept, does not have *apriori* meaning, and that it has to be gradually defined through translation processes. Thus, world class must be given meaning through action. They demonstrate how firms' past histories are mobilized and constituted as part of this process. History is interpreted and reinterpreted by organizational actors in strategic ways. Indeed, translation processes involve the bridging of boundaries that key stakeholders gain a voice in the emerging network, and in the process, define what is world class through their interactions. History, as an interpretation of the past, becomes a key resource that is drawn up on even as it is being made.

Pinch (chap. 14) uses the social construction of technology (SCOT) literature in his study of the early emergence of electronic music synthesizers, Why go to a Piano Store to Buy a Synthesizer: Path Dependence and the Social Construction of Technology. Pinch's follow-the-actor approach generates insight to facets associated with the emergence of a synthesizer. His study illustrates that, as with other technological fields in the making, the technology, customer preferences, and relevant social groups do not preexist. In this regard, it is interesting to note that even the inventors of the synthesizer, Moog and Buchla, had very different visions. Indeed, the new technology emerged through a coevolutionary process involving a heterogeneous set of objects and people. Pinch describes how the new type of sound generated by a synthesizer was perceived as "weird shit" because it deviated from the existing notions of what was considered to be music. Pinch's description provides an appreciation of the importance of time for this weird sound to become accepted by a larger social group. During this timeframe, many actors played a role in using synthesizers. Pinch's description is a story of path creation, one where reflective entrepreneurs mindfully tried to navigate a flow of events that they attempted to shape. Indeed, Pinch's study demonstrates the interactive nature of the three dimensions introduced earlier—objects, relevance structures, and time—within which actors are embedded, and from which they disembed.

The End of a New Beginning

Each chapter is richer than the descriptions we offer. Each contributor has taken a process view, one where it is important to accord some agency to humans in their abilities to shape the emergence of novelty in real time. Together, the chapters represent a mosaic of ideas that build a perspective of path creation as a process of mindful deviation. Indeed, the chapters are an inspiration to all of us encouraging us to engage in more research that will enrich our understanding of processes associated with the emergence, stabilization, and erosion of paths.

ACKNOWLEDGMENTS

We have benefitted from our discussions with Kristian Kreiner, Paul Hirsch, and Roger Dunbar. We have also benefited from input offered by participants at the Path Dependence and Creation workshop in Denmark, 1997; Path Creation workshop in Maastricht, 1999; Samples for the Future SCANCOR Conference in Stanford, 1998; and the Knowledge Development Workshop CISTEMA in Denmark, 1999. We gratefully acknowledge the generous support from CISTEMA.

REFERENCES

Amabile, T. M. (1996, Jan.). The motivation for creativity in organizations. *Harvard Business Review*, pp. 1–14.
Astley, G. W. (1985). The two ecologies: Population and community perspectives on organizational evolution. *Administrative Science Quarterly*, 30, 224–241.
Arthur, B. (1994), *Increasing returns and path dependence in the economy*. Ann Arbor: The University of Michigan Press.
Arthur, B. (1996, Jul–Aug). Increasing returns and the new world of business. *Harvard Business Review*, pp. 100–109.
Arthur, B. (1988) Self-reinforcing mechanisms in economics. In P. Anderson et al. (Eds.), *The economy as an evolving complex system*. Reading, MA: Addison-Wesley.
Baum, J. A. C., & Silverman, B. (in press). Complexity, attractors, and path dependence and creation in technological evolution. In R. Garud & P. Karnøe (Eds.), *Path dependence and creation*. Mahwah, NJ: Lawrence Erlbaum Associates.
Baum, J., & Singh, J. V. (1994). Organizational niches and the dynamics of organizational founding. *Organization Science*, 5 (4), 483–501.
Berger, P., & Luckmann T. (1967). *The social construction of reality*. London: Penguin.
Bijker, W. E., Hughes T. P., & Pinch. T. J. (1987). *The social construction of technological systems*. Cambridge, MA: MIT Press.
Blumer, H. (1969). *Symbolic interactionism: Perspective and method*. Englewood Cliffs, NJ. Prentice-Hall.
Boisvert, R. (1998). *John Dewey: Rethinking our time*. State University of New York Press.
Braun, E., & Macdonald S. (1978). *Revolution in miniature*. New York: Cambridge University Press.
Brown, J. S., & Duguid, P. (1991). Organizational learning and communities of practice: Toward a unified view of working, learning and innovation. *Organization Science*, 2, 40–57.
Brown, S. L., & Eisenhardt, K. M. (1998). *Competing on the edge: Strategy as structured chaos*. Boston: Harvard Business School Press.
Burgelman, R. A. (1983). A process model of internal corporate venturing in diversified major firms. *Administrative Science Quarterly*, 28, 223–224.
Callon, M. (1986). The sociology of an actor–network: The case of the electric vehicle. In M. Callon, J. Law, & A. Rip (Eds.), *Mapping the dynamics of science and technology* 19–34. Hampshire, London: Macmillian.
Christensen, C. M. (1997). *The innovator's dilemma: When new technologies cause great firms to fail*. Boston: Harvard Business School Press.
Coyne, W. E. (1996). Building a tradition of innovation, *The Fifth UK Innovation Lecture*. London: Department of Trade and Industry.
Cyert, R., & March, J. G. (1963). *A behavioral theory of the firm*. Englewood Cliffs, NJ: Prentice-Hall.
David, P. (1985). Clio and the economics of QWERTY. *Economic History*, 75, 227–332.
Dewey, J. (1934). *Art as experience*, p. 51. New York: Minton, Balch.
Dierickx, I., & Cool, K. (1989). Asset stock accumulation and sustainability of competitive advantage. *Management Science*, 35 (12), 1504–1511.

Dooley K. J., & Van De Ven, A. H. (1999). Explaining complex organizational dynamics. *Organization Science, 10*(3), 358–372.

Dosi, G. (1982). Technological paradigms and technological trajectories. *Research Policy, 11*, 147–162.

Dougherty, D. (1992, May). Interpretive barriers to successful product innovations in large firms. *Organization Science, 3*(2), 179–202.

Dunbar, R., Garud, R., & Raghuram, S. (1996). Deframing in strategic analyses. *Journal of Management Inquiry, 5*(1) 23–34.

Farrell, J., & Saloner, G. (1986). Installed base and compatibility: Innovation, product preannouncements and predation. *American Economic Review, 76*, 940–955.

Feldman, M., & March, J. G. (1981). Information in organizations as signal and symbol. *Administrative Science Quarterly, 26*, 171–186.

Fligstein, N. (1997). Social skill and institutional theory, *American Behavioral Scientist, 40*(4) 397–405.

Garud, R., & Ahlstrom, D. (1997). Technology assessment: A socio-cognitive perspective. *Journal of Engineering and Technology Management, 14*, 25–48.

Garud, R., & Jain, S. (1996). Technology embeddedness. In J. Baum & J. Dutton (Eds.), *Advances in Strategic Management, 13*, 389–408. Greenwich, CT: JAI.

Garud, R., & Karnøe, P. (1999). When complexity is a problem, bricolage may be a solution. *CBS/NYU working paper.*

Garud, R., & Kumaraswamy, A. (1995). Technological and organizational designs to achieve economies of substitution. *Strategic Management Journal*, Vol. 16, pp. 93–110.

Garud, R., & Nayyar, P. (1994). Transformative capacity: Continual structuring by inter-temporal technology transfer. *Strategic Management Journal, 15*, 365–385.

Garud, R., & Rappa, M. (1994). A socio-cognitive model of technology evolution. *Organization Science, 5*(3), 344–362.

Geertz, C. (1973). *The interpretation of cultures.* New York: Basic.

Ghemawat, P. (1991). *Commitment: The dynamic of strategy.* New York: The Free Press.

Giddens, A. (1979). *Central problems in social theory.* Los Angeles: University of California Press.

Giddens, A. (1984). *The Constitution of Society.* Berkeley and Los Angeles: California University Press.

Granovetter, M. (1985). Economic action and social structure: the problem of embeddedness, *American Journal of Sociology, 91*(3), 481–510.

Grove, A. S. (1996). *Only the paranoid survive.* New York: Doubleday.

Hannan, M. T., & Freeman, J. (1977). The population ecology of organizations. *American Journal of Sociology, 82*, 929–964.

Hayek, F. A. (1948). *Individualism and economic order.* University of Chicago Press.

Hedberg, B. (1981). How organizations learn and unlearn. In P. C. Nystrome & W. H. Starbuck (Eds.), *Handbook of Organizational Design,* (Vol. 1, 3–27). New York: Oxford University Press.

Hirsch, P. (1975). Organizational environments and institutional arenas. *Administrative Science Quarterly, 20*, 327–344.

Hirsch, P. M., & Lounsbury M. (1997). Ending the Family Quarrel, *American Behavioral Scientist, 40*(4), 406–418.

Hughes, T. P. (1983). *Networks of power,* Baltimore: The Johns Hopkins University Press.

Katz, M. L., & Shapiro, C. (1985). Network externalities, competition, and compatibility. *American Economic Review, 75*, 424–440.

Karnøe, P. (1993). *Approaches to innovation in modern wind energy technology: Technology policy, science, engineers and craft traditions. No. 334.* Stanford University: Center for Economic Policy Research.

Karnøe, P. (1996). The Social Process of Competence Building. *International Journal of Technology Management, 11*(7/8), 770–789.

Karnøe P., & Garud, R. (in press). Path creation and dependence in the Danish wind turbine field. In J. Porac & M. Ventresca, *The social construction of industries and markets.* Oxford: Pergamon.

Karnøe, P. (1999). The Business Systems Framework and the Danish SME's. In P. Karnøe, P. H. Kristensen, & P. H. Andersen (Eds.), *Mobilizing resources and generating competencies: The remarkable success of small and medium sized enterprises in the Danish business system.* Copenhagen Business School Press.

Karnøe, P., Kristensen, P. H., & Andersen, P. H. (Eds.). (1999). *Mobilizing resources and generating competencies: The remarkable success of small and medium sized enterprises in the Danish business system.* Copenhagen Business School Press.

Kenney, M., & Curry, J. (1999). Knowledge creation and temporality in the information economy. In J. Porac & R. Garud (Eds.), *Cognition, knowledge and organizations* 149–170. Stamford, CT: JAI.

Kirzner, I. M. (1992). *The meaning of market process.* London: Routledge.

Koestler, A. (1964). *The Act of creation.* New York: Macmillian.

Koppl, R., & Langlois, R. (1994). When do ideas matter? A study in the natural selection of social games. In *Advances in Austrian Economics* (Vol. 1, 81–104). Stamford, CT: JAI.

Kreiner, K., & Tryggestad, K. (1997). *The co-production of chips and society: Unpackaging packaged knowledge.* CBS Working Paper, Denmark.

Kumaraswamy, A. (1996). A real options perspective of firms' R&D investments. *Unpublished doctoral dissertation,* New York University.

Latour, B. (1987). *Science in Action: How to follow engineers and scientists through society.* Cambridge, MA: Harvard University Press.

Law, J. (1992). Notes on the theory of the actor–network: Ordering, strategy, and heterogeneity 379–393. *Systems Practice, 5*(4).

Leonard-Barton, D. (1992). Core capabilities and core rigidities: A paradox in managing new product development. *Strategic Management Journal, 13,* 111–126.

Levitt, B. & March, J. G. (1988). Organizational Learning. *Annual Review of Sociology, 14,* 319–340.

Liebowitz, S. J., & Margolis, S. E. (1990). The fable of the keys. *Journal of Law and Economics, 22,* 1–26.

Lindblom, C. E. (1959). The Science of 'Muddling Through.' *Public Administration Review, 19,* 79–88.

Lindhal, L. (1988, Jan.). Spence Silver: A scholar and a gentleman. *3M Today, 15*(1) 12–17.

Luehrman, T. (1998, Oct.). Strategy as a portfolio of real options. *Harvard Business Review,* 89–99.

March, J. G. (1971). The Technology of foolishness. *Civilokonomen, 4,* 4–12.

March, J. G. (1991a). Exploration and exploitation in organizational learning. *Organization Science, 2*(1), 71–87.

March. J. G. (1991b). Learning from experience in ecologies of organizations. *Organizational Science, 25,* 2–9.

March. J. G. (1994). *A primer on decision making.* New York: The Free Press.

March, J. G. (1997). Foreward. In R. Garud, P. Nayyar, & Z. Shapira (Eds.), *Technological innovation: Oversights and foresights.* Cambridge, England: Cambridge University Press.

March, J. G. (1998). Research on organizations: Hopes for the past and lessons from the future. Lecture at the SCANCOR Conference, *Samples for the Future,* September 20–22.

Masuch, M. (1985). Vicious cycles in organizations. *Administrative Science Quarterly, 30,* 14–33.

Mintzberg, H., Raisinghani, O., & Theoret, A. (1976). The structure of unstructured decision processes. *Administrative Science Quarterly, 21,* 246–275.

Mitchell, J. (1940). Women—The longest revolution. In *New Left Review* (Nov–Dec, 1966). London.

Molina, A. (1999). Understanding the role of the technical in the build-up of sociotechnical constituencies. *Technovation, 19,* 1–29.

Nayak, P. R., & Ketteringham, J. M. (1986). *Breakthroughs!* New York: Rawson.

Nelson, R., & Winter S. G. (1982). *An evolutionary theory of economic change.* Cambridge, MA: Harvard University Press.

Nonaka, I. (1994, Feb.). A Dynamic theory of organizational knowledge creation. *Organization Science, 5*(1), 14–37.

North, D. (1990). *Institutions, institutional change and economic performance.* Cambridge, Engalnd: Cambridge University Press.

Pettigrew, A. M. (1992). Character and significance of strategic process research. *Strategic Management Journal, 13,* 5–16.

Pettigrew, A. M. (1992). Character and significance of strategic process research. *Strategic Management Journal, 13*, 5–16.

Polley, D., & Van de Ven, A. H. (1995). Learning by discovery during innovation development. *International Journal of Technology Management, 11*(7/8), 871–882.

Porter A. L., Roper T., Mason T., Rossini F., Banks J., & Wiederholt, B. (1991). Forecasting and management of technology. New York: Wiley.

Peters T. J., & Waterman, R. H. (1982). *In search of excellence.* New York: Harper & Row.

Powell W. W., & DiMaggio P. J. (Eds.). (1991). *The new institutionalism in organizational analysis.* Chicago: The University of Chicago Press.

Quinn, J. B. (1978). Strategic Change: Logical Incrementalism, *Sloan Management Review, 20*(1), 7–21.

Rosenberg, N. (1994). Exploring the black box. Technology, economics, and history. Cambridge, NY: University Press.

Ruttan, V. W. (1979) Induced institutional innovation, *Agricultural Economics Research, 31*(3), 32–35.

Sastry, A. (1997). Problems and paradoxes in a theoretical model of punctuated organizational change. *Administrative Science Quarterly, 42*, 237–275.

Saxanian, A. (1994). *Regional advantage.* Cambridge, MA: Harvard University Press.

Schumpeter, J. A. (1934). *The theory of economic development.* Cambridge, MA: Harvard University Press.

Schumpeter, J. A. (1942). *Capitalism, socialism, and democracy.* New York: Harper & Row.

Schutz, A. (1973). *Collected papers I, The problem with social reality* (4th ed.). The Hague: Martinus Nijhoff.

Schutz, A., & Luckmann, T. (1973). *The Structures of the Life-World.* Evanston, IL: Northwestern University Press.

Stacey, R. (1993). *Strategic management and organizational dynamics.* London: Pitman.

Starbuck, W. (1996). Unlearning ineffective or obsolete technologies, *International Journal of Technology Management, 11*(7/8), 725–737.

Staw, B. M. (1976). Knee-deep in the big muddy: A study of escalating commitment to a chosen course of action. *Organizational Behavior and Human Performance, 16*, 27–44.

Stinchcombe, A. L. (1965). Social structure and organizations. In J. G. March (Ed.), *Handbook of Organizations* (pp. 153–193). Chicago: Rand McNally.

Teece, D. J. (1987). Profiting from technological innovation: Implications for integration, collaboration, licensing and public policy 185–219. In D. J. Teece (Ed.), *The competitive challenge: Strategies for industrial innovation and renewal.* Cambridge, MA: Ballinger.

Tsoukas, H. (1996). The firm as a distributed knowledge system: A constructionist approach. *Strategic Management Journal, 11*(25), 11–25.

Usher, A. (1954). *A history of mechanical enventions.* Cambridge, MA: Harvard University Press.

Van de Ven, A. H., & Garud, R. (1993). The co-evolution of technical and institutional events in the development of an innovation. In J. Baum & J. Singh (Eds.), *Evolutionary dynamics of organizations.* (pp. 425-443). New York: Oxford University Press.

Van de Ven, A. H, Polley D., Garud R., & Venkataraman S. (1999). *The innovation journey.* New York: Oxford University Press.

Ventresca, M., & Porac, J. (2000). *The social construction of industries and markets.* Oxford: Pergamon.

Weick, K. E. (1976). Educational organizations as loosely coupled systems. *Administrative Science Quarterly, 42*, 35–67.

Weick, K. E. (1979). *The social psychology of organizing,* second edition. Reading, MA: Addison-Wesley.

Weick, K. E. (1999). Improvisation. *Organization Science,* special issue.

White, E. B. (1957, Feb.). The Living Language, *The New Yorker.*

Whitley, R. D. (1992). *European Business Systems.* London: Sage.

I

Path Dependence
And Beyond

2

When and How Chance and Human Will Can Twist the Arms of Clio: An Essay on Path Dependence in a World of Irreversibilites

Andrea P. Bassanini
OCED, Paris, France

Giovanni Dosi
Santa Anna School for Advanced Studies, Pisa, Italy

The idea of path dependence, despite rather different uses (and misuses) in diverse disciplines, is nonetheless commonly linked with the idea that "history matters" in the interpretation of whatever phenomenon one would like to explain. Or, in other words, to understand why a certain entity has become what it is, or why a certain variable has acquired the value that one observes, one needs to bring into the picture, among the explanatory "causes," the past of that entity or the previous time path of that variable. The bottom line of such an intuitive notion is indeed that history matters precisely because another history—even holding the causal linkages of the analyzed system invariant—would have possibly yielded a different outcome.

Needless to say, the idea is very appealing and might in fact be a major building block of a new interpretative paradigm, emerging with respect to both natural and social sciences (see Prigogine & Stenger, 1984, for a discussion that is also a sort of epistemological manifesto). If anything, the intuitive appeal for most social scientists (except a few economists) is such that it is worth discovering the circumstances under which history does *not* matter.

Conversely, even when history seemingly does matter, what does it precisely mean? Does it all relate to some differences in the starting point of the dynamic process under investigation? Or, does it concern a series of events occurring, so to speak, along the historical path? And, how big should these events or the differences in initial conditions be in order to reshape the course of future outcomes?

At even deeper, and more philosophical, levels, these issues bear far-reaching implications in terms of *chance* (or *discretionary will*) and *necessity* (to paraphrase a famous book by Jacques Monod). That is, what are the degrees of discretionality that individual agents, organizations, or a collection of them, enjoy in shaping their own future? Is it that "freedom is just the consciousness of necessity"—as the philosopher Baruch Spinoza put it? Or, isn't it an implication of path dependence that agents may reset their own paths, albeit within the limits of their historically inherited constraints?

The importance and the difficulty of these background issues is among the motivating reasons of this essay, developing on previous works in economics on similar subjects (see, among others, Arthur, 1988, 1994; David, 1975, 1989, 1993, 1996; Dosi & Kogut, 1993; Dosi & Metcalfe, 1991; and Dosi & Orsenigo, 1988). In the hard task of providing within a single essay some reasonably coherent assessment of different notions of path dependence, in the following we start from some archetypical examples and interpretative categories. Building on them, we proceed to disentangle some of the sources of path dependence, different levels of descriptions, and different time scales at which path dependence might or might not emerge. Together, we shall attempt an admittedly conjectural assessment of the processes by which chance or discretionary human will might de-lock collective histories from particular paths. Notwithstanding some obvious bias in favor of economic and, more broadly, social examples, reference also to natural sciences helps in highlighting the challenging scope of the issues at stake.

PATH DEPENDENCE, IRREVERSIBILITIES, AND NOTIONALLY ALTERNATIVE OUTCOMES: SOME ILLUSTRATIVE EXAMPLES

Some basic conceptual categories and more mundane examples may be useful starting points. First, note that for history to matter, phenomena need to develop along an irreversible time arrow, and, together, the actual outcome must be only one of many possible alternative realizations.

There are plenty of simple examples even from natural history. Consider the formation of a planet system: That a gas *nebula* would eventually collapse and develop in a planet system is, as a first approximation, an almost sure event. The actual position of planets, which is related to the spatial configuration of initial agglomerations of dust particles, is just one of the almost infinite, possible dispositions.

The creation of the Lake La Niña in Peru—quite a remarkable event of hydrographic change given that it is now the second largest in South America for surface size—is an example of the cumulative effect of many meteorological events that are thoroughly changing the conditions in a once desertic territory.

Plenty of illustrations may be given from natural sciences: for example, a property of autocatalytic reactions is that they generally have multiple steady states. Even small perturbations of a system resting in an unstable steady state could lead to different (and unpredictable) configurations (see e.g., Prigogine, 1980).

Regarding biology, in the Darwinian evolutionary paradigm, there is a striking tension between selection and mutation: *Speciation* is an irreversible branching process where the followed path is the outcome of complicated dynamics of mutation and selection from possible alternative developments (see Gould, 1977).

Of course, social sciences are where one is likely to find that the interpretation of most phenomena also implies some account of the courses of history leading to them, this is so because, most often, the structures and constraints inherited from the past, together with human discretionality, select from alternative notional forms of social organization and paths of change.

Second, the proposition that "history matters" intuitively goes together with some sort of thought experiment (or counterfactual) which can be rarely undertaken through an actual experiment (at least in the social and biological domain).

Any scientist, or for that matter any individual, facing outcomes of contingencies that seem to be determined by particular coincidences of events and/or timing of choices, implicitly or explicitly try to rerun the "tape of history," attempting to disentangle the ways big or marginal variations in actions, exogenous events, or timings in the aforementioned, might have led to outcomes whose effects could not be washed away by the sheer passing of time.

History as a discipline is largely based on that method, from the conjectures on the macro historical effects exerted by the fascination with the nose of Cleopatra all the way to the causes and effects of the defeat of Napoleon at Waterloo. Here there are micro, random, events (the ravishing beauty of a lady who happened to be there at a particular time; the rain over Waterloo that prevented the full deployment of the French artillery; etc.). These phenomena, however, interact with more macroscopic ones: One could argue for example that, given the decadence of Hellenistic kingdoms, they would have fallen pray in any case to the Roman State; and that Cleopatra's nose influenced, if anything, only the modes and the timing (and one could argue the same for Napoleon's political enterprise).

The long-term impact of micro, normally irrelevant events has intuitively to do with resilience of some higher level collective structures to micro fluctuations. But, at this higher level of observation, a similar question may be posited: Namely, even assuming that Cleopatra's nose mattered, could the Roman State and Hellenistic Kingdoms have evolved, with positive probability, along paths wherein, on the contrary, no such random events mattered?

On these issues, see also the fascinating analysis by Cipella (1965) arguing that throughout the Middle Ages the survival of the whole Western European society against Eastern and Southern invaders has been due mostly to crucial but minor lucky events (such as the vicissitudes of Dougal Kings).

Third, these caricatural examples point indeed to the time-scale dimension of path dependence. Does it relate primarily to the distribution over time of events and to the time scale of processes which would yield, in the long run, an invariant outcome? Or, on the contrary, are long-term outcomes themselves affected by these sequences of apparently random micro events?

Symmetrically, *fourth*, even when one observes on some time scale, that history matters within rather narrow boundaries, seemingly frozen by much more inertial social structures and institutions, what can one infer about path dependence on longer time scales? Intuitively one has to refer back to the path dependence properties of these structures and institutions themselves. Indeed, by successive recursions on longer and longer time frames, one easily goes back to the questions of path dependence in the observed biological history, all the way to the initial cosmogony.

Nearer to our subject, instances where history matters at some levels are increasingly studied in economics and studies of technological change, innovation and diffusion are crowded by such examples, well beyond the celebrated QWERTY keyboard example (David, 1985, 1995, 1996; West, 1994).

Cowan (1990) argued that the developments of nuclear reactor technology toward light-water reactors (instead of, e.g., gas-cooled reactors) was due to a sequence of decisions that first favored a technology that could have been employed easily by military submarines and later, because the widespread employment in submarines, yielded technological improvements which could be quickly applied for civil uses. In short, particular historical conditions, related to macroconditions (the cold war) and microdecisions (in particular, those of a single Admiral of the U.S. Navy), were at the origin of the chosen path.

Cowan and Gunby (1996) showed that developments of chemical and biological pesticides in Texas was quite dependent on sequences of self-reinforcing events that were only partially triggered by differences in climatic conditions.

The 640K memory constraint on DOS-based software was not an outcome of any optimization exercise but rather the result of the hasty choices of the IBM designers to obtain their first generation of PCs. What would have been the software developments had there not been the premium placed on better ways to use high-level memory subject to such a constraint (see David, 1996).

JVC's VHS and Sony's Beta were commercialized approximately at the same time. According to many studies (see Cusumano, Milonadis & Rosenbloom, 1992; and Liebowitz & Margolis, 1994), none of the two standards has ever been perceived as unambiguously better, and despite their incompatibility, their features were more or less the same because of the common derivation from the U-matic design. For these reasons, the relevant decisions of consumers were likely to be sequential both at the individual and at the collective levels. Plausibly, a consumer chooses first whether or not to adopt a VCR, and then, once the adoption decision has been made, turns to the choice of VCR to purchase. The role of the installed base of the two technologies in this market most likely matters. There are strong increasing returns in design and production of VCR models (so that historically all firms specialized just in one single standard) on the supply side, and increasing returns externalities due to the availability of home video rental services on the demand side (Cusumano et al., 1992). Despite technical similarities between the two standards, preferences were strongly heterogeneous, due to brand name loyalty. The key to JVC's success seems to have been the exploitation of heterogeneity of preferences through Original Equipment Manufacturers (OEM) agreements with European firms characterized by well-established market positions in electronic durable goods. After the invasion of the European market and the consequent reorientation of the home video rentals market, VHS moved toward dominating also the other, quite segmented, Japanese and American markets. However, what would have happened had Sony not been misguided by its in-house productive capacity and engaged in the same market penetration strategy?

Beyond these simple patterns of competition among technologies, there are a few examples where the emergent monopoly of one technology lasts for a long while until it is discarded by new radical discoveries revitalizing the old technology. For instance, Islas (1997) showed how gas turbines got an opportunity to regain a dominant position in the market of thermal power stations for mass-production of electricity after barely surviving in the military aircraft market niche. Tell (1997) argued that the fact that direct current electric technology survived in the railway transportation market segment was key for the rediscovery of that technology as long distance electricity carrier many years later.

Clearly it is doubtful that the new winning technology could be considered just a new version of the old defeated competitor. The re-emergence of gas turbines is an archetypical case: this re-emergence occurred through the de-velopment of different types of combined cycles, whereas at the beginning the gas turbine was only a complementary component of an hybrid technology (Auer, 1960; Islas, 1997; Pfenninger & Yannakopoulos, 1975; White, 1956).

All these examples point at some critical history dependence in the sub-sequent collective selection of some dominant product design or even in the selection of the dominant knowledge bases and technological paradigms (Dosi, 1982). At the same time, some of these historical instances also illus-trate the possibility of de-locking from some apparently dominant techno-logical trajectories, due to changes in the knowledge bases—on different time scales—in related technological fields.

History looms large also at the broader level of country specific patterns of growth and specialization in international trade. Indeed, an enormous litera-ture, involving sociology, political science, and the political economy of growth, has convincingly emphasized the inertia and self-sustained reproduction of in-stitutions and organizational forms as determinants of growth of different na-tions, showing variegated patterns of catching up, falling behind and forging ahead (for some "stylized facts," see Abramovitz, 1985; Dosi, & Kogut, 1993; Dosi, Freeman and Fabiani, 1994a and Fagerberg, Verspagen, & von Tunzelmann, 1995). Still political and institutional lock-in is almost never com-plete, and what appears to be stable equilibria for a long period may be quickly disrupted by a sequence of strongly self-reinforcing, possibly surprising, events. This is the case, for instance, of the collapse of socialist regimes in Eastern Eu-rope,[1] but also the case of major technological discontinuities.

The nature of path dependent dynamics, the levels at which they might be detected and their sources are the topics of following sections.

WHAT IS PATH DEPENDENCE?

Clearly, path dependence must involve some irreversibility of the phe-nomena under consideration. At the very least, history must matter in some phenom- enologically defined short run. Call it a *weak* form of path depend-ence. Or in a *strong* form, history might affect also the long-term states the system will eventually obtain.

To illustrate more rigorously these points on irreversibility and path de-pendence (and the conceptual difference among them), let us start from some natural science examples.

In a closed environment all thermodynamic processes increase their en-tropy toward a maximum. Entropy can never be decreased.[2] In a sense, this

---

[1] For a different interpretation (although in our opinion short of satisfactory), see Caplin and Leahy (1994).
[2] See Prigogine (1980) for more details on irreversibility of thermodynamic reactions as compared to classical dynamic reactions.

is an archetypical example of an irreversible process, that is, a process that develops through history. However, despite its irreversibility, this canonical thermodynamical reaction is a typical example of processes where the effects of history are washed away as time goes by: All systems that have a single maximum of entropy, which is also a globally stable steady state (i.e., systems that admits a Lyapunov function over the whole phase space) converge toward a single asymptotic pattern. History matters and constrains the process during the transition, but the ending point is predictable and unique. If we heat the corner of a physically compact box, and we thermo-insulate it from the outside, we can observe an irreversible process of equalization of temperature inside the box. However the system converges toward the unique steady state where the expected value of temperature is the same everywhere in the box.

As an example from economic theory, consider the case of the neoclassical growth model à la Solow (Solow, 1956): Independent of any initial conditions, there is just one possible asymptotic steady state (implying, in that model, also the absence of per-capita output growth).

In this whole class of systems, there is an obvious irreversibility (one cannot "go back in time"). However, this implies convergence to an unchanged final destiny. In that sense there is no path dependence.

A general condition for history to matter in terms of asymptotic states that the system might attain is the existence of multiple equilibria reachable under different initial conditions. In turn, it is now well known that they are likely to appear with reference to the economic domain in the presence of some forms of dynamic increasing returns or collective externalities. Even sticking to otherwise very conventional assumptions on microbehaviors and collective interactions, it can be shown that positive feedbacks of some kind generally yield multiple growth trajectories, multiple specialization patterns in international trade, etc., depending on initial conditions.

PATH DEPENDENCE IN THE PRESENCE OF STABLE ATTRACTORS

In mathematical terms, the possibility of having many basins of attraction depends on the shape of the function describing its transition dynamics. Just as an example take $x(t)$ as the variable of interest and consider a function $f(.)$ relating its value in the future, for example, $x(t + 1)$, to its current value. Clearly the system is in a steady state only when

$$x(t + 1) = f(x(t)) = x(t).$$

Furthermore, a steady state x^* can be a (stable) attractor only if in a neighborhood of that point, the function $f(x) - x$ downcrosses the x axis, changing its value from positive to negative. In other terms, $x(t)$ should display a tendency to increase its value when it is below the (locally stable) attractor x^*, whereas the opposite holds when it is larger. Clearly if $f(.)$ is linear, just one steady state is admissible. In a one-dimensional space, this means that either the process displays a tendency to grow or collapse for ever or it tends to converge to the stable attractor for every initial condition. The possibility of either pattern depends on the nature of the attractor. However, similar properties are shared by functions that cross the 45° line no more than twice. For instance, all the concave and convex functions belong to this class. Note that the Solow model generates a concave transition function. The Kaldorian growth (Kaldor, 1957; Kaldor & Mirlees, 1962), on the other hand, has a linear transition. Both these cases are graphically represented in Fig. 2.1.

With more complex transition functions (for instance, in a growth model with a nonconvex production possibility set[3]), the possibility of

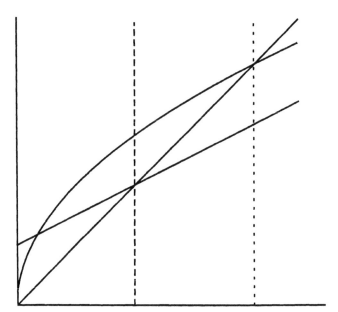

FIG. 2.1   Transition functions in the Solow and the Kaldor growth models. Dotted and dashed lines identify steady states.

---

[3]Simple treatments in the context of the growth literature can be found in Majumdar, Mitra, & Nyardo, 1989), King & Robson (1993), and Azariadis (1996). More complicated examples come from applications of chaos theory to economics (e.g., Brock & Malliaris, 1989, and Medio, 1992).

multiple locally stable attractors emerges. Consider the transition function depicted in Fig. 2.2.

A *fortiori*, multiple growth trajectories are likely to emerge in evolutionary models[4] sustained by the positive feedback structure linking, in probability, technological innovation, profitability, growth, and further innovations.

Another broad domain where a multiplicity of equilibria easily appears is the processes of selection, at the biological or economic levels, among heterogeneous entities, whenever there is some interaction in the contribution of various traits to the fitness (in biology) or to the competitiveness (in economics) of various entities. Consider the relationship in some biological environment between traits and fitness—that is, what is often called *fitness landscape*.

When the fitness contribution of every gene or trait is independent, the adaptation of biological systems in fixed environments occurs in an highly correlated Fujiyama single-peaked landscape (even when fitness contributions are randomly assigned). In such a case, whatever the mechanism of adaptation, provided that the fittest has an evolutive advantage, the system

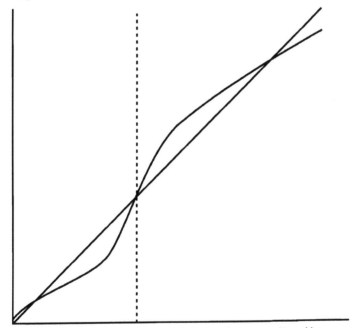

FIG. 2.2   Transition functions with two separate basins of Attraction. Dotted line separates basins.

---

[4]The classic reference for such type of models is Nelson and Winter (1982). An example where such multiplicity of growth paths is explicitly analyzed is in Dosi et al. (1994). A more detailed discussion of the relation between evolutionary processes and path dependence is in the chapter of this book by Ruttan (chap. 4, this volume), and, with a somewhat different view, in Dosi (1997).

converges from whatever initial condition toward the same maximum. The system belongs to the class exemplified by Fig. 2.1.

However, whenever *epistatic correlation* appears—for example, when the fitness contribution of each trait or gene depends also on other ones— even when the environment is fixed, the landscape tends to become rugged, that is highly nonlinear and multipeaked. A well known example is Kauffman's NK model (Kauffman, 1989, 1993; Kauffman & Levin, 1987).[5] Even more so, this picture extends to the case when each species landscapes are mutually interdependent (as in the so-called NKC models: See Kauffmann and Johnsen, 1991, and Bak, Flyvbjerg, & Lantrup, 1994). Both of these models belong to the class exemplified by Fig. 2.2.

Interestingly, there are many analogies between these models of biological evolution and the evolution of complex organizations (Levinthal, 1998a, 1998b). Heterogeneity of organizational patterns, as well as inertia in the face of changing environmental conditions, can emerge as properties of adaptative search in rugged fitness landscapes due to complementarities among organizational components. Contrary to contingency theories of organizations (Lawrence & Lorsch, 1967) claiming a high plasticity of organizational traits with respect to the requirements of changing competitive environments, Levinthal (1998b) convincingly argued that adaptation over rugged competitive landscapes may yield lock in onto different fitness peaks, even when the competitive conditions change. Hence both organizational variety at any point in time and organizational inertia over time.

SMALL AND BIG EVENTS

An economic system may be affected by big events (plagues, catastrophes, wars, major innovations, policy reforms, etc.), with very strong and persistent effects but usually very low frequence; and small events (weather conditions, incremental innovations, adoption choices, etc.), such that the analyst's filter cannot finely detect the linkage between single event and overall economic effects. Most of economic models take into account only the first type of events: In deterministic models with multiple steady states, history just selects the initial condition from which the attainable steady state is determined. As Azariadis (1996) asserted in his survey on multiple equilibria and growth:

> An alternative working hypothesis … takes the growth process of nations to be fundamentally the same except for differences in history, e.g., in the circumstances from

---

[5]Kauffman's analysis has mainly a statistical character: General properties of NK models are studied through random assignment of fitness contributions.

which the growth process begins. These are chiefly the starting stocks of human and physical capital, and the state of technology (p. 452).

For instance, in the modern theory of international trade, history looms very large but its representation is mainly reduced to initial conditions.[6] Similar considerations apply to deterministic equilibrium models of endogenous growth.[7]

Basically, a deterministic approach suffers from the limitation that it cannot represent the dynamic process that makes the whole unfolding history relevant: Everything historically relevant is over when the analyst's camcorder is switched on. And it doesn't help much to watch the crime scene when all the facts have already happened. Phenomena that are related to timing, potential repetition, and correlation of historical events are formally ruled out. All this notwithstanding, there is verbal acknowledgment of the dynamic importance of timing and repetition of events:

> Are the poor merely the victims of the circumstances in which they are initially placed by chance, environment or history? If the answer to the last question is yes, can a small or *temporary* improvement in the opportunities of persistently poor groups result in a large or permanent betterment of their lifetime income? (Azariadis, 1996, p. 451, italics added).

> Even temporary events, if they are strong enough, can permanently wrench an economy away from underdevelopment. If *temporary events lead to favorable initial conditions*, the economy continues to grow even without the stimulus of major additional innovations or other events similar to those that got the process started. (Becker et al., 1990, p. 33)

In general, the possibility of temporary events occurring along any historical path which have permanent effect on the future path itself, can hardly be treated within a deterministic framework. Conversely, a stochastic approach allows a more natural representation of both big and small events throughout the history of the system.[8]

---

[6]Examples are Krugman (1981, 1987, 1991b), Ethier (1982), and almost all the works in the Kaldorian tradition (e.g., Thirlwall, 1979; Fagerberg, 1988; Cimoli & Soete, 1992; Amable, 1993; McCombie & Thirlwall, 1994).

[7]In endogenous growth theory multiplicity of equilibria has generally more to do with coordination of decentralized decision making rather than with history (as shown by Benhabib & Perli, 1993, this is the case in Lucas (1988) and Romer (1990)). Still, as in the foregoing quotation, when a role for history is identified, this is reduced mainly to the selection of initial conditions (e.g., Becker et al., 1990; Azariadis & Drazen, 1990; Boldrin, 1992; Brezis et al., 1993; Cozzi, 1997; Boldrin & Levine, 1998).

[8]This statement should be qualified by considering also deterministic chaotic systems. Although forgoing remarks still apply on the fact that history is "squeezed" into initial conditions, chaotic systems imply sensitive dependence on initial conditions themselves, so that two systems starting with arbitrary close initial conditions may exhibit increasingly diverging trajectories.

Arthur (1989) models of emergence of technological lock-in through a sequence of individual adoption choices are a classical example of path dependence emerging through a sequence of small stochastic events. The standard story told by authors developing this class of competing technology dynamics models is the following (see Arthur, Ermoliev, & Kaniovski, 1987; Arthur, 1990; Cowan, 1990; Glaziev & Kaniovski, 1991; Dosi & Kaniovski, 1994; Kaniovski & Young, 1995).

Every period a new agent enters the market and chooses the technology that is best suited to its requirements, given its preferences, information structure, and the available technologies. Preferences are heterogeneous and a distribution of preferences in the population is given. Information and preferences determine a vector of payoff functions (whose dimension is equal to the number of available technologies) for every type of agent. Because of positive (negative) feedbacks, these functions depend on the number of previous adoptions. When an agent enters the market, it compares the values of these functions (given its preferences, the available information, and previous adoptions) and chooses the technology that yields the maximum perceived payoff. Which "type" of agent enters the market at any given time is a stochastic event whose probability depends on the distribution of types (e.g., of preferences) in the population. Because of positive (negative) feedbacks, the probability of adoption of a particular technology is an increasing (decreasing) function of the number of previous adoptions of that technology. These assumptions often lead to multiple asymptotic collective arrangements, none of which is certain at the beginning.

In essence, in deterministic (nonlinear) models, only initial conditions are the carriers of history, whereas in a stochastic framework it may well be that it is the whole sequence of events that determines which limit state is attained; and, conversely, from the same initial conditions, the system may evolve toward many different end states.

Continuing with our one dimensional graphic example we can illustrate this with Fig. 2.3. Imagine that there is an underlying stationary stochastic variable (e.g., weather conditions) that takes only two values that affect production differently (e.g., rain or sunshine). Figure 2.3 represents the transition functions relative to the different realizations of the stochastic variable in each instance. When the system lies to the right of the dashed line or to the left of the dotted line, only one steady state is attainable with positive probability. If initial conditions are between the two lines, the system initially fluctuates in this region, but eventually will trespass one of the lines and, with certainty, never cross it again. Both lines can be trespassed with positive probability. The dotted and dashed lines represents two absorbing

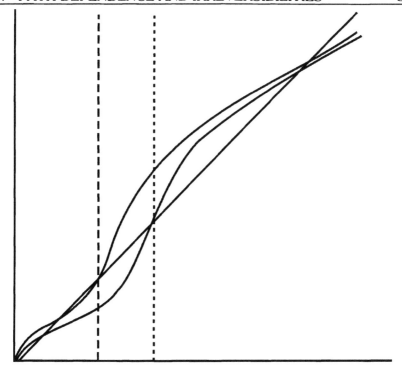

FIG. 2.3   Transition functions corresponding to two different values of the underlying stochastic variable. There are two separate basins of attraction. Dotted and dashed lines separate
basins.

barriers.[9] Obviously, still, there are some sets of initial conditions (metaphorically speaking, some sets of big events) that determine the long-run
outcome with certainty.[10]

Furthermore, notice that when the system fluctuates in the middle region, the nearer it gets to one barrier, the higher the probability to be absorbed by that barrier. The engine that is at work here is some kind of
positive feedback or self-reinforcing mechanism. Positive feedbacks (at least
of a local nature) are actually necessary to create local instability, and there-

[9]Let us clarify what we mean by *stochastic steady state* when the process admits more than one basin of
attraction (i.e., more than one probability distribution of limit states of the process conditional to events
at time 0). In fact, by definition, there is at most one asymptotic distribution for every initial state. A set of
states is said to be a *closed set* if no outside state can be reached with positive probability through a finite
sequence of steps from any state inside the set. We consider as stochastic steady states each invariant distribution defined for each irreducible subprocess with initial conditions inside each closed set, because
there is only one invariant distribution for every such set.

[10]But this need not to be the case when the state space is multi-dimensional (see Bassanini, 1997, and
David, 1975).

fore multiple long-run patterns that can be selected by history. Indeed, in our view, a good deal of economic processes driven by knowledge accumulation and innovation share these basic characteristics (for discussions, especially with reference to evolutionary models of economic change, see Dosi, 1997, and Nelson, 1995).

Let us recall the main points set forth in the previous subsections: Essentially a *path dependent phenomenon* is an irreversible dynamic process where there is a multiplicity of potential long-run outcomes, due to some kind of nonlinearity in the functions describing transitional dynamics. Furthermore, a complete description of the role of history can be accomplished if the whole sequence of relevant historical events is taken into account. From the point of view of formal modeling, this calls for a stochastic approach.

For expository reasons, until now we have considered only set-ups where dynamical systems eventually settle into a "resting" state. The foregoing examples consider situations wherein the asymptotic pattern is a steady state. This does not need to be the case. Actually, many examples, from the evolution of institutions, organizations, and technologies suggest a world wherein temporary resting states are "metastable" in the sense that, on longer time scale, they are persistently overcome by new developments leading to new "temporary" resting states. Therefore, we need to broaden the definition of path dependence to encompass the case where there is no convergence to any asymptotic behavior.

PATH DEPENDENCE WITHOUT ASYMPTOTICS

Consider the irregular motion performed by a pollen grain suspended in liquid. Its irregularity was first noted by the Scottish botanist Robert Brown in 1828. The motion was later explained by the random collision with the molecules of the liquid and described as a stochastic process that closely resembles a random walk (actually a random walk in continuous time and over a continuous state space: i.e. a Brownian motion). The trajectory followed by the pollen grain is an irreversible phenomenon and still, even if we could take the liquid surface as approximately infinite, history would not matter asymptotically. Two pollen grain would come arbitrarily close to each other in finite time, and an infinite number of times afterward; the position of a grain in a time instant is all that matters in order to know the probability distribution over future trajectories (i.e. the process has a Markovian nature); thus, differences in the two pollen grain's histories are eliminated in finite time.

Note that foregoing observation about the motion of particles over a surface does not apply to the motion of particles in a volume. It is a well known property of random walks in more than two dimensions that two realizations do not cross with certainty. In this latter case, we are actually facing a true instance of path dependence.

In terms of a general definition of path dependence, able to encompass cases of path dependence without asymptotics, we can say that a dynamical system is path dependent whenever the trajectories described by two possible realizations do not come arbitrarily close to each other infinitely many times with certainty. If the dynamical system is described by a Markovian process (such as a random walk), this statement reduces to the following: The trajectories described by two possible realizations do not cross with certainty.

Path dependence without any asymptotic lock-in is even more likely if one allows the endogenous change of the state space in which the dynamics are nested. This is, after all, the case of many biological, economic, and technological phenomena.

In biology, the evolution of species is the result of a branching process that often displays alternation of stages of convergence to a (approximately invariant) local fitness landscape followed by stages of temporary permanence in the neighborhood of local maxima until disruption of the equilibrium occurs and a new pattern of convergence toward different fitness peaks along a reshaped fitness landscape takes place. Simulation of the NKC model with local interactions (see Kauffman & Johnsen, 1991) gives an easy way to visualize this process. If interactions are local for every single species there is an alternation of stages of landscape stability and instability due to mutation of the interacting species. The ruggedness of the global landscape (that is the landscape referred to as "the whole bundle of genes," independent of their species), may generate rapidly diffusing "avalanches" of change, even after mutation of only one species. Representing graphically the spatial structure of interactions, regions of mutating species and regions of species in a "resting" state can be observed (see Kauffman & Johnsen, 1991).

The economics of innovation and diffusion of technologies are full of examples of apparent convergence to a dominant technology, intertwined with the arrival in the longer term of a new one which displaces the old. For instance, the competition between the bloomery and the puddle steel production processes, leading to the dominance of the latter around mid-19th century, was disrupted by the introduction (around 1860), and subsequent dominance of the Bessemer and later open hearth processes, followed by the introduction of two new competitors (electric and oxygen processes) during the second half of the 20th century (Gruebler, 1991; Nakicenovic & Gruebler, 1991). Such dynamics can be interpreted in terms of the theory of dominant designs (Abernathy & Utterback, 1978; Anderson & Tushman, 1990; Henderson & Clark, 1990; see Tushman & Murmann, 1998 for a comprehensive survey of the literature), or almost equivalently, in terms of technological paradigms yielding relatively ordered technological trajectories of more incremental change (Dosi, 1982, 1988).

Whatever interpretative lenses are chosen, however, the path de-pendence of such processes of innovation and diffusion entail some more subtle questions. At one extreme, one could argue that the whole pattern of emergence of a paradigm or design, its diffusion, dominance, time-consuming displacement by a new one, is in fact a story of the suc-cessive discovery and establishment of knowledge bases and products which are objectively better than on some technological and economic measures, than the one they are displacing and than any other one that could compete with them at that particular time: hence, of course, irre-versibility, but not much path dependence. Were it not for the persistent arrival of new (and superior) paradigms, the system would indeed lock into an asymptotic state, and it would do so independently of initial con-ditions and/or early stochastic fluctuations. In the aforementioned lan-guage, for any state of technological knowledge, the system displays a single peaked landscape.

Conversely, at the opposite extreme, a few scholars claim an overwhelm-ing driving role to history, especially in the form of social and political fac-tors. A good summary of this view (indeed still far from more radical "social constructionists") is by Tushman and Murmann (1998) who suggested that

> ... dominant designs are not driven by technical or economic superiority, but by sociopolitical/institutional processes of compromise and accommodation between communities of interest moderated by economic and technical constraints. The more complex the product, the more accentuated these institutional forces intrude in the emergence of a dominant design. (p. 260)

Thus, they imply evolutionary process over landscapes, which, at least in terms of technical efficiency, are for a good portion flat and, thus, lock or de-lock as a result of individual and collective wills and politics.

The view we tentatively suggest here does indeed share with the latter the appreciation of the importance of social factors in the selection among notionally alternative technological paradigms and archetypical artifacts, *when such alternative exists*, and especially in the early history of a technology (what in Dosi, 1982, is called the *pre-paradigmatic phase*). Here, in our view, is where path dependence primarily rests. Seen from the symmetric angle of social discretionality in governing the future course of events, it is the phase of early emergence of new technological paradigms that provides the pri-mary "window of opportunity" (as David, 1986, and Perez & Soete, 1988 put it), for social governance and choice.

However, we do share with the former view, first, the idea that often technical constraints might be overwhelmingly binding (that is, caricaturally, to repeat again a famous sentence by Keith Pavitt "We would never want to fly on an airplane that is only a social construction!"). Second, irrespectively of the drivers in the early process of selection (and also of the long-term notional opportunities of technologies that could have been but have not been chosen), we believe that technological learning does indeed display local (paradigm-specific) dynamic increasing returns to knowledge accumulation. Hence, as time goes by, the peaks in the landscape associated with the dominant technologies become higher and higher, so that, most often, major discontinuities are associated with the emergence of radically new knowledge bases (e.g. electricity v. steam power, semiconductors vs. thermoionic valves, etc.) which radically change the space of exploration for further advancement.

SOURCES OF PATH DEPENDENCE

As discussed in the foregoing section, some combinations of irreversibilities and nonlinearities are the essential determinants of path dependence. Let us expand upon the factor accounting for these irreversibilities and nonlinearities.[11]

In general, at *micro level*, some irreversibility condition emerges whenever the past irremediably influences the behavioral framework of the agents. For example, their choice set and payoff structures depend on time and past decisions; their problem-solving competencies, preferences and models of the world changes in history dependent fashion; or, much more trivially, irreversibility just takes the form of an unchanged decision algorithm that agents carry with them since their birth. Note that situations where the decision process of one or many agents is sequential typically generates some kind of irreversibility.

In general, we shall say that there is path dependence at an individual level whenever history influences irreversibly the choice set and the behavioral algorithms of the agents so that e.g. if the system at some future time, $t + \tau$, is suddenly reversed back to its macroscopic state at t, the microscopic identities and behaviors of the agents would remain irreversibly changed as compared to those present at t.

---

[11]More detailed discussions by one of us of these and related issues is in Dosi and Metcalfe (1991) and Dosi and Kogut (1993).

At *system level*, irreversibility entails, somewhat loosely speaking, a decreasing probability, as time goes by, of going back to a state that the system visited before, or switching over to a state that the system could have attained, had it had another history up to that time.

Nonlinearities, both at the local and at the global level, stem out of some kind of dynamic increasing returns, which puts into play some process of "cumulative causation" in the dynamics. Dynamic increasing returns can emerge on the supply side because of economies of scale, irreversibility of investment, asymmetry of information. More generally, they are likely to be a common property of learning and accumulation of technological capabilities with their typical features of locality and cumulativeness.[12] Widely studied phenomena such as learning by doing, learning by interacting and learning by using all entail positive feedbacks (see Arrow, 1962; Lundvall, 1993; and Rosenberg, 1982).

The dynamical interactions between knowledge accumulation, market expansion, and reduction of the (hedonic) price of goods are common features of diffusion processes of particular technologies and dominant designs, and they are indeed what drives specific technological trajectories. For example, think of the well documented dynamics of information-processing capabilities, speed and prices in microelectronics (Dosi, 1984 and Malerba, 1987). Or at the level of a single (dominant) artifact, consider the case of the Boeing 727, 737, and the 747, which have been on the jet aircraft market for decades, and have undergone constant modification of the design and improvement in structural soundness, wing design, payload capacity, and engine efficiency as they accumulated airline adoptions and hours of flight (Rosenberg, 1982). Similar observations can be made for many helicopter designs (Saviotti & Trickett, 1992).

Demand side positive feedbacks are equally important. Network externalities (see, e.g., Katz & Shapiro, 1994) have been receiving much attention in the last 2 decades. For example, telecommunication devices and networks, as a first approximation, do not tend to provide any utility per se but only as a function of the number of adopters of compatible technologies with whom the communication is possible (more formal analyses of this intuitive property can be found in Economides, 1996; Oren & Smith, 1981; Rohlfs, 1974). The user benefits of a particular hardware system depend on the availability of software. The quantity and variety of software may depend on the size of the market if these are increasing returns in software production. This is the case of VCRs, microprocessors, hi-fi devices and, in general, systems made of complementary products that need not be consumed in fixed proportions (Church & Gandal, 1993; Cusumano et al., 1992; Katz & Shapiro, 1985, 1994).

---

[12]Quite diverse but complementary empirical and theoretical arguments supporting the "cumulative causation" view, can be found in Myrdal (1957), Kaldor (1981), Zeckhauser (1968), Atkinson and Stiglitz (1969), David (1975), Nelson (1981, 1995), Dosi (1988), Levin, Cohen, & Mowery (1985, 1987), Antonelli (1995), Stoneman (1995), Arthur (1994), Freeman (1994), and Rosneberg (1976).

More generally, network externalities of some kind (on the demand and/or on the supply side) and the development of commonly shared standards all entail some sort of positive feedback, and, thus, also a potential source of path dependence. In this respect, the story of the dominance of the QWERTY keyboard, discussed by David (1985, 1996) is the most celebrated example. Moreover, on the demand side, all phenomena of endogenous evolution of preferences with social imitation, conformity and bandwagon effects are likely to involve nonlinear feedbacks between collective interactions and microbehaviors (Aversi, Dosi, Fagiolo, Meacci, & Olivetti, 1999; Bernheim, 1994; Brock & Durlauf, 1995, 1998; Dosi and Metcalfe, 1991; and Duesenberry, 1949).

Another quite general source of positive feedbacks is related to the emergence of social customs, conventions and collectively shared norms. Their development also implies the change in rewards and penalties facing individual decisions and the evolution of cognitive patterns and behavioral algorithms supporting these norms and customs. And with that goes a nonlinear self-reinforcing process.

Indeed, as argued by David (1995), *institutions are* one of the fundamental *carriers of history*. In fact, they carry history in several ways. First, they carry and inertially reproduce the birthmarks of their origin and tend to persist even beyond the point when the conditions that originally justified their existence, if any, cease to be there. Second, they generally contribute to structure the context wherein the processes of socialization and learning of the agents and their interactions take place. In that sense, one could say that institutions contribute to shape the fitness landscapes for individual economic actors and their changes over time.[13]

In brief, institutions bring to bear the whole constraining weight of past history on the possible scope of discretionary behaviors of individual agents; and relatedly, contribute to determine the set of possible worlds that collective dynamics attain, given the current structure of the system. At the same time, they also represent social technologies of coordination, as Nelson and Sampat (1998) argued. As such, alike technologies *stricto sensu*, they are also a source of path dependent opportunities for social learning.

LEVELS AT WHICH PATH DEPENDENCE MAY OCCUR

It should be clear from the foregoing discussion that we believe that path dependence is a rather common property at different level of observation—ranging from individual behaviors to business organizations, all the way to the whole economies—and within different domains—including techno-

---

[13]These telegraphic points are presented in more details in Dosi (1995), and Dosi and Coriat(1998); for germane arguments, see Granovetter (1985), Boyer (1996) and Nelson and Sampat (1998).

logical change, economic growth, and institutional dynamics. However it is important to emphasize that there need not be any isomorphism in the degrees of irreversibility and path dependence across different levels of observation (the issue is discussed at greater length in Dosi & Metcalfe, 1991). For sake of illustration, one may think of two extreme archetypes. In the first one, the behavior of heterogeneous micro entities is trivially path dependent in the sense that an initial condition (say, genes at birth) makes them repeat endlessly the same behavioral repertoire (e.g., in the language of game theory, they might deterministically play a single pure strategy). Still, the system dynamics might or might not be path dependent. This will be determined by the nature of the interactions and the related fitness landscape (cf. previous paragraphs). Indeed, in the simplest case of a selection dynamics on a single-peaked landscape, under relatively weak assumptions, it can be shown that the system converges to the unique evolutionary stable strategy equilibrium (see, for example, Foster & Young, 1990; Fudenberg & and Kreps, 1993). Hence, there is an irreversible system dynamics that is not path dependent. Conversely, at the other extreme, one may easily think of systems composed of agents who do not embody any long-lasting effect of their idiosyncratic histories on their own behavior, which nonetheless exhibit strong collective path dependence.

A conceptually distinct issue regards the aggregate properties of a collection of path dependent processes. Consider an economy with an increasing number of identical sectors, without intersectoral input/output linkages or intersectoral knowledge spillovers. Even if each sector displays path dependent dynamics, due to the law of large numbers, the limit of that economy exhibits ergodic patterns. In this case we are conceptually in the presence of a repeated sample of independent observations drawn from the same population. Whatever aggregating rule we employ, the asymptotic value of this variable can be known with certainty from the parameters of the model.

Ultimately the properties of the aggregate as compared to the properties of its constituent parts depends on the structure of interactions among the latter. In Bassanini and Dosi (2000) we examined the international diffusion of competing technologies, formalizing the idea that convergence to the same technology or standard depends on the relative weight and strength of international spillovers as compared to nationwide (or regional) externalities. Aggregate versus local path dependence seems therefore to depend on the structure of interactions. When local interactions are strong (i.e., every unit depends on each other with intensity above a certain threshold) path dependence at local level induces path dependence at global level. Still, no conclusive results have ever been provided to define a minimum interaction threshold below which no path dependent aggregate outcome is observable.

As an empirical illustration of the point, compare the case of the VCR market with that of computer keyboards. Historically, in the former, gaining leadership in the European market, with the consequent bias in the related home video market, was crucial to VHS to resolve in its favor the battle for leadership in the Japanese market as well (Cusumano et al., 1992). Keyboards tell a different story: although in the English-speaking world the QWERTY keyboard represents the standard, in the French-speaking world a slightly different version (the AZERTY keyboard) is by far the more adopted one. Clearly, geographical areas with the same language tend to be reflected in spillover clusters due to free "migration" of typists, similar training institutions, etc.

However, no conclusion can be drawn from the observation of the absence of a global monopoly of technology about the ergodicity or path dependence in the worldwide diffusion process. Putting another way, what would be the outcome, in terms of world market shares, of running the tape twice? In general, no strong result, to our knowledge, has been achieved in the relationship between structures of interactions and distributions of asymptotic states (a more detailed discussion can be found in Bassanini & Dosi, 1998).

LOCKING INTO AND ESCAPING
FROM PATH DEPENDENCIES

In a nutshell, the thrust of our foregoing argument is that, even confining ourselves to social phenomena, there are some very general sources of path dependence intimately associated with (a) the cumulative characteristics of knowledge accumulation, (b) the nature of organizations (in general, including business organizations) with their "epistatic correlations" in behavioral traits, mechanisms of coordination, routines, patterns of organizational learning, (c) the externalities and dynamic increasing returns that the process of economic growth most often entails, (d) the network of social relations path dependently constraining and shaping the action sets, decision algorithms, and preferences of agents, and more generally, (e) the very nature of institutions as "carriers of history."

As such, the ubiquitous presence of path dependence implies, as argued above, a view of socioeconomic phenomena deeply tainted by irreversibility and various forms of "lock-in" into particular organizational structures and/or trajectories of change. The tape cannot re-run twice (except in the *gedankenexperiment* of the analyst) and the only one history that actually occurred provides all the constraints, as well as all the opportunities, that social agents face at any particular time.

But "lock-ins" seldom have an absolute nature, and the unfolding of history, while closing more or less irremediably opportunities that were available but not seized upon at some past time is also a source of new "possible worlds." Hence, in some sense, there always are "windows of opportunities," using again David's terminology, which allows "de-locking" and escape from the tyranny of the past.

Some factors operating to this effect have been already discussed, when considering path dependence without asymptotics. In particular, first, at technological level, the emergence of new technological paradigms, we have seen, do represent a major source of "de-locking," which in turn often involves the emergence of a new set of business actors, a new knowledge base, new communities of practitioners (e.g. scientists, engineers, etc.), and even new forms of corporate organization.

Second, heterogeneity among agents and imperfect adaptation of agents with organizations and broader social networks do represent a persistent source of variety and in a sense a sort of insurance that the system will never completely lock into a trajectory or behavioral mode (more on this issue in Dosi & Coriat, 1998a). Furthermore, under some circumstances, non-average behaviors may yield symmetry-breaking effects on the distribution of traits in the population and yield macroscopic (i.e., system level) transitions (see e.g., Allen, 1988).

Third, the coevolutionary nature of many processes of socioeconomic adaptation is, as already argued, a source of lock-in but also entails a potential for de-locking and major discontinuities. Often, technologies, behavioral traits, and organizational forms are selected in multiple landscapes, and according to different criteria of "fitness." For example, as discussed in Dosi and Coriat (1998b), organizational routines are selected both in relation to their problem-solving efficacy and their ability to represent mechanisms of organizational governance and social control. That is, they can been seen as the outcome of adaptation in two different landscapes. In turn, increasing "misadaptation" in one of the multiple domains in which the fitness of routines, technologies, etc., are evaluated may entail far-reaching discontinuities and, so to speak, reopen the process of search also for new relatively path independent combinations.

Finally, fourth, entrenched path dependencies might be broken by "invasions": new organizational forms originally developed in other contexts that spread and become at least for a period, the new dominant paradigm. Think, in this respect, of the diffusion from the U.S. throughout the world of the M-form, "Fordist," corporation, or, more recently, the spreading of Japanese industrial practices, also implying organizational de-locking from older established organizational forms.

CONCLUSIONS

The attempt of accounting in a thorough and rigorous way for the role of history in socieconomic phenomena is fascinating enterprise that is only at its beginning. However, some lessons can already be drawn also on normative levels.

As Arthur (1989) put it, path dependent processes *may* (although not always) display properties of (a) suboptimality (in the sense that other ex ante attainable historical paths would have implied socially superior outcomes); (b) potential inflexibility (i.e., increasing lock-in, irremediable from the point of view of ex-post discretionary intervention by agents); and (c) ex ante unpredictability.

In such worlds, it is precisely in the phases of early "seeding" and development of path dependent processes that the scope of discretionary (individual and collective) choices is higher, while later on, the weight of the past history may well bind freedom to rather narrow boundaries. In essence, it is in the subtle relations among path creations, path dependencies, and the various forms of de-locking mechanisms just discussed that one sees the inevitable tension between freedom and necessity characteristic of many social phenomena. More detailed, historical and formal, studies of path dependent processes will allow the development of sorts of "theories of the possible worlds," defining the notional states that are attainable given all the weight of an irreversible past, and thus also, paraphrasing March (1991), determining the scope of what one may explore and what one has to exploit, or just inevitably swallow.

REFERENCES

Abernathy, W., & Utterback, J. (1978). Patterns of Industrial Innovation. *Technology Review, 80,* 40–47.

Abramovitz, M. (1986). Catching Up, Forging Ahead, and Falling Behind. *J. of Econ. History, 46,* 385–406.

Allen, P. M. (1988). *Evolution, Innovation and Economics,* in Dosi et al. (Eds.)(1988).

Amable, B. (1993). National Effects of Learning, International Specialization and Growth Paths. In D. Foray & C. Freeman (Eds.), *Technology and the Wealth of Nations.* London: Pinter.

Anderson, P., & Tushman, M. (1990). Technological Discontinuities and Dominant Designs. *Administrative Sci. Quart, 35,* 604–633.

Antonelli, C. (1995). *The Economics of Localized Technological Change and Industrial Dynamics.* Dordrecht: Kluwer.

Arrow, K. J. (1962). The Economic Implications of Learning by Doing. *Rev. Econ. Studies, 26,* 155–173.

Arthur, W. B. (1988). *Self-Reinforcing Mechanisms in Economics.* P. W. Anderson, K. Arrow & D. Pines (Eds.), *The Economy as an Evolving Complex System.* Reading, MA: Addison-Wesley.

Arthur, W. B. (1989). Competing Technologies, Increasing Returns and Lock-In by Historical Events. *Econ. J., 99,* 116–131.

Arthur, W. B. (1990). "Sylicon Valley" Locational Clusters: When Do Increasing Returns Imply Monopoly? *Math. Soc. Sci*, 19, 235–251.

Arthur, W. B. (1994). *Increasing Returns and Path-Dependence in the Economy*. Ann Arbor: University of Michigan Press.

Arthur, W. B., Ermoliev, Y., & Kaniovski, Y. (1987). Path-Dependent Processes and the Emergence of Macro-Structure. *Eur. J. of Operation Research*, 30, 294–303.

Atkinson, A. B., & Stiglitz, J. (1969). A New View of Technological Change. *Econ. J.*, 79, 573–578.

Auer, W. (1960). Exemples Pratique de Recuperation de Chaleur dans les Installations Combinees de Turbines a Gaz et a Vapeur. *Brown Boveri Journal*, 47, 800–825.

Aversi, R., Dosi, G., Fagiolo, G., Meacci, M., & Olivetti, C. (1999). Demand Dynamics with Socially Evolving Preferences. *Industrial and Corporate Change*, 8.

Azariadis, C. (1996). The Economics of Poverty Traps. Part One: Complete Markets. *J. of Economic Growth*, 1, 449–486.

Azariadis, C., & Drazen, A. (1990). Threshold Externalities in Economic Development. *Quart. J. Econ.*, 105, 501–526.

Bak, P., H. Flyvbjerg, & B. Lautrup (1994). Evolution and Co-Evolution in a Rugged Fitness Landscape. In C. G. Langton (Ed.), *Artificial Life III. SFI Studies in the Sciences of Complexity*. Redwood City, CA: Addison-Wesley.

Bassanini, A. P. (1997). Localized Technological Change and Path Dependent Growth. IIASA Interim Report IR-97-086.

Bassanini, A. P., & Dosi, G. (1998, unpublished manuscript). Competing Technologies, Technological Monopolies, and the Rate of Convergence to a Stable Market Structure. Pisa: S. Anna School of Advanced Studies.

Bassanini, A. P., & Dosi, G. (2000). Heterogeneous Agents, Complementarities, and Diffusion: Do Increasing Returns Imply Convergence to International Technological Monopolies? In D. Delli, M. Gatti, Gallegati & A. Kirman (Eds.), *Market Structure, Aggregation and Heterogeneity*. Berlin/New York: Springer Verlag.

Becker, G. S., Murphy, K. M., & Tamura, R. (1990). Human Capital, Fertility, and Economic Growth. *J. Pol. Econ.*, 98, S12–S37.

Benhabib, J., & R. Perli (1994). Uniqueness and Indeterminacy: On the Dynamics of Endogenous Growth. *J. Econ. Theory*, 63, 113–142.

Bernheim, B. D. (1994). A Theory of Conformity. *J. Pol. Econ.*, 102, 841–877.

Boldrin, M. (1992). Dynamic Externalities, Multiple Equilibria, and Growth. *J. Econ. Theory*, 58, 198–218.

Boldrin, M., & Levine, D. (1998, unpublished manuscript). *Economic Growth with Perfect Competition I: Homogeneous Agents*.

Boyer, R. *Elements for an Institutional Approach to Economics*, Paris, CEPREMAP (mimeo).

Brezis, E. S., & Krugman, P. (1997). Technology and the Life Cycle of Cities. *J. Econ. Growth*, 2, 369–383.

Brezis, E. S., Krugman, P., & Tsiddon, D. (1993). Leapfrogging in International Competition: A Theory of Cycles in National Technological Leadership. *Amer. Econ. Rev.*, 83, 1211–1219.

Brock, W. A., & S. N. Durlauf (1995). *Discrete Choice with Social Interactions I: Theory*. NBER Working Paper No. 5291.

Brock, W. A., & Durlauf, S. N. (forthcoming). A Formal Model of Theory Choice in Science. *Econ. Theory*.

Brock, W. A., & A. G. Malliaris (1989). *Differential Equations, Stability and Chaos in Dynamic Economics*. Amsterdam: North-Holland.

Brynjolfsson, E., & Kemerer, C. F. (1996). Network Externalities in Microcomputer Software: An Econometric Analysis of the Spreadsheet Market. *Management Science*, 42, 1627–1647.

Caplin, A., & Leahy, J. (1994). Business as Usual, Market Crashes, and Wisdom After the Fact. *Amer. Econ. Rev.*, 84, 548–565.

Church, J., & Gandal, N. (1993). Complementary Network Externalities and Technological Adoption. *Int. J. of Industrial Organization*, 11, 239–260.

Cimoli, M., & Soete, L. (1992). A Generalized Technology Gap Trade Model. *Economie Appliquee*, 45, 33–54.

Cipolla C. M. (1965), *Guns and Sails in the early Phase of European Expansion, 1400-1700*. London: Collins.

Cowan, R. (1990). Nuclear Power Reactors: A Study in Technological Lock-In. *J. of Econ. History, 50*, 541–567.

Cowan, R., & Gunby, P. (1996). Sprayed to Death: Path-Dependence: Lock-In and Pest Control Strategies. *Econ. J., 106*, 521–542.

Cozzi, G. (1997). Exploring Growth Trajectories. *J. Econ. Growth, 2*, 385–398.

Cusumano, M. A., Y. Milonadis, & R. S. Rosenbloom (1992). Strategic Maneuvering and Mass-Market Dynamics: The Triumph of VHS Over Beta. *Business History Rev, 66*, 51–94.

David, P. (1975). *Technical Choice, Innovation and Economic Growth*, Cambridge: Cambridge University Press.

David, P. (1985). Clio and the Economics of QWERTY. *Am Econ. Rev., Papers and Proceedings, 75*, 332–372.

David, P. (1993). Path-dependence and Predictability in Dynamic Systems with Local Network Externalities: A Paradigm for Historical Economics. In D. Foray & C. Freeman (Eds.), *Technology and the Wealth of Nations*. London: Pinter.

David, P. (1995). Are the Institutions the Carriers of History? *Structural Change and Economic Dynamics, 6*.

David, P. (1986). *Path-Dependence and the Quest for Historical Economics*, unpublished manuscript, Stanford: Stanford University.

Dosi, G. (1982). Technological Paradigms and Technological Trajectories. *Research Policy, 11*, 142–167.

Dosi, G. (1984). *Technical Change and Industrial Transformation*. London: MacMillan.

Dosi, G. (1988). Sources, Procedures and Microeconomic Effect of Innovation. *J. Econ. Lit., 26*, 1120–1171.

Dosi, G. (1995). Hierarchies, Market and Power: Some Foundational Issues on the Nature of Contemporary Economic Organization. *Industrial and Corporate Change, 4*, 1-19.

Dosi, G. (1997). Opportunities, Incentives and the Collective Patterns of Technological Change. *Econ. J., 107*, 1530–1547.

Dosi, G., & Coriat B. (1998a). The Institutional Embeddedness of Economic Change. An Assessment of the Evolutionist and Regulationist Research Programs. In K. Nielsen & B. Johnson (Eds.), *Institutions and Economic Change*. Aldershot: Elgar.

Dosi, G., & Coriat, B. (1998b). Learning How to Govern and Learning How to Solve Problems. On the Double Nature of Routines as Problem Solving and Governance Devices. In Chandler A.D., Hagitröm P. and Sölvell Ö. (Eds.), *The Dynamic Firm*. Oxford: Oxford University Press.

Dosi, G., Ermoliev Y., & Kaniovski, Y. (1994). Generalized Urn Schemes and Technological Dynamics. *J. Math. Econ, 23*, 1–19.

Dosi, G., Fabiani, S., Aversi, R., & Meacci, M. (1994). The Dynamics of International Differentiation. A Multi-Country Evolutionary Model. *Industrial and Corporate Change, 3*, 225-242.

Dosi, G., C. Freeman and S. Fabiani (1994a). The process of economic development. Introducing some stylized facts on technologies, firms and institutions. *Industrial and Corporate Change, 3*, 1-45.

Dosi, G., Freeman, C. R., Nelson, G., Silverberg, & L. Soete (1988). *Technical Change and Economic Theory*. London: Francis Pinter and New York: Columbia University Press.

Dosi, G., & Kaniovski, Y. (1994). On Badly Behaved Dynamics. *J. of Evolutionary Econ., 4*, 93–123.

Dosi, G., & Kogut, B. (1993). National Specificities and the Context of Change: The Co-evolution of Organization and Technology. In Kogut B. (Ed.), *Country Competitiveness—Technology and Re-organization of Work*. Oxford/New York: Oxford University Press.

Dosi, G., & J. S. Metcalfe. (1991). On Some Notions of Irreversibility in Economics. In P. P. Saviotti & J. S. Metcalfe (Eds.), *Evolutionary Theories of Economic and Technological Change*. Chur: Harwood.

Dosi, G. & Orsenigo, L. (1988). Coordination and Transformation: An Overview of Structures, Behaviors and Change. In Dosi et al.

Duesenberry, J. (1949) *Income, Savings and the Theory of Consumer Behavior*. Cambridge, MA, Harvard University Press.

Economides, N. (1996). The Economics of Networks. *Int. J. of Industrial Organization, 14*, 673–99,

Ethier, W. J. (1982). Decreasing Costs in International Trade and Frank Graham's Argument for Protection. *Econometrica*, 50, 1243–1268.

Fagerberg, J. (1988). International Competitiveness. *Econ. J.*, 98, 355-374.

Fagerberg, J., Verspagen, B., & von Tunzelmann, N. (Eds.). (1995). *The Dynamics of Technology, Trade, and Growth*. Aldershot: Elgar.

Farrell, J., & Saloner, G. (1985). Standardization, Compatibility and Innovation. *Rand J. of Econ.*, 16, 70–83.

Farrell, J., & Saloner, G. (1986). Installed Base and Compatibility: Innovation, Product Preannouncements, and Predation. *Amer. Econ. Rev*, 76, 940–955,

Freeman, C. (1992). *The Economics of Industrial Innovation*, 2nd ed. London: Pinter.

Foster, D., & Young , H. P. (1990). Stochastic Evolutionary Game Dynamics. *Theoretical Population Biology*, 38, 219–232.

Fudenberg, D., & D. Kreps (1993). Learning Mixed Equilibria. *Games Econ. Behavior*, 4, 320–367.

Gerschenkron, A. (1962). *Economic Backwardness in Historical Perspective*. Cambridge, MA: The Belknap Press of Harvard University Press.

Glaziev, S., & Kaniovski, Y. (1991). Diffusion of Innovations under Conditions of Uncertainty: A Stochastic Approach. In N. Nakicenovic & A. Gruebler (Eds.), *Diffusion of Technologies and Social Behavior*. Berlin: Springer.

Gould, S. J. (1977). *Ever Since Darwin*. New York: Norton.

Granovetter, M. (1985). Economic Action and Social Structure: The Problem of Embeddedness. *American Journal of Sociology*, 51, 481–510.

Gruebler, A. (1986). Diffusion: Long Term Patterns and Discontinuities. *Technological Forecasting and Social change*, 39, 159–180.

Henderson, R., & Clark, K. (1990). Architectural Innovation. *Administrative Sci. Quart*, 35, 9–30.

Islas J. (1997). Getting round the lock-in in electricity generating systems: The case of gas turbines. *Res. Policy*, 26, 49-66.

Kaldor, N. (1957). A Model of Economic Growth. *Econ. J.*, 68, 591–624.

Kaldor, N. (1981). The Role of Increasing Returns, Technical Progress and Cumulative Causation in the Theory of International Trade and Economic Growth. *Economie Appliquee*, 30, 11–30.

Kaldor, N., & Mirlees, J. A. (1962). A New Model of Economic Growth. *Rev. Econ. Studies*, 29, 174–190.

Kaniovski, Y., & Young, H. P. (1995). Learning Dynamics in Games with Stochastic Perturbations. *Games and Econ. Behavior*, 11, 330–363.

Katz, M. L., & Shapiro, C. (1985). Network Externalities, Competition, and Compatibility. *Amer. Econ. Rev.*, 75, 424–440.

Katz, M. L., & Shapiro, C. (1986b). Technology Adoption in the Presence of Network Externalities. *J. Pol. Econ*, 822–841.

Katz, M. L., & Shapiro, C. (1992). Product Introduction with Network Externalities. *J. of Industrial Econ.*, 40, 55–84.

Katz, M. L., & Shapiro, C. (1994). Systems Competition and Network Effects. *J. of Econ. Perspectives*, 8, 93–115.

Kauffman, S. A. (1989). Adaptation in Rugged Fitness Landscapes. In D. L. Stein (Ed.), *Lectures in the Science of Complexity, SFI Studies in the Sciences of Complexity* (vol. 1). Redwood City, CA: Addison–Wesley.

Kauffman, S. A. (1990). *Origins of Order*. Oxford: Oxford University Press.

Kauffman, S. A., & Johnsen, S. (1991). Co-Evolution to the Edge of Chaos. In C. G. Langton et al. (Eds.), *Artificial Life II, SFI Studies in the Sciences of Complexity*. Redwood City, CA: Addison-Wesley.

Kauffman, S. A., & S. Levin (1987). Towards a General Theory of Adaptive Walks on Rugged Landscapes. *J. Theoretical Biology*, 128, 11–45.

King, M., & Robson, M. (1993). A Dynamic Model of Investment and Endogenous Growth. *Scandinavian J. of Econ*, 95, 445–66.

Krugman, P. (1981). Trade, Accumulation, and Uneven Development. *J. Dev. Econ*, 8, 149–161.

Krugman, P. (1987). The Narrow Moving Band, the Dutch Disease, and the Competitive Consequences of Mrs. Thatcher. *J. Dev. Econ.*, 27, 41–55.

Krugman, P. (1991). *Geography and Trade.* Cambridge, MA: MIT Press.

Levin, R., Cohen, W., & Mowery, D. (1985). R&D Appropriability, Opportunity and Market Structure. *Am. Econ. Rev., Papers and Proceedings, 75,* 20–24.

Levinthal, D. Adaptation on Rugged Landscapes. *Management Science.*

Levinthal, D. (2000). Organizational Capabilities in Complex Worlds. In G. Dosi, R. Nelson & S. Winter (Eds.), *The Nature and Dynamics of Organizational Capabilities.* Oxford: Oxford University Press.

Liebowitz, S. J., & Margolis, S. E. (1994). Network Externality: An Uncommon Tragedy. *J. of Econ. Perspectives, 8,* 133–150.

Liebowitz, S. J., & S. E. Margolis (1995). Path-Dependence, Lock-In and History. *Journal of Law, Economics and Organization, 5,* 205–226.

Lucas, R. E. Jr. (1988). On the Mechanics of Economic Development. *J. Mon. Econ., 22,* 3–42.

Lundvall, B. A. (Ed.). (1992). *National Innovation Systems.* London: Pinter.

Majumdar, M., Mitra, T., & Nyarko, Y. (1989). Dynamic Optimization under Uncertainty: Non-Convex Feasible Sets. In G. R. Feiwel (Ed.), *Joan Robinson and Modern Economic Theory.* New York: New York University Press.

Malerba, F. (1987). *The Semiconductor Business.* Madison, WI: Wisconsin University Press.

March, J. (1991). Exploitation and Exploration. *Organization Science.*

McCombie, J. S. L., & A. P. Thirlwall (1994). *Economic Growth and the Balance-of-Payments Constraint.* New York: St. Martins's Press.

Medio, A. (1992). *Chaotic Dynamics.* Cambridge: Cambridge University Press.

Milgrom, P., & J. Roberts (1990). The Economics of Modern Manufacturing: Technology, Strategy, and Organization. *Amer. Econ. Rev., 80,* 511–528.

Myrdal, G. (1957). *Economic Theory and Underdeveloped Regions,* London: Duckworth.

Nakicenovic, N., & A Gruebler (1991). Long Waves, Technology Diffusion, and Substitution. *Review, 14,* 313–342.

Nelson, R. (1981). Research on Productivity Growth and Productivity Differences: Dead Ends and New Departures. *J. of Econ. Lit., 19,* 1029–1064.

Nelson, R. (1995). Recent Evolutionary Theorizing About Economic Change. *J. of Econ. Lit., 33,* 48–90.

Nelson, R., & B. Sampat 1998, (Unpublished manuscript). *Making Sense of Institutions as a Factor of Economic Growth,* New York: Columbia University.

Nelson, R., & S. G. Winter (1982). *An Evolutionary Theory of Economic Change.* Cambridge, MA: Belknap Press of Harvard University Press.

Oren, S., & S. Smith (1981). Critical Mass and Tariff Structure in Electronic Communications Markets. *Bell J. of Econ, 12,* 467–487.

Perez, C., & L. Soete (1975). Catching Up and Windows of Opportunity. In G. Dosi et al. (Eds.)(1988).

Pfenninger, H., & G. Yannakopoulos (1975). Centrales a Vapeur avec Chaudiere a Foyer Suralimenté. *Brown Boveri Journal, 62,* 285–308.

Prigogine, I. (1980). *From Being to Becoming.* New York: Freeman.

Prigogine, I., & I. Stengers (1984). *Order out of Chaos.* London: Heinemann.

Rohlfs, J., A. (1974). Theory of Interdependent Demand for a Communication Service. *Bell J. of Econ., 5,* 16–37.

Romer, P. (1990). Endogenous Technological Change. *J. Pol. Econ., 98,* S71–S102.

Rosenberg, N. (1976). *Perspectives on Technology.* Cambridge: Cambridge University Press.

Rosenberg, N. (1982). *Inside the Black Box.* Cambridge: Cambridge University Press.

Saviotti, P. P., & Trickett, A.(1992). The Evolution of Helicopter Technology, 1940–1986. *Economics of Innovation and New Technology, 2,* 111–130.

Solow, R. (1956). A Contribution to the Theory of Economic Growth. *Quart. J. Econ., 70,* 65–94.

Stoneman, P. (Ed.). (1995). *Handbook of the Economics of Innovation and Technological Change.* Oxford: Blackwell.

Tell, F. (1997, Unpublished manuscript). *Innovation, Abandonment and Technological Progress: Path-Dependence and Knowledge Management in the High Voltage Power Transmission Industry* Laxenburg, Austria: IIASA

Thirlwall, A. P. (1979). The Balance of Payments Constraint as an Explanation of International Growth Differences. *BNL Quarterly Review, 28*, 45–53.

Tushman, M. L., & J. P. Murmann (1998). Dominant Designs, Technology Cycles, and Organizational Outcomes. In B. Staw & L. L. Cummings (Eds.), *Research in Organizational Behavior.* Greenwich, CT: JAI.

Verdoorn, P. J. (1949). Fattori che Regolano lo Sviluppo della Produttivitá del Lavoro. *L'Industria, 1,* 3–10.

West, L. J. (1994). Keyboard Efficiencies Revisited: Direct Measures of Speed on the Dvorak and Standard Keyboards. *The Delta Pi Epsilon Journal, 36,* 49–61.

White, A. O. (1956). The Place of the Gas Turbine in Electric Power Generation. *Combustion, 27,* 49–56.

Zeckhauser, R. (1968). Optimality in a World of Progress and Learning. *Rev. Econ. Studies, 35,* 363–365.

3

Unpacking Path Dependence: Differential Valuations Accorded History Across Disciplines

Paul M. Hirsch
James J. Gillespie
Northwestern University

The concept of *path dependence*, originating in Arthur's (1989) and David's (1985) studies of how past actions affect specific technological trajectories, offers a welcome opportunity to reassess long-standing issues and debates, across and within disciplines, concerning the role and importance of history and temporality. Although the questions raised by this topic are not new, the combination of recent developments in the hard sciences (e.g., chaos theory) and the theoretical extensions of concepts in the social sciences (e.g., path creation), with their important application to technological trajectories, invite a broader inquiry. In this chapter, we assess what these new perspectives contribute to, and can learn from, disciplines and subfields that have long considered many of the same issues. We also consider the triadic reciprocal relationship between path creation, path dependence, and path destruction. In doing so, we suggest a more process-oriented examination of the paths accompanying knowledge vectors and technological trajectories.

Much of the chapter to follow provides a lay scholar's overview of the concept of path dependence and an exploration of how social science disciplines outside of economics (this concept's home base) have long discussed some of the issues raised and rediscovered in the literature on path dependence. This chapter "unpacks" the concept of path dependence, highlighting what is new and unique about its contribution, and in what areas it complements or takes issue with models and frameworks in economics and behavioral science. We encompass a variety of behavioral and social science disciplines: anthropology, economics, history, management, sociology, and political science. After reviewing the relation of path dependence to each discipline individually, we propose ways in which this concept can be expanded to encompass issues ignored so far, and ways it might, in turn, make contributions to the disciplines.

Within and between social science disciplines, what are the differential valuations accorded history? Across many relevant disciplines, we find historical issues discussed under a variety of titles and rubrics. The value assigned *historical inquiry* diverges greatly, ranging from *significant* (in either a positive or negative direction), to *none at all* when its relevance is denied from the outset. In the economic and social sciences, then, the answers to these questions are by no means settled. A lack of consensus still surrounds the answers to a controversial question like: "Does knowing about the early events and subsequent paths followed by a phenomenon enhance, detract from, or make absolutely no difference to the substance of our analyses and explanations of the present, or predictions for the future?" Each field and subfield's answer follows a conceptual filter and preferred paradigm, subscribed to by its home discipline. Ruttan (chap. 4, this volume) identified at least three major approaches to the examination of the sources of technological development: induced technical change, evolutionary theory, and path dependence. Even within the path dependence approach, which is the focus of our examination, one finds varying perspectives, with many unsettled issues.

We raise and address each of the following questions: (a) What will the focus on path creation and dependence contribute to the understanding of technological development beyond that already provided by studies in these basic disciplines? (b) What can these other disciplines contribute to our understanding of the concepts of path creation, dependence, and destruction? (c) What can path creation and dependence contribute to other disciplines, in terms of better understanding and explaining the role of history and temporality? In our view, these questions cannot be addressed by an insular examination focused only on the path dependence literature; a broader, more critical inquiry is required.

In this chapter, we thus (a) examine the relation of path creation and dependence to the larger intellectual contexts in which historical paths have been analyzed across the social sciences; (b) tease out how elements from long-standing debates and discussions of this topic can both inform, and be informed by, the more recent revival of interest stimulated by the research of scholars in other intellectual fields; (c) inquire into how findings about path dependence in one arena may be cycled back and contribute to other arenas; and (d) argue that most technological development is best conceptualized as continuous processes wherein creation, dependence and destruction of paths are highly interrelated.

OVERVIEW OF PATH DEPENDENCE

Path dependence is an interdisciplinary concept, first developed by economic historians that has acquired increasing currency among many social scientists. David (1985) defined path dependence: "A path-dependent sequence of economic changes is one in which important influences upon the eventual outcome can be exerted by temporally remote events, including happenings dominated by chance elements rather than systematic forces" (p. 332). Arthur (1989) delineated four self-reinforcing elements of path dependence: large fixed costs, learning effects, coordination effects, and adaptive expectations.

At critical historical junctures, two or more technologies offering comparable services compete for resources, but the first to establish a significant foothold will benefit from positive feedback mechanisms and become dominant. Footholds can persist because each technological trial provides experience that increases the payoffs to that technology in subsequent trials, while decreasing uncertainty about its efficacy. Which technology gains the initial foothold depends on diverse factors such as business connections, political climate, whims of early developers, and successful performances by prototypes. Path dependence dynamics become particularly compelling when two scientific and technological fields stand in close, complementary relationship. For example, the bacterial theory of diseases (i.e., bacteriology) had to be coupled with observational technology (i.e., the microscope) before additional scientific progress could be made; in this sense, two paths had to interact before the dependent path could be formed.

The path dependence literature makes a distinction between ergodicity and nonergodicity. *Ergodic* (i.e., non-path dependent) processes are those wherein outcomes are not importantly affected by the sequence of events. In contrast, *nonergodic* (i.e., path dependent) processes do not automatically converge to a predictable, well-specified distribution of outcomes. Whereas small events are flexibly averaged away under ergodocity, their consequences inflexibility persist under nonergodocity. Liebowitz and Margolis (1995) further divided path dependence according to first, second, and third degrees. With first degree path dependence, there is sensitivity to starting points but no necessary inefficiency. With second degree path dependence, there is sensitivity to starting points and inefficient outcomes that are costly to alter. With third degree path dependence, not only do intertemporal effects propagate error, the inefficient outcomes are avoidable. Liebowitz and Margolis (1995) argued that, "Third degree path dependence is the only form of path dependence that conflicts with the neoclassical model of relentlessly rational behavior leading to efficient, and therefore predictable, outcomes" (p. 207).

One of the basic arguments is whether path dependence operates in the real world (whether first, second, or third degree) or whether it is purely a theoretical artifact. Beyond this basic ontological issue, there are numerous other vexing issues focusing on the details of this concept. Path dependence emphasizes that, at key focal points, small decisions can have large future consequences. An interesting question is why path dependence depends only on "small events." Presumably, large events could just as well establish technological trajectories, but these cases would be less interesting because they are more obvious. What is considerably less intuitive is that major changes could follow from seemingly insignificant changes (see Bassanini & Dosi, chap. 2, this volume), for a discussion of big and small events.

Another issue concerns the predictability of path dependence. According to Arthur (1989), a path dependent outcome is "predictable if the small degree of uncertainty built in 'averages away' so that the observer has enough information to predetermine market shares accurately in the long-run" (p. 118). The predictability of future developments is sometimes very low but at other times very high, when the system becomes locked in. A key question is whether the critical or focal points are predictable: Could a researcher determine before hand which technology will win out? The answer would be *no* in those cases where path dependence is caused by insignificant and random events, which turns out to be a nontrivial percentage of cases. (By "chance" or "random" events, path dependence scholars usually mean those beyond the *ex ante* knowledge of the observer.) Arthur (1989) presented a model of technological competition in which transient and unpredictable factors exert great leverage over outcomes.

Another set of issues and questions concerns the reciprocal relationship between path creation and path dependence. A path must have an origin, a creation in some time and place. Yet, the literature on path dependence devotes relatively little attention to path creation. Before we can arrive at a complete understanding of the processes and outcomes driving technological development, the concepts of path creation and dependence must be integrated. An understanding based exclusively on path dependence could, both in theory and in practice, trap us, and thereby limit our abilities to maneuver into the future by reproducing the past (Karnoe & Garud, 1997).

Having noted that the past matters, it would be useful for path dependence scholars to also address how the present matters.

The path dependence literature contains many interesting examples and case studies. Rosenberg (1994) argued that the abundance of petroleum in the United States caused a whole set of path dependent phenomena that shifted U.S. industry to the use of oil and gas. Rockoff (1994) discussed how Akron, Ohio came to be the center of the American tire industry through a series of path creating factors, including the decision of Benjamin Franklin Goodrich to locate there. Other examples include: (a) the U.S. nuclear industry and why it is almost completely dominated by light-water reactors (Bupp & Derian, 1978), (b) the U.S. automobile industry and why it is dominated by petroleum powered cars (McLaughlin, 1954); (c) the triumph of the VHS videotaping technology over the Beta format (Arthur, 1990); and (d) the adoption of digital rather than analog for the internet (Edwards, 1996).

The best known example of path dependence involves the QWERTY keyboard. David (1985) argued that the inferior (at least judged by typing speeds) QWERTY keyboard came to dominate over the technologically superior Dvorak Simplified Keyboard for a variety of path dependent reasons. David posits that there are three basic factors that caused the QWERTY keyboard to "win": technical interrelatedness, economies of scale, and quasiirreversibility of investment. Liebowitz and Margolis (1990) offered a broad critique of David's position, arguing that QWERTY is not a true example of market failure. Whatever the merits of this critique, the QWERTY example continues to resonate with scholars of path dependence and observers of the field.

The diversity of frameworks modeling the role of history is illustrated by the richness of terminologies that have developed to address many sides of path creation and dependence. Our compilation and decoding of the language and interpretation of path dependence yields several related terms and descriptors from various disciplines: chaos theory, competency traps, cumulative causation, feedback loops, inertia, irreversibility, knowledge vectors, lags, lock-ins, lock-outs, nonerogodic processes, nonrational escalation of behavior, organizational commitment, momentum, punctuated equilibrium, stickiness, structuration, sunk costs, technological determinism, technological trajectories, threat rigidity, unintended consequences, and virtuous/vicious cycles.

The development of concepts analogous to path dependence is not restricted to the social sciences. *Chaos theory*, derived from physics and mathematics, emphasizes that minor changes in initial conditions can produce major critics effects. *Evolutionary biology* emphasizes that key events can cause development to proceed down one branch, while foreclosing other branches. Also in biology, *contingency theory* emphasizes both the irreversibility of natural selection and the unpredictability of fitness criteria. Path dependence similarly notes the irreversibility and unpredictability of technology development.

The proliferation in terminology involves more than just semantics. It shows that diverse disciplines offer potentially unique identifications, definitions, and explanations for the consequences past actions and events have for subsequent technological development. One cannot arrive at a truly integrated perspective on the economic, technical, and institutional forces that create and sustain technological paths without first integrating the diverse frameworks and terminology that different disciplines apply to the phenomenon. Much of this terminology does not find its way into the work of path dependence scholars. Few articles on path dependence make a serious attempt to integrate the terminology and existing constructs that various disciplines, particularly those in the social sciences, have generated to describe and explain the impact of history on the present and the future. In Table 3.1, we present a typological overview of the predominant valuations of history in different social sciences disciplines and fields. This table is not intended to be comprehensive. Our goal is simply to give the reader a general sense of the history related terminology that is invoked by various social sciences.

EXAMINATION OF SPECIFIC DISCIPLINES

Anthropology. The relationship between anthropology and history has always been close, reflecting the influence of evolutionary theory and ethnohistory. Prior to the 1970s, anthropologists used history to (a) provide historical background and preface to a larger study, (b) contrast one geographic area or social group with another in a different time period, or (c) serve as a data bank for examples. History was valued only for its ability to supply background, contrast, or data. These uses of history were ahistorical because they largely ignored the emergence, maintenance, and transformation of social groups, relationships, and institutions over time.

However, anthropology increasingly utilizes history in ways that are intellectually integral to research agendas, reflecting the growing consensus that all of cultural and social life has historicity. According to Kellogg (1994):

> Instead of static pictures of societies in an ethnographic present, anthropologists increasingly have sought to describe a dynamically changing world in which groups survive by making decisions, altering strategies, and changing, sometimes consciously and sometimes as an unexpected consequence of previous decisions. (p. 12)

This anthropological use of history shares similarities with the path dependence approach. Scholars of path dependence also seek to describe a dy-

TABLE 3.1

Typology of Predominant Valuations of History Across Disciplines

	Anthropology	Economics	History	Management	Political Science	Sociology
Use of History	Background; contrast social groups	Data source; validation of present theories; stimulus for new theories	Description, explanation, understanding, serve political agendas	Examples, case studies	Modeling long-term political change	Data source, theorizing about social processes, studying past social groups
Source of Decisions and Outcomes	Conscious decision making; unintended consequences	Conscious decision making; game theoretic consideration of another's behavior	Path dependence, unexpected consequences	Conscious decision making	Leadership, mass movements	Collective decision making
Key Terminology	Heritage, life states, patterns of change	Increasing returns, sunk costs, lags, stickiness, standards, coordination failure	Unintended consequences, technological determinism	Organizational commitment, competency traps	Escalation of actions	Structuration
Primary Domain	Cultures	Markets, firms, households	Events, intellectual and social processes	Organizations	Public sector, the polity	Societies
Goal of Actors	Survival of culture	Utility maximization	Social determined	Profit making	Power accumulation	Persistence of society
Logic of Action	Adaptation	Rational actor, bounded rationality	Social construction	Cost benefit	Indeterminate	Identity

namically changing world, emphasizing the importance of crucial decisions and strategies that determine which paths are taken (or forsaken). Both path dependence and this anthropological approach allow that outcomes can result from either conscious decision making or as an unexpected consequence of previous decisions.

Different areas of anthropology employ history in service of very different objectives; for example, with anthropologists oriented toward political economy using history largely as a source of longitudinal data and with anthropologists oriented toward culturalism using history as a fundamental explanatory building block. Nevertheless, history serves as a bridge between these factions within anthropology. It is interesting to contrast the anthropological approach with that of the discipline of history:

> [H]istorians and anthropologists tend to hold fundamentally different ideas about what history is and how it should be done. Anthropologists often use history to understand emergent social or cultural formations, while historians tend to be preoccupied with the pastness of the past. (Kellogg, 1994, p. 13). If fit into this comparison, path dependence falls closer to the anthropological approach than to the historical approach. Scholars of path dependence are not interested in the pastness of the past, for its own sake; they draw on history to explain current and emergent economic and technological formations.

Economics.   Neoclassical economists generally have been critical of path dependent theories. A key assumption of path dependence is increasing returns to scale, which runs counter to the paradigm of modern equilibrium economics. Also, the fundamental path dependent assumption of irreversibility is hostile to the neoclassical premise of smooth substitution possibilities. Until recent years, economists have not incorporated these assumptions of scale economies and irreversibility, making it more difficult to achieve conceptual integration with path dependence appraches.

Yet, the discipline of economics does use history for several purposes: advocacy, validation, and statistical power. Economists routinely play a prominent role in public policy debates. But

> Neither policymakers nor the general public are likely to understand nor be persuaded by the sort of mathematical–theoretical arguments and statistical proofs that are the normal fare of the professional journals. The economist is forced back upon the historical illustration to make this point (Rockoff, 1994, p. 63).

Economists also use history for comparative analysis and validation: The logic being that, if economic theory works well in diverse places and times, it can be applied to modern markets.

It is important to note that historical analysis is not synonymous with path dependence analysis, in the same way that *history* is not necessarily identical to *time*. Although economics often uses history without fully engaging it, the discipline has engaged issues closely related to path dependence (e.g., see the growing body of economic literature on standards and externalities). Some economists argue that most of the concepts involving path dependence can be captured under the traditional rubric of external and internal increasing returns to scale. They also argue that path dependence repeats many of the insights achieved by early economic scholars like Hayek (1939) and Marshall (1920). Because current data often lack sufficient variance, economists cannot identify underlying relationships using statistical techniques; to address this difficulty, lengthy time series data are employed. However, this use of history largely ignores the social context in which the data are generated. Economists utilize history primarily to illustrate the power of their own theories, often employing statistical time series without inquiring too closely into contextual reliability. Unfortunately, historians make few attempts to engage and challenge economists who venture into their domain of historical analysis.

History.   There are both similarities and differences between historical and path dependence approaches. If there is any one idea central to historical ways of thinking, it is that the order of things makes a difference and events that "are not single properties, or simple things, but complex conjunctures in which complex actors encounter complex structures" (Abbott, 1994, p. 101). The idea that the order of things matters is obviously central to path dependence too. But in comparison to history, path dependence focuses more on generalizable causes and less on the almost paralyzing complexity of any given situation.

The discipline of history focuses on complex forms of change, synthesis, and interaction within societies over time, and the basic questions focus on evolution and transformation. Path dependence focuses on factors that trigger change, particularly those of a technical nature, and how these changes interact with and are synthesized into the existing economic structure. For historians, any particular outcome is framed as the product of a long sequence of contingent events, where the outcome could have been very different even if events had been only slightly different. This is consistent with the path dependence approach that stresses the potential consequentiality of seemingly small changes.

The discipline of history emphasizes accident, contingency, and process, so most historians are willing to forsake generalization and theory building in service of more fine-grained, comprehensive detailing of events and outcomes. History utilizes an extensive factual mastery of specific contextual situations to study qualitatively and quantitatively diverse economic, political, and social actors, events, and institutions. "History remains fact centered, at least so far; theory plays much less a role than in the other social sciences (Monkkonen, 1994, p. 8)." In terms of taking context seriously, path dependence is more effective than traditional neoclassical economic models but less effective than historical approaches.

More often than not, historians deliberately reject the positivistic search for generalizations and regularities so typical of other social sciences. In fact, historians revel in the ambiguity that attends their study of the irregularities and special cases that other disciplines try to eliminate or statistically control for: *Ceteris paribus* (i.e., all things being equal) is explicitly rejected. "History might almost be defined as the discipline which deals with things which are never equal and cannot be supposed to be equal" (Hobsbawn, 1981, p. 633). Path dependence is similar to the extent that accidental and contingent factors play an important role, but historians do not try to explain these factors using an ahistorical and acontextual generalized model, whereas scholars of path dependence make every attempt to explain processes and outcomes using rational actor economic models. Although path dependence research seeks to identify crucial inflection points that alter trajectories of economic development, it ultimately returns to a focus on the equilibrium so characteristic of neoclassical models. Although history is equally concerned with critical historical junctures, it does not place emphasis on the establishment of equilibriums.

Management.   Although the concept of path dependence is typically employed to explain technological processes and outcomes, similar concepts also arise in the management literature. Cohen and Levinthal (1990) noted that organizations have an "absorptive capacity" that allows them to take advantage of future opportunities because past actions have built up a set of skills and knowledge. Levitt and March (1988) noted that nonrational dynamics such as competency traps and superstitious learning can establish "positive feedback loops" that cause a firm to continue using an inferior procedure even when given the option of switching to a superior procedure. Baum and Singh (1994) and other population ecologists further elaborated the consequences of such inertia.

Tushman and Rosenkopf (1992) emphasized that, in the absence of a radical disruption, standardization and dominant designs can sharply restrict the degree of future variation in organizations. Staw (1981) noted that for a variety of organizational and psychological reasons (e.g., external justification, norms of consistency, and self-justification) decision makers may tend to nonrationally escalate their commitment to a course of action. Closely related is the idea of "threat rigidity" (Staw, Sandelands, & Dutton, 1981) emphasizing that, when faced with challenges and threats related to poor performance, organizations tend to rigidly adhere to current strategies instead of engaging in failure induced change.

A number of management scholars have commented on the role of "momentum" (Amburgey & Miner, 1992; Kelley & Amburgey, 1991; Miller & Friesen, 1980). Based on well-established norms, procedures, practices, and routines, existing organizational actions and decision-making processes can build up a momentum that makes it extremely difficult to alter strategic trajectories. Ghemawat (1991) noted that "commitment" involves the tendency of earlier choices to constrain later ones: "I define commitment as the degree of difficulty of flip-flops. More precisely, a strategy embodies commitment to the extent that, if adopted, it is likely to persist" (p. 15). Thus, the management literature has focused broadly on how the investment of resources by a firm creates path dependent strategies.

Political Science.   The connection between history and political science is rather close and, as one scholar notes, "History is past politics and politics is present history (Robertson, 1994, p. 113)." However, as has been the case in other disciplines, political science has experienced a waxing and waning in the importance accorded the role of history. In the 1950s and 1960s, the behavioral revolution drove political science toward an emphasis on cross-sectional data and objectified empirical analysis, thereby displacing detailed historical analysis of singular actors and events. Since the 1960s, history has returned to political science in two forms: (a) historical behaviorists using longitudinal data to model long-term political change and (b) "new institutionalists" using history to examine institutions and institutionalism.

Behavioral political scientists turned to the study of history when it became necessary to address questions that could not be resolved by examining cross-sectional variation at a single point in time. Yet, despite the use of historical data, the orientation of behavioral political scientists remains strongly tied to functionalist and rational actor perspectives. In contrasts, new institutionalists assume that past decisions critically shape current and future institutional development, making this approach less acontextual than historical behavioral political science. In this regard, the approach of new institutionalists is very similar to that of path dependence scholars.

Sociology.   In the 1960s, Parsonianism was the dominant perspective in sociology. With his deemphasis on social change, Parsons (1960) was not particularly solicitous of historical approaches to sociological issues. The Chicago and Columbia Schools of sociology, with their emphasis on communities and deviance, and reliance on case studies and field observation, served in opposition to the Parsonian perspective. During this time, quantitative empiricism was another important perspective that emerged in sociology. The quantitative empirical approach included a prominent role for general linear models, as well as a reification and operationalization of causality through data-driven variables. Most recently, the "neoinstitutional" or "new" institutional perspective has emerged. It examines historical issues, but largely only insofar as these function to strengthen and solidify existing paths. For more attention to path creation and the dynamics of change, the field is more directed to "old" institutionalism (Hirsch, 1997; Powell & DiMaggio, 1991; Scott, 1996).

Parsonianism, the Chicago School, and quantitative empiricism (prior to event studies) are markedly different theoretical and methodological approaches to the study of social phenomena, but they share a general failure to incorporate history into their programs of study. In the 1970s, historical sociology emerged to address this deficiency, by directly criticizing the ahistorical, acontextual orientation of the discipline and instead stressing social change, drawing heavily on Marxist perspectives. Even within the field of historical sociology, there are as least three distinctive uses of history: as a source of data over time; as a foundation for theorizing about social processes; and as a means of intensively studying past social groups.

The use of history primarily as a source of cross-sectional data often produces research that displays lessened sensitivity to historical context and contingency. It is not clear that such an approach represents a true integration of history and sociology. Also, sociologists often theorize secondary historical accounts as the data sources, without being sufficiently critical of the potentially biased and artifactual nature of such data. Sociologists who use history for social process theorizing focus on accident and contingency, in this regard being similar in approach to the discipline of history (Zald, 1990).

One can also view the role of history in sociology from the perspective of narrativism, a methodological and theoretical approach that portrays social phenomenon as a chronologically ordered sequence of complex and contingent events. Sociological narrativists reject rational choice and causally oriented theories in favor of approaches that emphasize storytelling and the importance of identity. The particular and the specific take priority over the collective and the statistical. The narrative approach effectively deals with situations sufficiently complex that use of functionalist data reduction statistical techniques would result in important information being discarded. Although the narrative approach does not reify causal analysis, neither does it completely abandon attempts to derive more abstract explanations (Merton, 1967, 1993). Weber emphasized the need to search for "ideal type" narratives as a method for generalizing.

Much like path dependence, narrativism is effective at incorporating instances where small, contingent events have significant ramifications. Narrativists also stress the path dependent nature of social processes. However, sociological narrativists routinely bring the actor and human agency back into explanations of social phenomena, whereas path dependent research rarely dwells on the role of individual actors. Because of its deep tradition in history and sociology, the narrative approach offers a promising link between the two disciplines and path dependence.

PROCESSES OF PATH DEPENDENCE, DESTRUCTION, AND CREATION

Examining the path dependence literature alongside the literature from multiple disciplines and fields considerably enriches our analysis of the path dependent process. It also suggests the need to incorporate path destruction and path creation processes into our analysis. For example, anthropology and history both examine the past with a level of detail and complexity that path dependence scholars would be well served to emulate. Political science focuses attention on the contested and political nature of technological development. Sociology reminds us that all technological development is embedded within broader cultural and societal frameworks. Management emphasizes the human agency and purposefulness that lies behind many technological innovations. Each discipline has a comparative advantage in studying technological development that path dependence cannot fully match, given its interdisciplinary nature. Yet, an enriched understanding of path-related processes, based on disciplinary and interdisciplinary research, generates insights, raises questions, and adds complexity. Based on our reading of the disciplinary literature, as well as on selected contributions to this volume, we now discuss some of these issues.

Path dependence processes. It is important to recognize the coevolutionary nature of path dependence. Most studies of path dependence focus exclusively on the technological dimensions. The *coevolutionary perspective* emphasizes that technical innovation is a process rather than an outcome. This process involves coevolution of social, cognitive, and institutional dimensions along with the technological dimension (Baum & Singh, 1994; Garud, Nayyar, & Shapira, 1997). There is an ongoing reciprocal relationship between technology and environment that both enables and constrains development of new technologies (Van de Ven & Garud, 1989, 1993). Coevolution has been described as involving the contingent shaping of one arena by activities and events in another arena (Karnoe & Garud, 1997). Given that multiple arenas (e.g., social, cognitive, and institutional) impinge on the technological arena, coevolution involving path dependence might be better termed multiple evolution or multivariate evolution.

This raises issues less well examined by the coevolutionary literature: What determines the extent to which the technical arena is loosely or tightly coupled (Weick, 1976) to another arena? What are the multiple levels of analysis on which co-evolutionary path dependence operates? What is the relationship between geographically distant arenas versus geographically local arenas? For example, at a macro level, the technical arena may be loosely coupled to a nation-level institutional arena (e.g., the U.S. patent system), but at a micro local level, the technical arena may be tightly coupled to a local-level social arena (e.g., the network of Silicon Valley computer scientists and electrical engineers). Path dependent co-evolution can involve loose or tight coupling, macro or micro levels, and distant or local arenas. Also, as Bassanini and Dosi (chap. 2, this volume) point out, a phenomenon can be path dependent at one level but path independent at another. These distinctions concerning level and location, and combinations and permutations thereof, are underexamined by scholars of path creation and dependence.

The economic geography (Storper & Walker, 1989) and flexible specialization literatures (Piore & Sabel, 1984) offer potential insights on the geographic dimensions of path dependence. Chapter 5 of this volume by Kenney and von Burg provides insights on how the coupling of technology production regimes to different local, geographic arenas (Silicon Valley and Route 128) produces divergence in economic success. In conjunction with earlier work (Florida & Kenney, 1997; Robertson & Langlois, 1995), the Kenney and Burg examination indicates that the social structure of innovation prevailing in a particular region constitutes a coevolving arena with the technology.

Although progress is being made on understanding the geographic dimensions of coevolution, other issues await more detailed examination. What is the relationship between the individual entrepreneur and path dependence? If much of the earlier literature on entrepreneurship overly attended to the role of individual entrepreneurs and their personal characteristics, the path dependence literature does not focus enough on the role of individual actors. This is ironic given the tendency of path dependence to "romance" small events and thereby ignore or deny larger social and institutional contexts.

The sociocognitive model of technology evolution does incorporate multiple levels of analysis: "[T]here are two processes that unfold simultaneously during the evolution of a technology. One is a process of inversion at the micro-level of individual cognition. The second is a process of institutionalization at the macro-level of shared cognition (Garud & Rappa,

1994, p. 359)." When considering these multiple levels of analysis, an unaddressed issue is what occurs when one system evolves more or less slowly than another system? Or when a system evolves more or less slowly than its parts? The coevolution perspective seems to assume that the rate of system-to-system (i.e., interpath) or system-to-parts (i.e., intrapath) evolution is comparable, but they could be very dissimilar.

Some paths are likely to exhibit more dependency and stability than other paths. Most technological systems consist of numerous integrated components, and some of these paths will be embedded within other paths:

> [A]lthough the QWERTY keyboard configuration has remained constant, other aspects of typewriting, such as the shift to electric typewriters and then computers, have changed dramatically. General theory is necessary to explain why some aspects of typewriter technology seem to be strongly path dependent and others do not. (Kiser, 1996, p. 263)

The other social science disciplines can assist in supplying this general theory. For example, psychology can reveal the general microbehavioral dynamics behind the tendency toward inertia in human decision making. Sociology and political science can reveal the general social and political dynamics surrounding resistance to change in technological systems.

This differential interpath dependency may be one of the keys to successful technological development: Stable technological trahectories are usually identifiable only because enough of the rest of the world is stable. If too many environmental arenas are in flux, there may not be enough stability to sustain technological progress. If not enough environmental arenas are in flux, excessive stability may stifle technological progress. Yet, in terms of stability, to the extent that path dependent outcomes are suboptimal, they do contain the seeds for their own destruction.

In terms of maximizing technological development, the "just embeddedness" perspective (Garud & Jain, 1996) offers insights into the optimal degree of stability and tight coupling versus instability and loose coupling. *Embeddedness* refers to the contingent nature of economic activity by virtue of its situation in broader arenas: cognitive, cultural, social, political, and institutional (Uzzi, 1997; Zukin & DiMaggio, 1990). The just embeddedness perspective addresses questions such as how to balance the beneficial coordinating effects with the detrimental constraining effects of institutional standards. In other words, what is the optimal fit between the coevolutionary technical and institutional paths? Garud and Jain (1996) offer just embeddedness as the solution that enables without constraining: "By 'just' embedded we mean that standards and the processes associated with them provide the coordination required to carry out technical activities in the present, and, at the same time, not constrain the migration of the technology to new functionalities in the future" (p. 39). Garud and Jain examine technical–institutional just embeddedness. Future research could explore other types of interpath just embeddedness (e.g., technical–cognitive, technical–social, or even technical–technical). Garud and Jain focus on the enabling and constraining effects of standards. Evaluation routines, which are themselves institutionalized, play a crucial role in determining which standards will or will not become institutionalized (Garud & Rappa, 1994). Within economics, there is a growing literature on standards and network externalities, much of it focused on straightforward technical or institutional features. However, there are standards that serve to coordinate human behavior rather than connect technology, which is not surprising given that routines and standards are types of social institutions (Langlois, 1986). Incorporating the work of Garud and Jain (1996), Langlois and Savage (this volume) note that the balancing of enablement and constraint is particularly interesting with regard to the interaction of public routines (i.e., standards) with the private routines of individuals and organizations. Thus, routines and standards also operate on multiple levels of analysis.

Path Destruction Processes.   In comparison to path dependence and path creation, scholars of economics, management, and technology have devoted relatively little attention to path destruction. Yet, we need to talk about path destruction because the transition from path dependence to path creation is typically contentious. Because power and politics play a crucial role in path competition and path destruction, nonmarket actors often play a crucial role. For example, Rao and Singh examine the crucial role nonmarket actors such as auto clubs, antispeed vigilante organizations, and engineering professions played in the early development of the automobile industry. The medical industry offers a potentially rich example of how professional organizations, principally those led by physicians, strongly influenced the development of standards (Langlois & Savage).

Path competition is not restricted to technological markets. If we define technology as consisting of artifacts, beliefs, and evaluation routines (Garud & Rappa, 1994), there is potential competition along all these dimensions. Similarly, there is typically path competition in the cognitive, political, social, and institutional environments. An underexamined direction for research on technological development is closer scrutiny of path destruction processes.

Path Creation Processes.   Technological development necessarily involves path creation. However, path dependence seems to leave little room for action, thus making it less attractive for those interested in change and voluntaristic action. David and Bunn's (1988) work on Edison and electricity technology shows that conflict can be an integral component of path dependence dynamics. And in theory, there is no reason why path dependence cannot be intentionally created, as well as unintentionally emergent. A great deal of financial and political capital may be expended in a purposeful effort to establish a particular technological path (Granovetter, 1992; Mokyr, 1990).

Much of the early literature on path dependence did a limited job of theorizing about how firms or industries could escape technological lock-in, but scholars are beginning to address this theoretical deficiency (Garud et al., 1997). As discussed earlier, the coevolutionary perspective offers a promising theoretical framework in which to model the continual, reciprocal interaction between technology and the broader environment (Garud & Rappa, 1994; Tushman & Rosenkopf, 1992; Van de Ven & Garud, 1993). Also, Garud and Nayyar (1994) offered the concept of *transformative capacity* to describe the ability of firms to exploit technological opportunities inside the firm. As such, the concept is an complement to the notion of absorptive capacity, which concerns utilization of external technology (Cohen & Levinthal, 1990).

Transformative capacity reminds us that path creation does not necessarily mean creating technology from entirely unknown, external elements. Many firms, particularly those in research and development (R&D) intensive industries, have storehouses of technical knowledge (i.e., latent path dependencies awaiting sociocognitive and technological enactment). Many revolutionary technological outcomes are accomplished using evolutionary strands (Garud, Nayyar, & Shapira, 1997), through processes of "cumulative synthesis" (Usher, 1954) and "bisociation" (Koestler, 1964). In part, path creation involves the bringing together of previously disparate technological path dependencies. It also involves combining latent paths (i.e., knowledge-in-waiting from the technical storehouses) with one or more existing paths.

Although in theory path dependence can be used to explain both efficient and inefficient outcomes, in practice it is typically employed to explain inefficiency. Why is the bulk of attention devoted to cases where path dependence leads to suboptimal market outcomes? Perhaps because these cases offered the most vivid challenges to neoclassical orthodoxy concerning the efficiency of markets. The growing literature on path creation might be termed *efficient path dependency* (highlighting the contrast to the previous focus on inefficient path dependency).

We know that "humans can not only respond to the world or shape it in ways that match their competencies, but they also have the ability to abandon their past and create a new future" (Garud et al., 1997, p. 346). This perspective allows scholars of path creation and dependence to generate prescriptive advice for business managers, public policy officials, and other key decision makers concerning how to strategically create pathways of technological development that will maximize firm and/or society utility (e.g., constructive technology assessment). Standing alone, without the conceptual companionship of path creation, path dependence places excessive emphasis on suboptimality and on maintaining internal consistency, thereby constraining future options and encouraging the nonrational escalation of commitment.

Many scholars describe path creation and technological development as fundamentally sociocognitive (Garud & Rappa, 1994), largely the product of social construction (Berger & Luckman, 1967; Bijker, Hughes, & Pinch, 1987). Innovation itself is a negotiated activity, where ideas, markets, and institutions coevolve. In such a world, technological fits do not just happen, but must be tailored (Garud, Nayyar, & Shapira, 1997). The Porac, Rosa, and Saxon chapter (chap. 8, this volume) provides a compelling case study showing how markets are socially constructed and fundamentally knowledge based, with the central economic competition involving ideas, not resources. And because of interpretative flexibility, any given technology may be given different meaning by different groups.

Much of the path dependence literature focuses on the developmental stage of technology. Path creation reminds scholars of the need to focus on the market for technology also, as this importantly affects development. "In understanding how a path dependency or technical trajectory develops it is important to focus upon the use to which a new technology is put" (Pinch, chap. 14, this volume). Product markets are enacted, so one needs to examine the socio-cognitive dynamics driving creation of markets for particular technologies. Examination of the demand side of technology more clearly shows how a revolutionary conceptual system can have a ripple effect through multiple product markets. For example, development of the minivan market helped change the way Americans think about the car/truck conceptual boundary, cognitively paving the way for the booming market in sport-utility vehicles (Porac, Rosa, Spanjol, Saxon, & Nielsen, chap. 8, this volume).

Rao and Singh (chap. 9, this volume) note that an important dimension of technological innovation is creation of new organizational forms. Technology is not freestanding; it requires institutional and organizational carriers for complete integration into the sociocognitive fabric of a culture. For example, in the absence of aggressive marketing and creative framing by established firms with a vested interest, it is unlikely that the sport utility vehicle could have attained its current popularity. In the case of radical technological innovation, a substantially new organizational form may be required to serve as carrier. This type of organizational speciation plays an indispensable role in path creation and in the evolution of organizational diversity (Rao & Singh, chap. 9, this volume). Looking at multiple levels of analysis, it is important to incorporate the role of individual institutional entrepreneurs. These are the individuals who are the driving forces behind creation of new organizational forms (e.g., the pivotal role public planner Robert Moses played in metropolitan New York City by creating a new set of independent governmental authorities to build bridges, highways, and other infrastructure).

CONCLUSION

Path dependence deserves credit for bringing history back into analysis of economic and technological development, stimulating economists and other social scientists to address the limitations of their largely ahistorical models. Path dependence points out the paradox of researchers using history as a data bank and source of economic variance, even while simultaneously assuming that history does not fundamentally differ from, and has no implications for, the present. In recent decades, across the social sciences, one observes a trend toward static analysis and theoretical abstraction. This has produced equilibrium analysis in economics, flourishing of the rational model actor in political science, and various versions of functionalism/structuralism in anthropology and sociology. This trend has come at the expense of the more contextual, historical, and evolutionary perspectives that dominated the social sciences at the turn of the century. Although the basic concepts advanced by path dependence may not be entirely (or even mostly) novel, they have been revived and presented in a compelling manner that crystallizes the important implications of history for subsequent events, particularly technological development.

In this chapter, we suggested ways in which path dependence can build on the extensive examination of historical and temporal consequences that are a longstanding component of other social science disciplines. In addition, we noted ways in which the social sciences can learn from path dependence research. By unpacking the concept of path dependence, we intended to raise important theoretical questions about the concept. Does path dependence contain any valued added over existing theories of innovation and social changes? Is path dependence truly an interdisciplinary field, or just a subset of economics? Hopefully the reader will be led to reflect on these more profound disciplinary issues, in addition to substantive considerations of technological path dependence, destruction, and creation.

Given the possibility of human agency, no path is ever entirely dependent; and given the cumulative nature of R&D, no path is ever really new. Thus, the past is not completely fixed, and the future is not completely open.

To some extent, we should think in terms of "malleable pasts" and "fixed futures." The processes of path dependence, destruction, and creation involve a continual, reciprocal interaction between past, present, and future. However, the way we follow paths may be very different from the way we destroy paths or from the way we create paths. More detailed, comparative, process-oriented analysis is needed of these different path types. This will permit the further unpacking of path dependence and related concepts.

A final emergent goal of this chapter is to argue for a "path as process" approach to understanding technological development. Like Karnoe and Garud (1997), we believe technological progress "must be understood as a constant process of path dependence and creation that occurs as actors in different arenas reproduce, enact and negotiate with one another" (p. 6). Innovation is an ongoing, never-ending transformative process rather than a one-time outcome from which we are unlikely to move on or escape. An important issue for future research concerns the generative processes underlying path creation, dependence, and destruction. Specifically, are these processes different? If beliefs are the generative forces that set path-creation processes in motion (Garud & Rappa, 1994), where are the forces driving path dependence and destruction? Both the coevolutionary perspective and the sociocognitive model of technological evolution emphasize that innovation is a process rather than an outcome, and these theoretical frameworks have made an important contribution toward the effort to understand this complex process. Yet as organization scholars (and other social scientists) devote increasing attention to the processes that underlie the emergence of new technological fields, and the new communities, industries, and regions these fields spawn, more attention will need to be devoted to comparing the generative forces underlying path creation, dependence, and destruction, particularly as regards the use, production, and regulation of these technologies.

ACKNOWLEDGMENTS

We thank James Campbell, Robert Launay, Joel Mokyr, Kathleen Thelen, and especially, Marc Ventresca for their valuable comments on all or parts of this chapter.

REFERENCES

Abbott, A. (1994). History and sociology. In E. Monkkonen (Ed.), *Engaging the past: The uses of history across the social sciences* (pp. 77–112). Durham, NC: Duke University Press.

Amburgey, T. L., & Miner, A. S. (1992). Strategic momentum: The effects of repetitive, positional, and contextual momentum on merger activity. *Strategic Management Journal, 13*, 335–348.

Arthur, W. B. (1989). Competing technologies, increasing returns, and lock-in by historical events. *Economic Journal, 99*, 116–131.

Arthur, W. B. (1990). Competing technologies and lock-in by historical small events: The dynamics of allocation under increasing returns. International Institute for Applied Systems Analysis (Paper No. WP-83-92). Laxenburg, Austria.

Baum, J. A., & Singh, J. V. (Eds.).(1994). Evolutionary dynamics of organizations.

Berger, P., & Luckmann, T. (1967). The social construction of reality. London: Penguin.

Bijker, W. (1987). The Social Construction of Bakelite: Toward a Theory of Invention. In W. E., Bijker, T. P. Hughes & T.J. Pinch (Eds.), The Social Construction of Technological Systems. Cambridge, MA: MIT Press (pp. 159-187).

Bupp, I., & Derian, J. (1978). Light water: How the nuclear dream dissolved. New York: Basic Books.

Cohen, W. M., & Levinthal, D. A. (1990). Absorptive capacity: A new perspective on learning and innovation. Administrative Science Quarterly, 35, 128–152.

David, P. (1985). Clio and the economics of QWERTY. American Economic Review Proceedings, 75, 332–337.

Edwards, P. N. (1996). The closed world: Computers and the politics of discourse in cold war America. Cambridge, MA: MIT Press.

Florida, R., & Kenney, M. (1997). Financiers of innovation: Venture capital, technological change and industrial development. Princeton, NJ: Princeton University Press.

Garud, R., & Jain, S. (1996). The embeddedness of technological systems. Advances in Strategic Management, 13, 389–408.

Garud, R., & Nayyar, P. R. (1994). Transformative capacity: Continual structuring by intertemporal technology transfer. Strategic Management Journal, 15, 365–385.

Garud, R., Nayyar, P., & Shapira, Z. (1997). Beating the odds: Towards a theory of technological innovation. In R. Garud, P. Nayyar, & Z. Shapira (Eds.), Technological innovation: Oversights and foresights (pp. 20–40). Cambridge, England: Cambridge University Press.

Garud, R., & Rappa, M. A. (1994). A sociocognitive model of technology evolution: The case of cochlear implants. Organization Science, 5, 344–362.

Ghemawat, P. (1991). Commitment: The dynamic of strategy New York: Free Press.

Granovetter, M. (1992). Economic institutions as social constructions—a framework for analysis. Acta Sociologica, 35, 3–11.

Hirsch, P. M. (1997). Sociology without social structure: Neoinstitutional theory meets brave new world. American Journal of Sociology, 102, 1702–1723.

Hobsbawm, E. J. (1981). The contribution of history to social science. International Social Science, 33, 624–640.

Karnoe, P., & Garud, R. (Forthcoming). Path creation and dependence in the Danish wind turbine field. In J. Porac & M. Ventresca (Eds.), The social construction of industries and markets. Tarrytown, NY: Pergamon.

Kelley, D., & Amburgey, T. L. (1991). Organizational inertia and momentum: A dynamic model of strategic change. Academy of Management Journal, 34, 591–612.

Kellogg, S. (1994). Ten years of historical research and writing by anthropologists, 1980–1990. In E. Monkkonen (Ed.), Engaging the past: The uses of history across the social sciences (pp. 9–47). Durham, NC: Duke University Press.

Kiser, E. (1996). The revival of narrative in historical sociology: What rational choice theory can contribute. Politics & Society, 24, 249–271.

Koestler, A. (1964). The act of creation. New York: MacMillan.

Langlois, R. N. (1986). The new institutional economics: An introductory essay. In R. N. Langlois (Ed.), Economics as a process: Essays in the new institutional economics (pp. 1–25). New York: Cambridge University Press.

Levitt, B., & March, J. G. (1988). Organizational learning. Annual Review of Sociology, 14, 319–340.

Liebowitz, S. J., & Margolis, S. E. (1990). The fable of the keys. Journal of Law & Economics, 22, 1–26.

Liebowitz, S. J., & Margolis, S. E. (1995). Path dependence, lock-in, and history. Journal of Law, Economics, & Organization, 11, 205–226.

McLaughlin, C. (1954). The Stanley steamer: A study in unsuccessful innovation. *Explorations in Entrepreneurial History, 7*, 37–47.

Merton, R. (1967). *On theoretical sociology: Five essays old and new.* New York: Free Press.

Merton, R. (1993). *On the shoulder of giants: A Shandean postcript* (3rd ed.). Chicago: University of Chicago Press.

Miller, D., & Friesen, P. H. (1980). Momentum and revolution in organizational adaptation. *Academy of Management Journal, 22*, 591–614.

Mokyr, J. (1990). *The level of riches: Technological creativity and economic progress.* New York: Oxford University Press.

Monkkonen, E. H. (1994). Introduction. In E. Monkkonen, (Ed.), *Engaging the past: The uses of history across the social sciences* (pp. 1–8). Durham, NC: Duke University Press.

Piore, M., & Sabel, C. (1984). *The second industrial divide.* New York: Basic Books.

Powell, W. W., & DiMaggio, P. (1991). *The new institutionalism in organizational analysis.* Chicago: University of Chicago Press.

Robertson, D. B. (1994). History, behaviorialism, and the return to institutionalism in American political science. In E. Monkkonen (Ed.), *Engaging the past: The uses of history across the social sciences* (pp. 113–153). Durham, NC: Duke University Press.

Robertson, P., & Langlois, R. (1995). Innovation, networks and vertical integration. *Research Policy, 24*(4), 543–562.

Rockoff, H. (1994). History and economics. In E. Monkkonen, (Ed.), *Engaging the past: The uses of history across the social sciences* (pp. 48–76). Durham, NC: Duke University Press.

Rosenberg, N. (1994). *Exploring the black box.* Cambridge, England: Cambridge University Press.

Scott, R. W. (1995). *Institutions and organizations.* Thousand Oaks, CA: Sage.

Staw, B. M. (1981). The escalation of commitment to a course of action. *Academy of Management Review, 6*, 577–587.

Staw, B. M., Sandelands, L., & Dutton, J. E. (1981). Threat rigidity cycles in organizational behavior. *Administrative Science Quarterly, 26*, 501–524.

Storper, M., & Walker, R. (1989). *The capitalist imperative: Territory, technology, and industrial growth.* London: Basil Blackwell.

Tushman, M. L., & Rosenkopf, L. (1992). Organizational determinants of technological change: Toward a sociology of technical evolution. In L. L. Cummings & B. M. Staw (Eds.), *Research in Organizational Behavior, 14*, 311–347. Greenwich, CT: JAI.

Usher, A. P. (1954). *A history of mechanical inventions.* Cambridge, MA: Harvard University Press.

Uzzi, B. (1997). Social structure and competition in interfirm networks: The paradox of embeddedness. *Administrative Science Quarterly, 42*, 35–67.

Van de Ven, A. H., & Garud, R. (1989). A framework for understanding the emergence of new industries. In R. Rosenbloom & R. Burgelman (Eds.), *Research in technological innovation, management and policy* (pp. 195–225). Greenwich, CT: JAI.

Weick, K. (1976). Educational organizations as loosely coupled systems. *Administrative Science Quarterly, 21*, 1–19.

Zald, M. N. (1990). History, sociology, and theories of organization. In J.E. Jackson (Ed.), *Institutions in American society: Essays in market, political, & social organization* (pp. 81–108). Ann Arbor: University of Michigan Press.

Zukin, S., & DiMaggio, P. (1990). *Structures of capital: The social organization of the economy.* New York: Cambridge University Press.

4

Sources of Technical Change: Induced Innovation, Evolutionary Theory, and Path Dependence

Vernon W. Ruttan
University of Minnesota

This is an appropriate time to take stock, as economists, of our understanding of the determinants of the rate and direction of technical change. The 1960s through the 1980s were very productive of new theory and empirical insight into the process of technical change. In the 1960s and 1970s, major attention focused on the implications of changes in demand and in relative factor prices. In the late 1970s and early 1980s, attention shifted to evolutionary models inspired by a revival of interest in Schumpeter's insight into the sources of economic development. Since the early 1980s these have been complemented by the development of historically grounded "path dependent" models of technical change.

Each of these models has contributed substantial insight into the generation and choice of new technology. It appears to me, however, that each research agenda is approaching a dead-end. In this chapter, I argue that the three models—induced, evolutionary and path dependent—represent elements of a more general theory. This chapter reviews the development of the three models to identify their complementarity and to suggest how they might be incorporated into a more general theory.

INDUCED TECHNICAL CHANGE

There are at least three major traditions of research that have attempted to confront the impact of change in the economic environment on the rate and direction of technical change. The "demand pull" tradition has emphasized the relative importance of market demand on the supply of knowledge in inducing advances in technology. There has also been a longstanding debate among economic historians about the extent to which differences in English and American technology during the 19th century were influenced by relative factor endowments and prices. A third tradition stems from attempts by economic theorists to understand the apparent stability in factor shares in the American economy during the 20th century despite the very large substitution of capital for labor. At a more microlevel, there is a large literature in the fields of agricultural and resource economics on the role of differences and changes in relative factor endowments on the direction of technical change.

Demand Pull and the Rate of Technical Change

Schumpeter, whose writings have been exceptionally important in influencing the way economists think about technical change, made a sharp distinction between invention (and the inventor) and innovation (and the innovator): "Innovation is possible without anything we should identify as invention, and invention does not necessarily induce innovation but produces itself ... no economically relevant effect at all" (Schumpeter, 1934, p. 84).

The Chicago sociologist, Gilfillan, viewed invention as proceeding under the stress of necessity with the individual innovator being an instrument of luck and process (Gilfillan, 1935).

In his now classic study of the invention and diffusion of hybrid maize, Griliches demonstrated the role of demand in determining the timing and location of invention (Griliches, 1957). Schmookler, in a massive study of patent statistics for inventions in four industries (railroads, agricultural machinery, paper, and petroleum), concluded that demand was more important in stimulating inventive activity than advances in the state of knowledge (Schmookler, 1962, 1966). The Griliches–Schmookler demand induced model received further support from papers by Lucas (1967) and Ben-Zion and Ruttan (1975, 1978) that showed technical change to be responsive to aggregate demand. In the mid-1960s, Vernon (1966, 1979) introduced a demand pull model to interpret the initial invention and diffusion of consumer durable technologies—such as automobiles, television, refrigerators and washing machines—in the United States rather than in other developed countries. His interpretation came just as the United States was about to lose its dominance in several of these technologies to Japan.

Arguments about the priority of the role of demand side forces and supply side forces, such as advances in knowledge, in inducing advances in technology were intensified in the late 1960s. A study conducted by the Office of the Director of Defense Research and Engineering purported to show that the significant "research events" contributing to the development of 20 major weapons systems were predominantly motivated by military need rather than disinterested scientific inquiry. This view was challenged in studies commissioned by the National Science Foundation that, not unexpectedly, found that science events were of much greater importance as a source of technical change (Thirtle & Ruttan, 1987).

In a review of the "demand pull–supply push" controversy, Mowery and Rosenberg (1979) argued that much of the research purporting to show that technical change has been demand induced is seriously flawed. They insist that the concept of demand employed in many of the studies has been so broad or imprecise as to embrace virtually all possible determinants. Rosenberg also insists that the demand pull perspective has ignored "the whole thrust of modern science and the manner in which the growth of specialized knowledge has shaped and enlarged man's technological capacities" (Rosenberg, 1974, p. 93). Research conducted from a demand pull perspective appears to have atrophied since the late 1970s, partly as a result of the Rosenberg criticism.

Careful industry studies, such as the study of innovation in the chemical industry by Walsh, suggest that both "supply and demand factors play an important role in innovation and in the life cycles of industries, but the relationship between the two varies with time and the maturity of the industrial sector concerned" (Walsh, 1984, p. 233). A rigorous econometrics study by Scherer (1982) that simultaneously tests both the demand induced and supply push hypotheses across a broad range of industries confirms the earlier Schmookler finding of strong association between capital goods investment and invention. But Scherer found a weaker association between demand pull and industrial materials inventions. He also found that the introduction of an index of technological opportunity based on the richness of an industry's knowledge base added significantly to the power of his model to explain differences in the level of inventive activity among industries.

It should no longer be necessary to insist that basic research is the cornucopia from which all inventive activity must flow to conclude that investment in the generation of scientific and technical knowledge can open up new possibilities for technical change. Nor should it be necessary to demonstrate that advances in knowledge, inventive activity, and technical change flow automatically from changes in demand to conclude that changes in demand represent a powerful inducement for the allocation of research resources.

Factor Endowments and the Direction of Technical Change

Modern interest in the effect of factor endowments on the direction of technical change dates to the early 1960s. Hicks (1932, 1963) had earlier suggested:

> The real reason for the predominance of labor saving inventions is surely that which was hinted at in our discussion of substitution. A change in the relative prices of the factors of production is itself a spur to innovation and to inventions of a particular kind—directed at economizing the use of a factor which has become relatively expensive (pp. 124–125).

Hicks suggestion received implicit assent but little attention until the early 1960s. In his work on the theory of wages, Rothschild repeated the Hicks' argument (Rothschild, 1956). In a book on economic growth, Fellner argued that firms with some degree of monopsony power had an incentive to make "improvements" that economized on the progressively more expensive factors of production and that expectations of future changes in relative factor prices would be sufficient to induce even firms operating in a purely competitive environment to seek improvements that would save the more expensive factors (Fellner, 1956; see also Fellner 1961, 1962).

An intense dialogue around the issue of induced innovation by economic theorists in the 1960s and early 1970s was triggered by Salters' explicit criticism of the Hicks' induced technical change hypothesis. Salter insisted "at competitive equilibrium each factor is being paid its marginal value product; therefore all factors are equally expensive to firms" (Salter, 1960, p. 16). He went on to argue that "the entrepreneur is interested in reducing costs in total, not particular costs or capital costs. When labor costs rise any advance that reduces total cost is welcome, and whether this is achieved by saving labor or saving capital is irrelevant" (Salter, 1960, pp. 43–44; see also Blaug, 1963). It is difficult to understand why Salters' criticism attracted so much attention except that students of economic growth were increasingly puzzled about why, in the presence of substantial capital deepening in the U.S. economy, factor shares to labor and capital had appeared to remain relatively stable. The differential growth rates of labor and capital in the U.S. economy were regarded as too large to be explained by simple substitution along a neoclassical production function.

The Growth Theoretic Model.   The debates about induced technical change centered on two alternative models—a growth theoretic approach and a microeconomic version. The most formally developed version was the growth theoretic approach introduced by Kennedy (1964). The Kennedy article initiated an extended debate on the theoretical foundations and the implications of incorporating the process of induced technical change into the theory of economic growth (Drandakis & Phelps, 1966; Kennedy, 1966; Samuelson, 1965, 1966; Wan, 1971).

In the Kennedy model the initial conditions included: given factor prices, an exogenously given budget for research and development, and a fundamental trade-off (a transformation function) between the rate of reduction in labor requirements and the reduction of capital requirements. The model assumes a production function with factor augmenting technical change. Kennedy cast his analysis in terms of the effect of changes in relative factor shares rather than changes in relative factor prices on bias in invention because of the growth theory implications.

The following example from Binswanger (1973) represents an intuitive interpretation of the Kennedy model.

> Suppose it is equally expensive to develop either a new technology that will reduce labor requirements by 10 percent or one that will reduce capital requirements by 10 percent. If the capital share is equal to the labor share, entrepreneurs will be indifferent between the two courses of action.... The outcomes of both choices will be neutral technical change. If, however, the labor share is 60 percent, all entrepreneurs will choose the labor reducing version. If the elasticity of substitution is less than one, this will go on until the labor and capital shares again become equal, provided the induced bias in technical change does not alter the (fundamental) trade-off relationship between technical changes that reduce labor requirements on the one hand, or capital requirements on the other. (p. 32)

The Kennedy variant of induced innovation was subsequently incorporated into neoclassical growth theory (Wan, 1971). Nordhaus (1973) noted,

> Until recently, only Harrod-neutral (or purely labor augmenting) technological change could be introduced into neoclassical growth without leading to bizarre results. Neoclassical growth models were "saved" from such restrictiveness by the introduction of the theory of induced innovation. Under the usual neoclassical assumptions and, in addition, when the innovation possibility curve takes the form assumed by Kennedy and Samuelson the system settles down into a balanced growth path exactly like that of the labor-augmenting case. (p. 209)

By the early 1970s, the growth theoretic approach to induce technical change was under severe attack (David, 1975; Nordhaus, 1969, 1973; Wan, 1971). Nordhaus (1973) noted that in the Kennedy model, no resources are allocated to inventive activity. A valid theory "of induced innovation requires at least two productive activities; production and invention. If there is no invention then the theory of induced innovation is just a disguised case of growth theory with exogenous technological change" (p. 210). He further notes that the Kennedy innovation possibility frontier (IPF) implies that the rate of capital augmenting technological change is independent of the level of labor augmentation. Thus, as technological change accumulates there is no effect on the trade-off between labor and capital augmenting technological change (Nordhaus, 1973). He insisted that the model is "too defective to be used in serious economic analysis." (Nordhaus, 1973, p. 208). The growth theoretic version of induced innovation has never recovered from the criticism of its inadequate microeconomic foundation.[1]

The Microeconomic Model.   A second approach to induced innovation, built directly on Hicksian microeconomic foundations, was developed by Ahmad (1966). His criticism of the growth theoretic approach initiated a vigorous exchange (Ahmad, 1967a, 1967b; Fellner, 1967; Kennedy, 1967). In his 1973 critique, Nordhaus mentioned that Ahmad was the only person to attempt to formulate the theory of induced technical change along microeconomic lines but he did not comment explicitly on the Ahmad paper or on the subsequent exchange.[2]

In his model, Ahmad employed the concept of a historic *innovation possibility curve* (IPC). At a given time, there exists a set of potential production processes, determined by the basic state of knowledge, available to be developed. Each process in the set is characterized by an isoquant with rather narrow possibilities for substitution. Each process in the set requires that resources be devoted to research and development before the process can actively be employed in production. The IPC is the envelope of all unit isoquants of the subset of those potential processes that the entrepreneur might develop with a given amount of research and development expenditure.

Assume that I_t is the unit isoquant describing a technological process available in time t and that IPC_t is the corresponding IPC (Fig. 4.1). Given the relative factor prices described by line P_tP_t, I_t is the cost minimizing technology. Once I_t is developed, the remainder of the IPC becomes irrelevant because, for period t + 1, the IPC shifts inward to some IPC_{t+1}. This occurs because it would take the same R & D resources to go from I_t to any other technique on IPC_t as to go from I_t to any technique on IPC_{t+1}. If fac-

---

[1]Another reason for the decline in interest among economic theorists was the difficulty, pointed out by Diamond, McFadden, and Rodriquez (1978) in simultaneously measuring the bias of technical change and the elasticity of substitution between factors. This problem had, however, already been solved (Binswanger, 1974b; Binswanger & Ruttan, 1978). For a more recent discussion see Haltmaier (1986).

[2]It is interesting to speculate on what the future course of induced innovation theory might have been if the Ahmad article had, as it might have, appeared first. The initial drafts of the articles were written while Kennedy was teaching at the University of the West Indies (Kingston) and Ahmad was teaching at the University of Khartoum (Sudan). Ahmad submitted his article to the *Economic Journal* in 1963. Kennedy served as a reviewer of the Ahmad article. His article, which was published in 1964, was originally written as a comment on the Ahmad article. Ahmad's article was rewritten, resubmitted, and published in 1966 (Ruttan & Hayami, 1994).

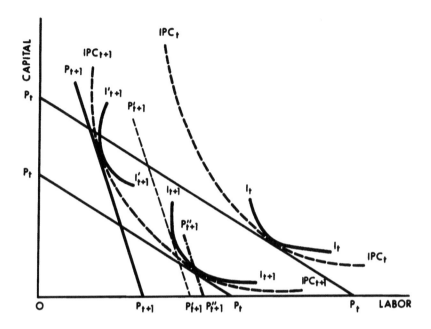

FIG. 4.1. Ahmad's induced innovation model. Adapted from, "On The Theory of Induced Invention," by S. Ahmad. Copyright 1966 by *Economic Journal*, 76, p. Adapted with permission.

tor prices remain unchanged and technical change is neutral, the new unit isoquant will be I_{t+1} on IPC_{t+1}. If, however, factor prices change to $P_{t+1}P_{t+1}$, then it is no longer optimal to develop I_{t+1}. Instead, a technological process corresponding to some It_{t+1} becomes optimal. In the graph, $P_{t+1}P_{t+1}$ corresponds to a rise in the relative price of labor. If the IPC has shifted neutrally, It_{t+1} will be relatively labor saving in comparison to I_t.

Ahmad's graphical exposition is useful as an illustration of the induced innovation process of a one period microeconomic model in which a firm or a research institute has a fixed exogenous budget constraint. When research budgets are no longer fixed, a mathematical exposition is more convenient (Binswanger, 1978). In a multiperiod model, the shift from I_t to It_{t+1} would occur in a series of steps in response to incremental shifts from P_t to P_{t+1}. One way of describing this process would be to appeal to *learning by doing* and *learning by using* concepts (Arrow, 1962; Rosenberg, 1982a).

Dialogue With Data

The initial dialogues about the logic of the Kennedy–Samuelson–Weizsäcker growth theoretic and the Hicks–Ahmad microeconomic approaches to induced technical change were conducted within the confines of the standard two factor (labor and capital) neoclassical model. Among economic historians, there has been a continuing debate about the role of land abundance on the direction of technical change in the industrial sector. Among agricultural economists, there has emerged a large literature on the bias of technical change along mechanical (labor saving) and biological (land saving) directions.

Habakkuk (1962) argued that the ratio of land to labor, which was higher in the United States than in Britain, raised real wages in American agriculture and thereby increased the cost of labor to manufacturers. Habakkuk argued, in effect, that in the 19th century, the higher U.S. wage rates resulted not only in the substitution of capital for labor (more capital), but induced technical changes (better capital) biased in a labor saving direction (James & Skinner, 1985). The issue became controversial among economic historians even before they became fully sensitive to the emerging theoretical debates of the 1960s around the issue of induced technical change or the earlier empirical work by Hayami & Ruttan (1970, 1971, 1985).

The criticisms of the Rothbard–Habakkuk labor scarcity theses by Temin (1966) and the debates that his criticism engendered (Ames & Rosenberg, 1968; David, 1973, 1975; Fogel, 1967) focused primarily on the issue of the impact of land abundance on the substitution of capital for labor—the "more capital" rather than the "better capital" part of the thesis. David (1975) argued that economic historians "steered away from serious re-evaluation of the proposition about the rate and bias of innovation, precisely because standard economic analysis was thought to offer less reliable guidance there than on questions of the choice of alternative known techniques of production" (p. 31).

David insisted that the argument could not be resolved without a more intensive mining of the historical evidence. But recourse to measurement could not be expected to get very far without a theoretically grounded definition of an operational concept that distinguishes between choice of technology and technical change and between bias in the direction of technical change and the rate of technical change. David (1975) argued that this can be done by embracing "the concept of a concave, downward sloping 'innovation-possibility frontier' ... along the lines of the neoclassical theory of induced technical progress due to Kennedy, Weizsäcker and Samuelson" (p. 32). He then went on to argue along the same lines as Wan (1971) and Nordhaus (1973) that the particular pattern of changes in macroproduction relationships observed in the United States could not be rationalized within the framework of a stable innovation-possibility frontier. "While shifts of the innovation-possibility frontier are entirely conceivable, the necessity of accepting their occurrence in this context signifies a practical failure of the underlying theoretical construct. For the latter treats the position of the frontier as established autonomously for each economy, and has no explanation to offer for it" (David, 1975, p. 33).

David also insisted that bias in the direction of technical change could only be understood by building a theory of induced innovation on microeconomic foundations consistent with engineering and agronomic practice. To David this also meant abandoning both neoclassical growth theory and the neoclassical theory of the firm. Furthermore, it would be necessary to incorporate the intimate evolutionary connection "between factor prices, the choice of technique and the rate and direction of global technical change" (David, 1975, p. 61).

In attempting to develop a nonneoclassical evolutionary and historical approach to induced innovation, David introduced the concepts of (a) linear fixed-coefficient processes or techniques from activity analyses (which he credits to Chenery), (b) a latent set of potential processes that could be designed with the currently existing state of knowledge (which he attributed to Salter). He then added (c) localized learning which directs technical change toward the origin along a specific process ray (that he attributes to Stiglitz); and (d) a probablistic learning process that is bounded by transition probabilities that depend on the firm's initial technical state (David, 1975). The model of the search process appears to have been inspired by the Nelson–Winter evolutionary model (see next section). He insisted, however, that his transition probabilities, in which past states—the firms initial myopic selection of a technical process—influences the future course of development, is "clearly non-Markovian" (David, 1975, p. 81).

David differentiated his approach from neoclassical production theory by suggesting that substitution may involve an element of innovation. This is similar to the mechanism that Ahmad (1966) and Hayami and Ruttan (1970) had earlier employed to account for the shift in the IPC (or in David's terms, the FPF). It should be viewed an extension rather than an alternative to the neoclassical model. He also differentiated the mechanism that accounts for the evolutionary nature of technical change from that form employed by Nelson and Winter.

David then turned to the technical relationships among natural resources, labor and capital. He argued, drawing on the work of Ames and Rosenberg (1968) and his own earlier work (David, 1966), that in the mid-19th century mechanical technology and land were complements:

The relevant fundamental production functions for the various branches of industry and in agriculture did not possess the property of being separable in the raw materials and natural resource inputs; instead the relative capital intensive techniques ... were also relatively resource using (David, 1975, p. 88).

Greater availability of natural resources facilitates the substitution of capital for labor.

Thus, even if the same labor capital price ratios had faced producers in Britain and America, the comparatively greater availability of natural resources would have suggested to some American producers the design and to others the selection of more capital intensive methods.... In America the on-going capital formation spurred by the greater possibilities of jointly substituting natural resources and capital for labor may well have been responsible for driving up the price of labor from the demand side" (David, 1975, pp. 89–90).

The formal introduction of the role of relative resource abundance (or scarcity) clearly represents an important extension as compared to the traditional two-factor (labor and capital) neoclassical models. But the primary significance is that David opened the door, and identified most of the elements, of what has since become known as the path dependent model of technical change (David, 1975).

There are substantial differences in the extent to which the several induced technical change models have been tested against empirical data. The demand induced model was developed in close association with empirical studies and was not subjected to formal modeling or theoretical critique until fairly late (Lucas, 1967; Mowery & Rosenberg, 1979). The growth theoretic version of factor induced technical change has been peculiarly unproductive of empirical research. The only test against empirical data seems to have been by Fellner. Fellner (1961) interpreted his results as indicating that, except during periods of very rapid increase in capital intensity, and hence rapidly rising demand for labor, the induced labor saving bias was sufficient to prevent the labor share from rising.

The microeconomic version of factor induced technical change has in contrast, been highly productive in stimulating a wide body of applied research. The first formal test based directly on microeconomic foundations was the Hayami–Ruttan test against the historical experience of agricultural development in the United States and Japan (Hayami & Ruttan, 1970).[3] It seemed apparent that neither the enormous differences in land–labor ratios between the two countries or the changes in each country

---

[3] At the time the article was written, Hayami and Ruttan were familiar with the growth theoretic literature by Fellner, Kennedy and Samuelson, but not with the Ahmad article and his subsequent exchange with Fellner and Kennedy. The inspiration for the 1970 Hayami–Ruttan paper was the historical observations about the development of British and American technology by Habakkuk (1962). See Ruttan and Hayami (1994).

over time could be explained by simple factor substitution. Hayami and Ruttan employed a four-factor model in which (a) land and mechanical power were regarded as complements and land and labor as substitutes, and (b) fertilizer and land infrastructure were regarded as complements and fertilizer and land as substitutes.

The process of advance in mechanical technology in the Hayami–Ruttan model are illustrated in the left hand panel of Fig. 4.2. PCI_0 represents the innovation possibility curve (IPC) in time zero; it is the envelope of less elastic unit isoquants that correspond, for example, to different types of harvesting machinery. The relationship between land and power is complementary. Land cum power is substituted for labor in response to a change in the wage rate relative to an index of land and power prices. The change in the price ratio from P_0 to P_1 induces the invention of labor saving machinery—say a combine for a reaper.

The process of advance in biological technology is illustrated in the right hand panel of Fig. 4.2. Here IPC_0 represents an IPC that is an envelope of relatively inelastic land-fertilizer isoquants such as L_0. When the fertilizer–land price ratio declines from P_0 to P_{1a} new technology—a more fertilizer responsive crop variety—represented by I_1 is developed along IPC_0. Since the substitution of fertilizer for land is facilitated by investment in land and water development the relationship between new fertilizer responsive varieties and land infrastructure is complementary.

In Fig. 4.2 the impact of advances in mechanical and biological technology on factor ratios are treated as if they are completely separable. This is clearly an oversimplification. It is not essential to the Hayami–Ruttan induced technical change model that changes in the land-labor ratio be a direct response to the price of land relative to the wage rate (Thirtle & Ruttan, 1987).

The econometric tests conducted by Hayami and Ruttan suggested that the enormous changes in factor proportions that occurred during the process of agricultural development in the two countries "represents a process of dynamic factor substitutions accompanying changes in the production function induced by changes in relative factor prices" (Hayami & Ruttan, 1970, p. 1135).

The initial Hayami–Ruttan article and the further exposition in their book on agricultural development (1971, 1971/1985) became the inspiration for a large number of empirical tests of the microeconomic version of the induced technical change hypothesis in the agricultural and natural resource sector. Binswanger (1974a, 1974b) advanced the methodology for measuring technical change bias with many factors of production. In a 1987 literature review, Thirtle and Ruttan (1987) listed 29 empirical studies of induced technical change in agriculture. Most of the studies draw their inspiration from the initial study by Hayami and Ruttan (1970). Thirtle and Ruttan also list 38 empirical

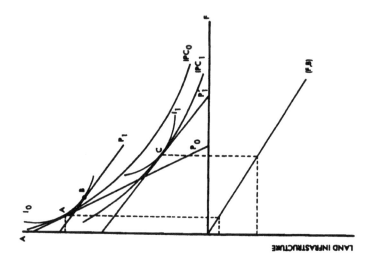

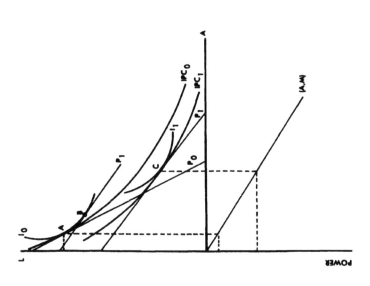

FIG. 4.2. Induced Technical change in Agriculture. Adapted from *Agricultural Development: An International Perspective*, (p. 91) by Y. Hayami and V. W. Ruttan, 1994, (Baltimore: Johns Hopkins University Press). Reprinted with permission.

studies in the industrial sector. The initial studies of biased technical progress change in industry typically did not involve direct tests of the induced technical change hypotheses. By the late 1970s and early 1980s, however, a substantial number of studies, some stimulated by the rise in energy prices in the 1970s, involved direct tests of the induced technical change hypotheses. Within the industrial sector, the evidence is strongest in the natural resource and raw material using industries (Jorgenson & Fraumeni, 1981; Jorgenson and Wilcoxen, 1993; Wright, 1990). As of the mid-1980s, the evidence of tests of the induced technical change hypotheses in agriculture, both in the United States and abroad, was sufficient to support the view that changes (and sometimes differences) in relative factor endowments and prices exert a substantial impact on the direction of technical change.[4]

EVOLUTIONARY THEORY

The modern revival of interest by economists in an evolutionary theory of technical change derives largely from a series of articles by Nelson and Winter in the mid-1970s (Nelson & Winter, 1973, 1974, 1975, 1977; Nelson, Winter, & Schuette, 1976).[5] These articles in turn served as a basis for the highly acclaimed book, *Evolutionary Theory of Economic Change* (Nelson & Winter, 1982). The theory advanced by Nelson and Winter has been identified by the authors as Schumpeterian in its interpretation of the process of economic change. In much of the literature that has drawn its inspiration from Nelson and Winter, "evolutionary" and "Schumpeterian" have been used as interchangeable.[6] The second cornerstone of the Nelson–Winter model is the behavioral theory of the firm in which profit maximizing behav-

---

[4]Olmstead and Rhode (1993) criticized the Hayami and Ruttan work on both conceptual and empirical grounds. At the conceptual level, they find confusion between the relative factor "change variant" that is used in explaining productivity growth over time within a given country, and the "level variant" of the model that is used in analysis of international productivity differences. They also argue, on the basis of regional tests in the United States, that the induced technical change model holds only for the central grain growing regions. In a later paper (Olmstead & Rhode, 1995) using state level data, they found somewhat stronger support for the induced technical change hypothesis. For further criticism and a defense, see Koppel (1994).

[5]Nelson and Winter identify Alchian (1950) and Penrose (1952) as representing direct intellectual antecedents of their work. For the theoretical foundations of the Nelson–Winter collaboration, see Winter (1971). For the historical and philosophical foundations, see Elster (1983) and Langolis and Everett (1994). Witt (1993) assembled many of the most important articles in the field of evolutionary economics in a collection of readings, *Evolutionary Economics*. For a review of recent evolutionary thought about economic change, see Nelson (1995).

[6]The Nelson–Winter model departs in its treatment of the linkage between invention and innovation. For Schumpeter, there was no necessary link between invention and innovation (Ruttan, 1959). Nelson and Winter (1982) employ the term evolutionary metaphorically—We emphatically disavow any intention to pursue biological analogies for their own sake (p. 11). Nelson and Winter regard their approach as closer to Lamankianism than Mendelianism. Yet their description of the evolutionary process of firm behavior and technical change as a Markov process, and their use of the Markov mechanism in their simulation, is analogous to the Mendelian model.

ior is replaced by decision rules that are applied routinely over an extended period of time (Cyert & March, 1963; Simon, 1955, 1959).

The Nelson–Winter evolutionary model, particularly chapters 9–11, jettisons much of what they consider to be the excess baggage of the neoclassical microeconomic model—"the global objective function, the well defined choice set, and the maximizing choice rationalization of firm's actions. And we see 'decision rules' as very close conceptual relatives of production 'techniques' whereas orthodoxy sees these things as very different" (Nelson & Winter, 1982, p. 14). The production function and all other regular and predictable behavior patterns of the firm is replaced by the concept of *routine*—

> a term that includes characteristics that range from well-specified technical routines for producing things, procedures for hiring and firing, ordering new inventory, or stepping up production of items in high demand to policies regarding investment, research and development (R&D), or advertising, and business strategies about product diversification and overseas investment (Nelson & Winter, 1982, p. 14).

The distinction between factor substitution and shifts in the production function is also abandoned. The two fundamental mechanisms in the Nelson–Winter models are the *search* for better techniques and the *selection* of firms by the market (Elster, 1983, p. 14). In their models the microeconomics of innovation are represented as "a stochastic process dependent on the search routines of individual firms" (Dosi, Giannetti, and Toninelli, 1992, p. 10). The activities leading to technical changes are characterized by local search for technical innovations, imitation of the practices of other firms, and satisfying economic behavior.

In their initial models, search by the firm for new technology, whether generated internally by R & D or transferred from suppliers or competitors, is set in motion when profits fall below a certain threshold. The models assume that in this search the firms draw samples from a distribution of input–output coefficients (Fig. 4.3). If A is the present input combination, then potential input coefficients are distributed around it such that there is a much greater probability of finding a point close to A then if finding one far away. Search is local. Once the firm finds a point B, it makes a profitability check. If costs are lower at B than at A, the firm adopts the point B and stops searching. Otherwise, search continues. Thus, the technology described by the point $B°$ input–output and factor ratios will be accepted if labor is relatively inexpensive, that is, if relative prices are described by line CD. But if labor is relatively expensive, as described by C¢D¢ the firm will reject the $B°$ technology and continue to search for another technology until it finds another point, say $B°¢$ The technology at point $B°¢$ will be labor saving relative to that at $B°$.

The stochastic technology search process is built into a model with many competing firms. All profits above a "normal" dividend (investors are satisficers rather than optimizers) are reinvested so that successful firms grow faster than the unsuccessful ones. The capital stock of the economy is determined by the total investment by all firms. Labor supply is elastic to the firm.

Simulation runs rather than formal analysis or tests against historical experience are employed to demonstrate the plausibility of the models. The simulations start from an initial point where all firms are equal. The model determines endogenously the output of the economy, the wage-rental rate, and the capital accumulation rates. Nelson and Winter have used a series of variations in their basic model to explain how changes in market structure influences the rate of technical change, the direction of technical change, and the importance of imitation and innovation.

When firms check the profitability of alternative techniques that their search processes uncover, a higher wage rate will cause certain techniques to fail the more profitable tests that would have passed at a lower wage rate, and enable others to pass the test that would have failed at a lower wage rate. The latter will be capital intensive relative to the former. Thus a higher wage rate

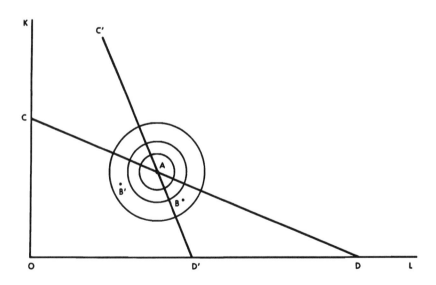

FIG. 4.3.   Sampling and selection of new input-output coefficients. "From Factor Price Change and Factor Substitution in an Evolutionary Model" By R. R. Nelson and S. G. Winter, 1975, *Bell Journal of Economics*, 6, p. 472. Copyright 1975 by. Adapted with permission.

nudges firms to move in a capital-intensive direction compared with that in which they would have gone. Also, the effect of a higher wage rate is to make all technologies less profitable (assuming, as in their model, a constant cost of capital) but the cost increase is proportionately greatest for those that involve a low capital–labor ratio. Because firms with high capital labor ratios are less adversely affected by high wage rates then those with low capital–labor ratios, capital intensive firms will tend to expand relatively to labor-intensive ones. For both of these reasons, a higher wage rate will tend to increase capital-intensity relative to what would have been obtained" (Nelson & Winter, 1974, p. 900). The responsiveness of the capital labor ratio to changes in relative factor prices is rather striking because, except for the profitability check, search (or research) outcomes are random (Nelson & Winter, 1982), and the inducement mechanism comes about through competition, survival and growth rather than through efforts to maximize profits.

The early Nelson–Winter models were criticized for the "dumb manager" assumption in which the search (or research) process is triggered only when profits fall below a threshold level. For example, "Here we assume that firms with positive capacity do not search if they are making positive or zero profits; they satisfice on their prevailing routines." (Nelson & Winter, 1982, p. 149). An implication is that an increase in demand for the product of an industry can lead to a reduction in research effort. This was hardly consistent with either historical evidence (Schmookler, 1966) or with a Schumpeterian perspective. The restriction was relaxed in the second round of Nelson and Winter models by the explicit introduction of directed research. As the wage/rental ratio rises research effort is allocated to sampling the spectrum of capital intensive techniques (Nelson & Winter, 1975, 1977).

Winter devoted considerable attention to extensions of the initial Nelson–Winter models. In a 1984 article, for example, he abandons the assumption of the level playing field in which the initial conditions were the same for all firms. The basic model is augmented by a model that includes entirely new firms. Winter uses this expanded model to explore the growth path of two industrial regimes. One is an "entrepreneurial regime" that he identifies with Schumpeter's *The Theory of Economic Development* (1911, 1934; originally published in German). The second is a "routinized regime" that he identifies with Schumpeter's *Capitalism, Socialism and Democracy* (1950). The entrepreneurial regime model is designed so that innovations are primarily associated with the entry of new firms. In the routinized regime, innovations are primarily the result of internal R & D by established firms. Several suggestions for further extension of the Nelson and Winter models, to include the creation of new industries, interaction among industries, and product innovation and imitation, for example, have been summarized and extended by Andersen (1994).

It is important to clarify the role of historical process in the Nelson–Winter evolutionary models. The condition of the industry in each time period shapes its condition in the following period.

> Some economic processes are conceived as working very fast, driving some of the model variables to (temporary) equilibrium values within a single period (or in a continuous time model, instantaneously). In both the entrepreneurial and routinized Schumpeterian models, for example, a short-run equilibrium price of output is established in every time period. Slower working processes of investment and of technological and organizational change, operate to modify the data of the short-run equilibrium system from period to period (or from instant to instant). The directions taken by these slower processes of change are directly influenced by the values taken by the subset of variables that are equilibrated in the individual period or instant. (Winter, 1984, p. 290)

Two questions that I find difficult to resolve is why there have been so few efforts by other scholars to (a) advance the Nelson–Winter methodology[7] or, (b) to test the correspondence between the plausible results of the Nelson–Winter simulations against the historical experience of particular firms or industries.[8] Simulation is capable of generating a wide range of plausible behavior. But the hypothesis generated by the simulations have seldom been subjected to rigorous empirical tests. The closest they or others come to empirical testing is the demonstration that it is possible to generate plausible economy wide growth paths or changes in marketshare.

Path Dependence

The argument that technical change is "path dependent" was vigorously advanced, by Arthur and colleagues in the late 1970s and early 1980s (Arthur 1989, 1994; Arthur, Ermoliev, and Kaniovski, 1987.)[9] In the mid and

---

[7]For a useful interpretation and extension, see Anderson (1994). Anderson's work is particularly helpful in clarifying the poorly documented computational steps of the Nelson–Winter models. Anderson supplements the mathematical notation employed by Nelson and Winter by an algorithmically oriented programming notation. An appendix, "Algorithmic Nelson and Winter Models" (p. 198–219) is particularly useful.

[8]Since the mid-1970s there has emerged a large body of empirical research on technical change that can be categorized as broadly Schumpeterian or evolutionary in inspiration (see the review by Freeman, 1994). The point I am making, however, is quite different. There has been very little effort to use the simulation models to generate hypothesis about the process of technology development and then to either identify historical counterparts or to test the outcomes against historical experience in a rigorous manner. The one exception with which I am familiar is the Evenson–Kislev (1975, pp. 140–155) stochastic model of technological discovery. The model was used to interpret the stages in sugar cane varietal development. The Evenson–Kislev model did not, however, draw directly on the Nelson–Winter stochastic model.

[9]Arthur encountered unusual delay before his work was accepted in leading economics journal. His 1986 Economic Journal paper was initially submitted to the American Economic Review in 1983. It was rejected by the American Economic Review twice and by the Quarterly Journal of Economics twice and accepted by the Economic Journal only after an appeal. By the time the paper was finally accepted in the Economic Journal, referees were noting that the path dependence idea was already recognized in the literature (Goss and Sheperd, 1994).

late 1980s, David presented the results of a series of historical studies of the typewriter keyboard, the electric light and power supply industries, and others, that served to buttress the plausibility of the path dependence perspective (David, 1985, 1986, 1993; David & Bunn, 1988). The emphasis on path dependence in David's more recent work represents, as noted earlier, an extension of his earlier research on the relationship between labor scarcity and modernization in 19th century America (David, 1975). This earlier work was strongly influenced by Arrow's article on learning by doing (Arrow, 1962) and by Habakkuk's historical research on British and American technology in the 19th century (Habakkuk, 1962).

The effect of the work by Arthur and his colleagues has been to emphasize the importance of increasing returns to scale as a source of technological lock-in. In some nonlinear dynamic systems positive feedbacks (Polya processes) may cause certain patterns or structures that emerge to be self-reinforcing. Such systems tend to be sensitive to early dynamical fluctuations. Often there is a multiplicity of patterns that are candidates for long-term self-reinforcement. The accumulation of small events early on 'pushes' the dynamics of technical choice into the orbit of one of these and thus 'selects' the structure that the system eventually locks into" (Arthur, Ermoliev, & Kaniovski, 1987, p. 294).

The authors provide an intuitive example: Think of an urn of an infinite capacity.

> Starting with one red and one white ball in the urn, add a ball each time, indefinitely, according to the following rule. Choose a ball in the urn at random and replace it; if it is red, add a red; if it is white, add a white. Obviously this process has increments that are path dependent—at any time the probabilities that the next ball added is red exactly equals the proportion red.... Polya proved in 1931 that in a scheme like this the proportion of red balls does tend to a limit X_1 and with probability one. But X is a random variable uniformly distributed between 0 and 1. (Arthur et al., 1987, p. 259).

Thus in an industry characterized by increasing returns, small historical or chance events that give one of several technologies an initial advantage can (but need not) "drive the adoption process into developing a technology that has inferior long run potential" (Arthur, 1989, p. 117)

The historical small events that result in path dependence are "outside the *ex ante* knowledge of the observer—beyond the resolving power of his model or abstraction of the situation" (Arthur, 1989, p. 118). Arthur employs a series of progressively complex models to simulate situations in which several technologies compete for adoption by a large number of economic agents. Agents have full knowledge of the technology and returns functions but not of the events that determines entry and choice of technology by other agents. His analyses is carried out for three technological regimes (constant, increasing, and diminishing returns) with respect to four properties of the paths of technical change (predictable, flexible, ergodic, path efficient).[10] The only unknown is the set of historical events that determine the sequence in which the agents make their choices. The question he attempts to answer is whether the fluctuations in the order of choice will make a difference in final adoption shares.

For further elaboration of the conditions for path dependence, see Bassanini and Dosi (chap. 2, this volume).

Arthur's simulations reinforce the importance of increasing returns as a necessary condition for technological lock-in.

> Under constant and diminishing returns the evolution of the market reflects only *a-priori* endowments, preferences, and transformation possibilities; small events cannot sway the outcome .... Under increasing returns, by contrast, many outcomes are possible. Insignificant circumstances become magnified by positive feedbacks to "tip" the system into the actual outcome "selected." The small events in history become important .... (Arthur, 1989, p. 127)

The network externalities are important not only because of their impact on the direction or path of technology development, but because they represent a source of market failure—welfare losses that cannot be resolved by normal market process—and hence call for public intervention (Arthur, 1994).

In *Technical Choice*, David (1975) characterized his work as an evolutionary alternative to neoclassical theory. As noted earlier, he explicitly rejected the Fellner and Kennedy versions of the induced technical change approach to the analysis of factor bias. He also rejected the early work of Nelson and Winter as being "fundamentally neo-classical-inspired" (David, 1975, p. 76).[11] But he shares the Nelson and Winter view that the neoclassical model is excessively restrictive because factor substitution typically involves not simply a movement along a given production function but an element of innovation leading to a shift in the function itself. He does as-

---

[10]A process is *predictable* if the small degree of uncertainty built in "averages away" so that the observer has enough knowledge to pre-determine market shares accurately in the long run; *flexible* if a subsidy or tax adjustment to one of the technologies' returns can always influence future market choice; *ergodic* (not path dependent) if different sequences of historical events lead to the same market outcome with probability one; ... and *path efficient* if at all times equal development (equal adoption) of the technology that is behind in adoption would not have paid off better. (Arthur, 1989, pp. 118, 199)

[11]For a further comparison of the David and Nelson–Winter evolutionary approaches, see Elster (1983). Elster notes that David regards the Nelson–Winter model as evolutionary, but ahistorical. In his view, it differs from the neoclassical model only in its conception of microeconomic behavior. It is ahistorical because, in David's view, the Markovian-like transition probabilities depend only on the current state and not on earlier state of the system. Elster rejects David's criticism of Nelson and Winter on the basis that for the past to have a causal influence on the present it must be "mediated by a chain of locally causal links" (Elster, 1983, p. 157). Thus, because all the history that is relevant to the prediction of the future is contained in the state description, if the present state is known, prediction cannot be improved by considering the past history of the system.

sume that the firm has knowledge of available (or potentially available) alternative technologies and chooses rationally among them.

David's early analysis of factor bias was remarkably similar to the Hicks–Ahmad–Hayami–Ruttan interpretation of the process of induced technical change. And, in spite of his emphasis in *Technical Choice* that the future development of the system depends not only on the present state but also on the way the present state evolved, I find his research on path dependence in the 1980s a distinct departure from his research on factor bias in the 1970s.

In his research in the mid- and late 1980s, David employs historical analysis of a series of technical changes—the typewriter keyboard, the electric light, and power supply industries—to buttress the plausibility of the path dependence perspective. His already classic paper on the economics of QWERTY (the first six letters on the left of the topmost row of letters on the typewriter and now the computer keyboard) explored why an inefficient (from today's perspective) typewriter keyboard was introduced and why it has persisted.[12] David's answer is that an innovation in typing method, touch typing, gave rise to three features "which were crucially important in causing QWERTY to become 'locked in' as the dominant keyboard arrangement. These features were *technical inter-relatedness, economics of scale*, and *quasi-irreversibility* of investment" (David, 1985, p. 334). *Technical interrelatedness* refers to the need for system compatibility—in this case the linkage between the design of the typewriter keyboard and a typist's memory of a particular keyboard arrangement. *Scale economics* referred to the decline in user cost of the QWERTY system (or any other system) as it gained in acceptance relative to other systems. The *quasiirreversibility of investments* is the result of the acquisition of specific touch typing skills (the "software"). These characteristics are sometimes bundled under the rubric of positive "network externalities."

As David has drawn increasingly on Arthur's path dependence model, it has biased his research even further in the direction of interpreting the QWERTY-like phenomenon in dynamic systems characterized by network externalities and path dependent technical change as a dominant paradigm for the history of technology (David, 1993).[13] This paradigm would seem par-

---

[12]Liebowitz and Margolis (1990, 1994) argued that David's version of the history of the market's rejection of the supposedly more efficient Dvorak keyboard represents bad history. Given the available knowledge and experience at the time QWERTY became dominant, it represented a rational choice of technology. For a response to the Liebowitz and Margolis criticism, see David (1997).

[13]Both Arthur and David emphasize the role of network externalities in locking in inferior technological trajectories. In their review of the creation and growth of Silicon Valley, Kenney and von Burg (chap. 5, this volume) note that while the evolution of Silicon Valley venture capital has a path dependent history, "it is difficult to imagine a more efficient system for formulating high technology startups."

ticularly apt at a time when the impact of scale economics on productivity growth has been rediscovered and embodied in a "new growth economics" literature (Barro & Sala-i Martin, 1995; Lucas, 1988; Romer, 1986).[14] But Arthur's results suggest some caution. "Increasing returns, if they are bounded, are in general not sufficient to guarantee eventual monopoly by a single technology" (Arthur, 1989, p. 126). And there is substantial empirical evidence that scale economies, which often depend on prior technical change, are typically bounded by the state of technology (Levin, 1977, pp. 208–221).

Both induced innovation and evolutionary theory suggest that as scale economies are exhausted (and profits decline) the pressure of growth in demand will focus scientific and technical effort on breaking the new technological barriers. Superior technologies that have lost out as a result of chance events in the first round of technical competition have frequently turned out to be successful as the industry developed.[15] And induced technical change theory suggests that research efforts will be directed to removing the constants on growth resulting from technological constraints or inelastic (or scarce) factor supplies.[16]

The transition from coal to petroleum-based feedstocks in the heavy organic chemical industry is a particularly dramatic example. From the 1870s through the 1930s, German leadership in the organic chemical industry was based on coal-based technology. Beginning in the 1920s, with the rapid growth in demand for gasoline for automobiles and trucks in the United States, a large and inexpensive supply of olefins became available as a by-product of petroleum refining. By the end of World War II, the U.S. chemical industry had shifted rapidly to petroleum-based feedstocks. In Germany this transition, impeded by skills, education and attitudes that had been developed under a coal-based industrial regime, was delayed by more than a decade. By the 1960s, however, Germany was making a rapid transi-

---

[14]Scale economies have become the new "black box" of contemporary growth theory. It is hard to believe that much of the productivity growth that is presumably accounted for by scale economies is not the disequilibrium effect of prior technical change (Landau & Rosenberg, 1992; Liebowitz & Margolis, 1994)

[15]See, for example, the exceedingly careful study of technological substitution in the case of Cochlear implants by Van de Ven and Garud (1993). The Cochlear implant is a biomedical invention that enables hearing by profoundly deaf people. The industry is characterized by the conditions that David and Arthur identify with technological lock-in. Yet in spite of initial commercial dominance the "single channel" technology was completely replaced by the "multiple channel" technology. For other cases, see Foray and Grubler (1990), Cheng and Van de Ven (1994) and Liebowitz and Margolis (1992, 1995).

[16]The development of semiconductor technology as a replacement for vacuum tubes for amplifying, rectifying, and modulating electrical signals is an example of a shift in technological trajectories induced by technological constraints (Dosi, 1984). The development of fertilizer responsive crop varieties represents an example of a shift in technological trajectories induced by changes in resource endowments (Hayami & Ruttan 1985). The emergence of the gas turbine from a niche technology to an important source of electric power generation since the early 1980s was induced, in part, by the exhaustion of scale in steam turbine generation (Islas, 1997).

tion to the petroleum-based feedstock path of technical change in heavy organic chemicals (Grant, Patterson, & Whitston, 1988; Stokes, 1994).

TOWARD A MORE GENERAL THEORY?

In this section I first summarize my assessment of the strengths and limitations of each of the three models of technical innovation. I then outline the elements of a more general theory. I would like to make clear to the reader my particular historical and epistemological bias: Departures from neoclassical microeconomic theory, when successful, are eventually seen as extensions and become incorporated into neoclassical theory. Thus, for example, the microeconomic version of induced technical change can now be viewed as an extension of, rather than a departure from, the neoclassical theory of the firm.[17]

Assessment

One common theme pervading the three approaches to understanding sources of technological change is the disagreement with the assumption in neoclassical growth models that a common production function is available to all countries regardless of human capital, resources or institutional endowments. It should now be obvious that differences in productivity levels and rates of growth cannot be overcome by the simple transfer of capital and technology. The asymmetries between firms and between countries in resource endowments and in scientific and technological capabilities are not easily overcome. The technologies that are capable of becoming the most productive sources of growth are often location specific (Kenney & von Burg, chap. 5, this volume). A second common theme is an emphasis on microfoundations. This emphasis on microfoundations is common to the approaches that have abandoned neoclassical microeconomics as well as to those that have attempted to extend neoclassical theory (Dosi, personal communication, February 15, 1995).

The major limitation of the growth theoretic version of the induced innovation model is the implausibility of the innovation possibility function (IPF). The shape of the IPF is independent of the bias in the path of technical change. As technical change progresses, there is no effect on the "funda-

---

[17]Nelson and Winter attempt to confront this problem by arguing that there are two alternative views of neoclassical theory. One is the more rigorous "literal" view. The other is termed "the "tendency" view. Applied economists with a primary interest in interpreting economic history or behavior tend to employ the tendency view while theorists who are more concerned with the formal properties employ a more literal interpretation. They identify evolutionary theory with the tendency view (Nelson & Winter, 1975).

mental" trade-off between labor and capital augmenting technical change. Thus, as Nordhaus (1973) noted, the growth theoretic approach to induced innovation fails to rescue growth theory from treating technical change as exogenous. It has been unproductive of empirical research and is no longer viewed as an important contribution to growth theory.

The major limitation of the microeconomic version is that its internal mechanism—the learning, search, and formal R & D processes—remain inside a black box. The model is driven by exogenous changes in the economic environment in which the firm (or public research agency) finds itself. The microeconomic model has, nevertheless, been productive of a substantial body of empirical research and has helped to clarify the historical process of technical change, particularly at the industry and sector level both within and across countries.

The strength of the evolutionary model is precisely in the area where the microeconomic induced innovation model is weakest. It builds on the behavioral theory of the firm in an attempt to provide a more realistic description of the internal workings of the black box. The Nelson–Winter evolutionary approach has not, however, become a productive source of empirical research. The results of the various simulations are defended as plausible in terms of the stylized facts of industrial organization and of firm, sector and macroeconomic growth. It is possible that the reason for the lack of empirical testing is that the simulation methodology lends itself to the easy proliferation of plausible results. At present, the evolutionary approach must be regarded as a "point of view" rather than as a theory (Arrow, 1995).

The strengths of the path dependence model lie in the insistence of its practitioners on the importance of the sequence of specific microlevel historical events.[18] In this view, current choices of techniques become the link through which prevailing economic conditions may influence the future dimensions of technology and knowledge (David, 1975). However, the concept of technological lock-in, at least in the hands of its more rigorous practitioners, applies only to network technologies characterized by increasing returns to scale. In industries with constant or decreasing returns to scale, historical lock-in does not apply.

There can be no question that technical change is path dependent in the sense that it evolves from earlier technological development. In spite of somewhat similar motivation the path dependent literature has not consciously drawn on the Nelson–Winter work for inspiration (Arthur, personal communication, October 17, 1996). It is necessary to go beyond

---

[18]This emphasis creates an important opportunity to incorporate the contributions of noneconomists, particularly of historians and other social scientists along with those of economists into a more comprehensive understanding of the sources of technical change (Kenney & von Burg, chap. 5, this volume.)

the present path dependent models, however, to examine the forces responsible for changes in the rate and direction of technical change. But there is little discussion of how firms or industries escape from lock in. What happens when the scale economies resulting from an earlier change in technology have been exhausted and the industry enters a constant or decreasing returns stage? At this point in time, it seems apparent that changes in relative factor prices would, with some lag, have the effect of bending or biasing the path of technical change along the lines suggested by the theory of induced technical change. Similarly a new radical innovation may, at this stage, both increase the rate and modify the direction of technical change.

The study of technical change in the semiconductor industry by Dosi (1984) represents a useful illustration of the potential value of a more general model. The Dosi study is particularly rich in its depth of technical insight. At a rhetorical level, Dosi identifies his methodology with the Nelson–Winter evolutionary approach. In practice, however, he utilizes an eclectic combination of induced innovation, evolutionary, and path dependence interpretations of the process of semiconductor technology development. A more rigorous approach to the development of a general theory of the sources of technical change will be required to bridge the three "island empires."

Integration of Factor and Demand-Induced Technical Change Models

A first step toward developing a more general theory of technical change is to integrate the factor induced model and the demand induced models (Ruttan & Hayami, 1994). Binswanger, drawing on Nordhaus (1969) and Kamien and Schwartz (1969) sketched the outlines of a more general model that makes both the rate and direction of technical change endogenous (Binswanger, 1978).

If one assumes decreasing marginal productivity of research resources in applied research and technology development and, in addition, incorporates the effects of changes in product demand, then growth (decline) in product demand would increase (decrease) the optimum level of search and research expenditure. The larger research budget, induced by growth in product demand, increases the rate at which the metaproduction function shifts inward toward the origin. Even when the initial path of technological development is generated by "technology push," factor market forces often act to modify the path of technical change. Differential elasticities of factor supply result in changes in relative factor prices and direct research effort to save increasingly scarce factor supplies. The result is a nonneutral shift in both the neoclassical and the metaproduction functions.[19]

More recently, Christian has elaborated the Binswanger model and analyzed more formally the innovators decision to conduct research and development directed toward process innovation (Christian, 1993). As yet, however, there has been no attempt to implement empirically an integrated factor and demand induced innovation model.

Integration of Path Dependence and Induced Technical Change Models

A second step would be the integration of the induced technical change and the path dependent models. As noted earlier, David has pointed to the persistent failure to replace the inefficient QWERTY layout of the typewriter and computer keyboards with the more efficient DSK keyboard. Wright (1990) suggested that the historical resource intensity of American industry, based on domestic resource abundance, has been an important factor in weakening the capacity of American industry to adapt to a world where lower transportation costs and more open trading systems have reduced the traditional advantage of United States based firms. If this perspective is correct, Japan's industrial success may be attributed to its historical resource scarcity.

The difference in perspective seems to hinge on how the elasticity of substitution changes over time in response to changes in resource endowments or relative factor prices. David (1975) showed how localized induced technical change can lead to path dependent technical change. The effect of localization is to lower the elasticity of substitution and lock-in the trajectory of technical change. As relative factor prices continue to shift, however, it is hard to believe technological competition would not result in a bending of the path of technical change in the direction implied by changing factor endowments. The path dependent and the induced innovation models are appropriately viewed as complementary rather than as alternative interpretations of the forces that influence the direction of technical change.

The path dependent model will remain incomplete, however, until it is more fully integrated with the microeconomic version of the induced technical change model and with the Nelson–Winter evolutionary model. Development of an industry seldom proceeds indefinitely along an initially

---

[19]In Binswanger (1978a) both the production function and the metaproduction function are neoclassical. David (personal communication, October 24, 1997) noted that if the advances in technology are highly "localized," in the sense suggested by Atkinson and Stiglitz (1969), the neoclassical assumption is inappropriate. See also David (1975) and Antonelli (1995).

selected process ray (Landes, 1994). As technical progress slows down or scale economies erode, a shift in relative factor prices can be expected to induce an intensified search for technologies along a ray that is more consistent with contemporary factor prices.[20]

Integration of Induced Innovation, Path Dependence and Trade Theory

A third step would be the integration of induced technical change, path dependence, and international trade theory. Relative factor endowments play an important role in both the Hecksher–Ohlin approach to trade theory and the theory of induced technical change. Under the Hecksher–Ohlin assumptions each country exports its abundant factor-intensive commodity (Hamilton & Soderstrom, 1981). Induced technical change acts to make the scare factor (or its substitutes) more abundant. Except for an early article by Chipman (1970) and a more recent articles by Hamilton and Soderstrom (1981), and Davidson (1979), the relationship between the theory of induced technical change and international trade theory has remained almost completely unexplored. To the extent that trade can release the constraints of factor endowments on growth, the theory of induced technical change loses part of its power to explain the direction of bias in productivity growth. Conversely, to the extent that technical change can release the constraints on growth resulting from inelastic factor supplies, the power of the differential factor endowments explanation for trade is weakened. The revival of interest in growth theory combined with recent developments in the theory of international trade are opening up opportunities to explore the sources, rate and direction of technical change more fruitfully (Grossman & Helpman, 1991; Srinivasan, 1985).

The research agenda implied by the discussion in this section should proceed at two levels. At the theoretical level, research should be directed to the development of a succession of more fully integrated models of the sources of technical change. At the empirical level, research should be directed to more comprehensive testing of the induced, evolutionary, and path dependent models. There is substantial empirical literature on induced technical change in agricultural economics and economic history. But only limited efforts have been made to test the evolutionary and path dependence models against historical experience.

---

[20]See, for example, the patterns of factor substitution in the transition in primary energy sources and transportation infrastructure (Grubler & Nakicenovic, 1988; Nakicenovic, 1991).

INDUCED TECHNICAL CHANGE AND ENDOGENOUS GROWTH MODELS

Since the late 1980s, students of economic growth have been engaged in a re-evaluation of neoclassical growth models. One result of this re-examination has been the emergence of a new generation of endogenous growth models.

The major focus of the new "macroendogenous" growth models is to attribute differences in growth performance among countries to endogenous factors such as investment in human capital, learning by doing, scale economies, and technical change (Lucas, 1988; Romer, 1986, 1994). In the initial Romer–Lucas framework, the accumulation of human capital adds to the productivity of the person in whom it is embodied.[21] But the general level of productivity rises by more than can be accounted for or captured by the person or firm that makes each particular investment. Gains in scale economies are enhanced by the integration into multinational trading systems of economies that are human capital intensive (Grossman & Helpman, 1991).

The new growth literature has yet to incorporate the richness and depth of understanding of the sources of technical change that the three traditions reviewed in this chapter have achieved (Bardhan, 1995; Ruttan, 1998). Like the older neoclassical growth literature its focus is on the proximate sources of growth rather than the sources of technical change. A major challenge for the future is to integrate the insights about endogenous growth gained from the theoretical and empirical research conducted within the induced technical change, the evolutionary, and path dependence theories, with new insights into the relationship between human capital, scale, and trade opened up by the macroendogenous growth models.

ACKNOWLEDGMENTS

I am indebted to Esben Sloth-Anderson, W. Brian Arthur, Erhard Bruderer, Jason E. Christian, Paul A. David, Jerry Donato, Giovanni Dosi, Laura McCann, Richard Nelson, Nathan Rosenberg, Tugrul Temel, Michael A. Trueblood, Andrew Van de Ven, and Sidney Winter for comments on an earlier draft of this paper. Earlier versions of this paper have been presented in seminars at the International Institute of Applied Systems Analysis (IIASA), at the University of Minnesota Economic Development Center and at the Hong Kong University of Science and Technology. The research on which the paper is based was supported, in part, by a grant from the Alfred P. Sloan Foundation. Abbreviated versions of this paper have appeared in Ruttan (1996, 1997). For critical reviews, see Dosi (1997) and Wright (1997).

REFERENCES

Ahmad, S. (1966). On the theory of induced innovation. *Economic Journal, 76,* 344–357.
Ahmad, S. (1976a). A rejoinder to Professor Kennedy. *Economic Journal, 77,* 960–963.
Ahmad, S. (1967b). Reply to Professor Fellner. *Economic Journal, 77,* 662–664.
Alchian, A. A. (1950). Uncertainty, evolution and economic theory. *Journal of Political Economy,* 58, 211–222.
Ames, E., & Rosenberg, N. (1968). The Enfield Arsenal in theory and history. *Economic Journal,* 78, 730–733
Anderson, E. S. (1994). Evolutionary economics: Post-Schumpeterian contributions. London: Pinter.
Antonelli, C. (1995). *The economics of localized technological changes and industrial dynamics.* Dordrecht, The Netherlands: Kluwer Academic Publishers.
Arrow, K. (1962). The economic implications of learning by doing. *Review of Economic Studies, 29,* 155–173.
Arrow, K. (1995). Viewpoint. *Science 267,* 1617.
Arthur, W. B. (1989). Competing technologies, increasing returns, and lock-in by historical events. *The Economic Journal, 99,* 116–131.
Arthur, W. B. *Increasing returns and path dependence in the economy.* Ann Arbor: The University of Michigan Press.
Arthur, W. B., Ermoliev, Y. M., & Kaniovski, Y. M. (1987). Path dependence processes and the emergence of macro-structure. *European Journal of Operational Research, 30,* 294–303.
Atkinson, A. B., & Stiglitz J. E. (1969). A new view of technological change. *Economic Journal, 79,* 573–578.
Bardhan, P. (1995). The contribution of endogenous growth theory to the analysis of development problems: An assessment. In J. Behrman & T. N. Srinivason (Eds.), *Handbook of Development Economics* (Vol. 3, pp. 2984–2998). New York.
Barro, R. J., & Sala-i-Martin, X. (1995). *Economic Growth.* New York: McGraw Hill.
Ben-Zion, U., & Ruttan, V. W. (1975). Money in the production function: An interpretation of empirical results. *Review of Economics and Statistics, 57,* 246–247.
Ben-Zion, U., & Ruttan, V. W. (1978). Aggregate demand and the rate of technical change. In H. P. Binswanger, & V. W. Ruttan (Eds.), *Induced innovation: Technology institutions and development* (pp. 261–275). Baltimore: Johns Hopkins University Press.
Binswanger, H. P. (1973). The measurement of biased efficiency gains in U.S. and Japanese agriculture to test the induced innovation hypothesis. Unpublished PhD dissertation, North Carolina State University, Raleigh.
Binswanger, H. P. (1974a). A cost function approach to the measurement of elasticities of factor demand and elasticities of substitution. *American Journal of Agricultural Economics, 56,* 377–386.
Binswanger, H. P. (1974b). The measurement of technical change biases with many factors of production. *American Economic Review, 64,* 964–976.
Binswanger, H. P. (1978a). Induced technical change: Evolution of thought. In H. P. Binswanger, & V. W. Ruttan, (Eds.), *Induced innovation: Technology, institutions and development* (pp. 13–43). Baltimore: The Johns Hopkins University Press.
Binswanger, H. P. (1978b). The micro economics of induced technical change. In H. P. Binswanger & V. W. Ruttan, (Eds.), *Induced innovation: Technology, institutions and development* (pp. 91–127). Baltimore: The Johns Hopkins University Press.
Binswanger, H. P., & Ruttan, V. W. (1978). *Induced innovation: Technology, institutions and development.* Baltimore: The Johns Hopkins University Press.
Blaug, M. (1963). A survey of the theory of process-innovation. *Economica, 63,* 13–32.
Cheng, Y.-T., & Van de Ven, A. H. (1994). Learning the innovation journey: Order out of chaos? (Discussion paper No. 208). University of Minnesota Strategic Management Center, Minneapolis, Minnesota.

Chipman, J. S. (1970). Induced technical change and patterns of international trade. In R. Vernon (Ed.), *The technology factor in international trade* (pp. 95–127). New York: Columbia University Press.

Christian, J. E. (1993). The simple microeconomics of induced innovation (Mineograph). Department of Agricultural Economics, University of California, Davis.

Cyert, R.M., & March, J. G. (1963). *A behavioral theory of the firm.* Englewood Cliffs, NJ: Prentice Hall.

David, P. A. (1966). Mechanization of reaping in the ante-bellum midwest. In H. Rosovsky (Ed.), *Industrialization in two systems: Essays in honor of Alexander Gerscherkron* (pp. 3–39). New York: John Wiley.

David, P. A. (1973). Labor scarcity and the problem of technological practice and progress in the nineteenth century (Research paper No. 297). Harvard Institute of Economic Research, Cambridge, Massachusetts.

David, P. A. (1975). *Technical choice, innovation and economic growth.* Cambridge: Cambridge University Press.

David, P. A. (1985). Clio and the economics of QWERTY. *American Economic Review, 76,* 332–337.

David, P. A. (1986). Understanding the Economics of QWERTY: The necessity of history. In W. N. Parker, (Ed.), *Economic history and the modern economist* (pp.). New York: Basil Blackwell.

David, P. A. (1993). Path dependence and predictability in dynamic systems with local network externalities: A paradigm for historical economics. In D. Foray & C. Freeman (Eds.), *Technology and the wealth of nations: The dynamics of contracted advantage* (pp. 208–231). London: Pinter Publishers.

David, P. A. (1997). Path dependence and the quest for historical economics: One more chorus of the ballad of QWERTY (Mimeograph). All Souls College, Oxford University, England. David, P. A., & Bunn, J. A. (1988). The economics of gateway technologies and network evolution: Lessons from electricity supply history. *Information Economics and Policy, 3,* 165–202.

Davidson, W. H. (1979). Factor endowment, innovation and international trade theory. *Kyklos, 32,* 764–773.

Diamond, P., McFadden, D., & Rodriguez, M. (1978). Measurement of factor substitution and bias of technological change. In M. Fuss & D. McFadden (Eds.), *Production economics: A dual approach to theory and applications* (Vol 2, pp. 125–147). Amsterdam: North Holland.

Dosi, G. *Technical change and industrial transformation.* New York: St. Martins Press.

Dosi, G., Giannetti, R., & Toninelli, P.A. (Eds.). (1992). *Technology and enterprise in a historical perspective.* Oxford: Oxford University Press.

Drandakis, E. M., & Phelps, E. S. (1966). A model of induced invention, growth and distribution. *The Economic Journal, 76,* 823–840.

Elster, J. (1983). *Explaining technical change.* Cambridge: Cambridge University Press.

Evenson, R. E., & Kislev, Y. (1975). *Agricultural research and productivity.* New Haven: Yale University Press.

Fellner, W. (1956). *Trends and cycles in economic activity.* New York: Henry Holt.

Fellner, W. (1961). Two Propositions in the Theory of Induced Innovations. *Economic Journal, 71,* 305–308.

Fellner, W. (1962). Does the market direct the relative factor saving effects of technological progress? In R. R. Nelson (Ed.), *The rate and direction of inventive activity: Economic and social factors* (pp.) Princeton: Princeton University Press for the National Bureau of Economic Research.

Fellner, W. (1967). Comment on the induced bias. *Economic Journal, 77,* 664–665.

Fogel, R. W. (1967). The specification problem in economic history. *Journal of Economic History, 27,* 283–308.

Freeman, C. (1982). The economics of industrial innovation (2nd ed.). Cambridge, MA: MIT Press.

Freeman, C. (1994). The economics of technical change. *Cambridge Journal of Economics, 18,* 463–514.

Foray, D., & Grubler, A. (1990). Morphological analysis, diffusion and lock-out of technologies: Ferrous casting in France and the FRG. *Research Policy, 19,* 535–550.

Gans, J. S., & Shepherd, G. B. (1994). How are the mighty fallen: Rejected classic articles by leading economists, *Journal of Economic Perspectives* 1994, 8, #1: 165–179.

Garud, R., & Rappa, M. (1994) A socio-cognitive model of technology evolution: The case of cochlear implants. *Organization Science, 5,* 344–362.

Gilfillan, S. C. (1935). *The sociology of invention.* Chicago: Folliett.

Grant, W., Patterson, W., & Whitston, C. (1988) Government and the chemical industry: A comparative study of Britain and West Germany. Oxford: Oxford University Press.

Griliches, Z. (1957). Hybrid corn: An exploration in the economics of technological change. *Econometrica, 25,* 501–522.

Grubler, A., & Nakicenovic, N. (1988). The dynamic evolution of methane technologies. In T. H. Lee, H. R. Linden, D. A. Dryfus, D.A., & T. Vasco (Eds.), *The methane age* (pp. 13–44). Dordrecht: Kluwer Academic Publishers.

Habakkuk, H. J. (1962). *American and British technology in the nineteenth century: The search for labor saving inventions.* Cambridge, England: Cambridge University Press.

Haltmaier, J. (1986). Induced innovation and productivity growth: An empirical analysis (Special studies paper No. 220). Washington, D.C.: Federal Reserve Board.

Hamilton, C., & Soderstrom, H. T. (1981). Technology and international trade: A Heckscher-Ohlin approach. In S. Grassman & E. Londberg (Eds.), *The world economic order: Past and prospects* (pp. 198–229). London: Macmillan.

Hayami, Y., & Ruttan, V. W. (1970). Factor prices and technical change in agricultural development: The United States and Japan, 1880–1960. *Journal of Political Economy, 78,* 1115–1141.

Hayami, Y., & Ruttan, V. W. (1985). *Agricultural development: An international perspective.* Baltimore: The Johns Hopkins University Press. (Original work published in 1971.)

Hicks, J. (1963). *The theory of wages* (2nd ed.). London: Macmillan. (Original work published in 1932.)

Islas, J. (1997). Getting round the lock-in in electricity generating systems: The examples of the gas turbine. *Research Policy, 26,* 49–66.

James, J. A., & Skinner, J. S. (1985). Resolution of the labor-scarcity paradox. *Journal of Economic History, 45,* 513–539.

Jorgenson, D. W., & Fraumeni, B. M. (1981). Relative prices and technical change. In E. R. Bendt & B. Fields, (Eds.), *Modeling and measuring national resource substitution* (pp. 17–47). Cambridge, MA: MIT Press.

Jorgenson, D. W., & Wilcoxen, P. J. (1993). Energy, the environment, and economic growth. In A. V. Kneese & J. L. Sweeney (Eds.), *Handbook of resource and energy economics* (Vol. 3, pp. 1267–1349). Amsterdam: Elsevier Science Publishers.

Kamien, M. I., & Schwartz, N. L. (1969). Optimal induced technical change, *Econometrica, 36,* 1–17.

Kennedy, C. (1964). Induced bias in innovation and the theory of distribution. *The Economic Journal, 74,* 541–547.

Kennedy, C. (1966). Samuelson on induced innovation. *Review of Economics and Statistics, 48,* 442–444.

Kennedy, C. (1967). On the theory of induced innovation—A reply. *Economic Journal, 77,* 958–960

Koppel B. M. (Ed.) (1995). *Induced innovation theory and international agricultural development.* Baltimore, MD: The Johns Hopkins University Press.

Landau, R., & Rosenberg, N. (1992). Successful commercialization in the chemical process industries. In N. Rosenberg, R. Landau, & D. C. Mowery (Eds.), Technology and the wealth of nations (pp. 73–119). Stanford, CA: Stanford University Press.

Landes, D. S. (1994). What room for accident in history?: Explaining big changes by small events. *Economic History Review, 47,* 637–656.

Langolis, R. N., & Everett, M. J. (1994). What is evolutionary economics? In L. Magnuson (Ed.), Evolutionary and neo-Schumpeterian approaches to economics (pp. 11–47). Dordrecht: Kluwer.

Levin, R. C. (1977). Technical change and optimal scale: Some evidence and implications. *Southern Economic Journal, 44*, 208–221.

Liebowitz, S. J., & Margolis, S. E. (1990). The fable of the keys. *Journal of Law and Economics, 33*, 1–25.

Liebowitz, S. J., & Margolis, S. E. (1992). Market processes and the selection of standards (Working paper). School of Management, University of Texas at Dallas.

Liebowitz, S. J., & Margolis, S. E. (1994). Network externality: An uncommon tragedy. *Journal of Economic Perspectives, 8*(2), 133–150.

Liebowitz, S. J., & Margolis, S. E. (1995). Path dependence, lock-in, and history. *Journal of Law, Economics, and Organization, 11*, 205–226.

Lucas, R. E., Jr. (1967). Tests of a capital-theoretic model of technological change. *Review of Economic Studies, 34*, 175–180.

Lucas, R. E., Jr. (1988). On the mechanics of economic development. *Journal of Monetary Economics, 22*, 3–42.

Mowery, D. C., & Rosenberg, N. (1979). The influence of market demand upon innovation: A critical review of some recent empirical studies. *Research Policy, 8*, 103–153.

Nakicenovic, N. (1991). Diffusion of pervasive systems: A case of transport infrastructures. *Technological Forecasting and Social Change, 39*, 181–219.

Nelson, R. R. (1995). Recent evolutionary theorizing about economic change. *Journal of Economic Literature, 33*, 48–90.

Nelson, R. R., & Winter, S. G. (1973). Toward an evolutionary theory of economic capabilities. *American Economic Review, 63*, 440–449.

Nelson, R. R., & Winter, S. G. (1974). Neoclassical vs. Evolutionary theories of economic growth: Critique and prospects. *Economic Journal, 84*, 886–905.

Nelson, R. R., & Winter, S. G. (1975). Factor price changes and factor substitution in an evolutionary model. *Bell Journal of Economics, 6*, 466–486.

Nelson, R. R., Winter, S. G., & Schuette, H. L. (1976). Technical change in an evolutionary model. *Quarterly Journal of Economics, 40*, 90–118.

Nelson, R. R., & Winter, S. G. (1977). Simulation of Schumpeterian competition. *American Economic Review, 67*, 271–276.

Nelson, R. R., & Winter, S. G. (1982). *An evolutionary theory of economic change.* Cambridge, MA: Harvard University Press.

Nordhaus, W. D. *Invention, growth and welfare: A theoretical treatment of technical change.* Cambridge, MA: The MIT Press.

Nordhaus, W. D. (1973). Some skeptical thoughts on the theory of induced innovation. *Quarterly Journal of Economics, 87*, 208–219.

Olmstead, A. L., & Rhode, P. (1993) Induced innovation in American agricultures: A reconsideration. *Journal of Political Economy, 101*, 100–118.

Olmstead, A. L., & Rhode, P. (1995). Induced innovation in American agriculture: An econometric analysis (Mineograph). Institute of Government Affairs, University of California, Davis.

Penrose, E. T. (1952). Biological analogies to the theory of the firm. *American Economic Review, 42*, 804–819.

Romer, P. M. (1986). Increasing returns and long-run growth. *Journal of Political Economy. 94*, 1002–1037.

Romer. P. M. (1994). Idea gaps and object gaps in economic development. *Journal of Monetary Economics, 8*, 543–573.

Rosenberg, N. (1982a). Learning by using. In N. Rosenberg (Ed.), *Inside the black box: Technology and economics* (pp. 120–140). Cambridge, MA: Cambridge University Press.

Rosenberg, N. (1982b). Science, invention and economic growth. *Economic Journal*, 90–108.

Rothschild, K. (1956). *The theory of wages.* Oxford: Blackwell Publishers.

Ruttan, V. W., Usher, & Schumpeter, J. A. (1959). On invention, innovation, and technological change. *Quarterly Journal of Economics, 73,* 596–606.

Ruttan, V. W. (1996). Induced innovation and path dependence: A reassessment with respect to agricultural development and the environment. *Technological Forecasting and Social Change, 53,* 41–60.

Ruttan, V. W. (1997). Induced innovation, evolutionary theory and path dependence: Sources of technical change, *Economic Journal, 107,* 1520–1529.

Ruttan, V. W. (1998). The new growth theory and development economics: A survey. *Journal of Development Studies, 1998, 35:* 1–26.

Ruttan, V. W., & Hayami, Y. (1994). Induced innovation theory and agricultural development: A personal account. In B. Koppel (Ed.), *Induced innovation theory and international agricultural development: A reassessment* (pp. 22–36). Baltimore, MD: The Johns Hopkins University Press.

Salter, W. E. G. (1960). *Productivity and technical change.* (2nd ed.). Cambridge, England: Cambridge University Press.

Samuelson, Paul A. (1965). A theory of induced innovation along Kennedy–Weizsäcker lines. *Review of Economics and Statistics, 47,* 343–356.

Samuelson, P. A. (1966). Rejoinder: Agreements, disagreements, doubts and the case of induced Harrod-neutral technical change. *Review of Economics and Statistics, 48,* 444–448.

Scherer, F. M. (1982). Demand pull and technological inventions: Schmookler revisited. *Journal of Industrial Economics, 30,* 225–237.

Schmookler, J. (1962). Determinants of industrial invention. In R. R. Nelson (Ed.), *The rate of direction of inventive activity: Economic and social factors.* Princeton, NJ: Princeton University Press.

Schmookler, J. (1966). *Invention and economic growth.* Cambridge, MA: Harvard University Press.

Schumpeter, J. A. (1934). *The theory of economic development.* Cambridge, MA: Harvard University Press. (Original work published in German in 1911.)

Schumpeter, J. A. (1939). *Business cycles* (Vols. 1 & 2). New York: McGraw Hill.

Simon, H. A. (1955). A behavioral model of rational choice. *Quarterly Journal of Economics, 69,* 99–118.

Simon, H. A. (1959). Theories of decision making in economics. *American Economic Review, 49,* 253–283.

Srinivasan, T. N. (1985). Long run growth theories and empirics: Anything new? In I. Takatoshi & A. O. Krueger (Eds.), Growth theories in light of the East Asian experience (pp. 37–70). Chicago, IL: University of Chicago Press.

Stokes, R. G. (1994). Opting for oil: The political economy of technical change in the West German chemical industry, 1945–1961. Cambridge, England: Cambridge University Press.

Temin, P. (1966). Labor scarcity and the problems of American industrial efficiency. *Journal of Economic History, 26,* 277–298.

Thirtle, C. G., & Ruttan, V. W. (1987). *The role of demand and supply in the generation and diffusion of technical change.* London: Harwood Academic Publishers.

Van de Ven, A., & Garud, R. (1993). Innovation and industry development: The case of cochlear implants. In R. Burgelman & R. Rosenbloom (Eds.), *Research on technological innovation, management and policy* (Vol. 5, pp. 1–46) Greenwich, CT: JAI Press.

Vernon, R. (1966). International investment and international trade in the product cycle. *Quarterly Journal of Economics, 80,* 190–207.

Vernon, R. (1979). The product cycle hypothesis in a new international environment. *Oxford Bulletin of Economics and Statistics, 40,* 255–267.

Wan, H. Y., Jr., (1971). *Economic growth.* New York: Harcourt Brace Javonovich.

Walsh, V. (1984). Invention and innovation in the chemical industry: Demand–pull or discovery–push? *Research Policy, 13,* 211–234.

Winter, S. G. (1971). Satisficing, selection and the innovating remnant. *Quarterly Journal of Economics, 85,* 237–261.

Winter, S. G. (1984). Schumpeterian competition in alternative technological regimes. *Journal of Economic Behavior and Organization, 5,* 287–320.

Witt, U. (1993). *Evolutionary economics.* Aldersbat, England: Edward Elgar Publishing.

Wright, G. (1990). The origins of American industrial success, 1889–1940. *The American Economic Review, 80,* 651–667.

Wright, G. (1997). Toward a historical approach to technological change. *Economic Journal, 107,* 1560–1566.

II

From Path Dependence to Path Creation

5

Paths and Regions: The Creation and Growth of Silicon Valley

Martin Kenney
University of California, Davis
Berkeley Roundtable on the International Economy

Urs von Burg
University of St. Gallen, St. Gallen, Switzerland

For the last 40 years, a geographic and mental space in the San Francisco Bay Area now known as Silicon Valley has been the birthplace of many of the largest and fastest growing electronics firms in the world and a number of new industrial sectors.[1] The technologies these firms commercialized have had a significant impact on many aspects of social and economic life. To facilitate the commercialization, a set of institutions evolved in Silicon Valley to nurture the new firms; these were established to exploit the potential for rapid growth stemming from electronics innovations.

The observation that technologies and places have histories and that these histories matter is, by itself, unremarkable (see Bassanini & Dosi, chap. 2, this volume). If, to invert Voltaire's (1999) Dr. Pangloss, it is accepted that the current situation is not necessarily the best of all possible worlds; then we would leave the world of microeconomics and enter a world of struggle, strategy, and serendipity, in which human beings working alone

---

[1]This region is also the home of a large number of medical equipment firms and the largest concentration of biotechnology firms in the United States. These startups derived critical benefits from proximity to institutions such as venture capital described later in this chapter.

and in groups, create novelty, while being conditioned by their history. In this world, the theoretical models of microeconomics lose power and can only be accepted as partially valid, at best; more often, are irrelevant. Freed from the simplistic, totalizing, ahistorical model of microeconomics, we are regrettably confronted with complexity.

Path dependency directs inquiry toward the ways in which today's realities are based on yesterday's events. Silicon Valley is, in many ways, an ideal case for examining the strengths and weaknesses of path dependent explanations. The concepts from path dependency literature provide a useful departure point for understanding the creation and evolution of Silicon Valley. For one thing, Arthur (1994) specifically referred to Silicon Valley as an example of path dependent industrial clustering due to agglomeration effects. Implicit, but not articulated and examined, is the idea that paths are created by human actors operating in time (Karnøe & Garud, 1998). Throughout this chapter, our evaluation of the applicability of path dependent arguments for explaining the dynamics of Silicon Valley integrates the creative dimension of path development.

Particularly important for understanding Silicon Valley is linking the opportunities technological evolution provided with the creation of institutions and even specific regional industrial cultures.[2] In effect, path dependency is intimately related with path creation. Silicon Valley's institutions and cultures can be understood as a set of evolving, path dependent routines nurturing specific combinations of extrafirm industrial patterns within its circumscribed region (Foray, 1991; Storper & Walker, 1988). Regional growth cannot be reduced to either business or technical developments; rather, technology and institutions dialectically create an unfolding path (Hirsch & Gillespie, chap. 3, this volume). Cook and Seely-Brown (chap. 1, this volume) use the "generative dance" as a metaphor for describing this dialectic.[3]

This chapter considers how the concept of *path dependency* can be extended to thinking about the creation and evolution of regions. The second section reviews the previous explanations for the growth and development of Silicon Valley. It also introduces our argument that to understand Silicon Valley, it is helpful to see it as two separate economies. The first economy consists of existing firms, whereas the second economy consists of the institutions that have evolved to nurture new startups. The third section examines the genesis of Silicon Valley as a high-technology region. The fourth section shows how it

---

[2]See Rao and Singh (chap. 9, this volume) for a discussion of the institutionalization of particular technologies. Especially interesting is their discussion of the biotechnology industry, which was a clear beneficiary of the Economy Two, which will be discussed.

[3]*Ex poste facto*, the outcomes of this generative dance can be seen as a path or even a technological trajectory (on technological trajectories, see Dosi, 1984).

was the spin-off pattern developed among the semiconductor firms that catalyzed the industry, which was central to the creation of the institutions that now exist in Silicon Valley. The fifth section singles out venture capital as one critical institution for the development of Silicon Valley. The conclusion discusses the strengths and weaknesses of the path dependent perspective for explaining the creation and subsequent evolution of Silicon Valley.

PATH DEPENDENCY

The concept of path dependency was developed by economists such as Arthur (1988) and David (1986) to describe the phenomenon they noticed of apparently inferior technologies dominating market spaces (for a more detailed discussion, see Hirsch & Gillespie, chap. 3, this volume). Arthur (1994) developed abstract mathematical models showing how features such as increasing returns could create winner-take-all outcomes. David (1986, 1990, 1997, 1999), in a series of historical articles, demonstrated how this occurred in the adoption of specific technologies. They found that under certain conditions, early decisions reverberate through history, closing alternative paths and validating a single path. The implication is that history matters and outcomes need not be rational or optimal.

Though there have been a number of critiques of the claims of path dependency,[4] this chapter considers path dependency as a significant contribution precisely because it problematizes the present. Path dependence accepts that small events can have very large later impacts. What is significant in this stance is that it permits these small events to be precipitated by noneconomic events. This is a critical opening for explanations not dependent on simple short-term profit maximizing. In a path dependent world, social constructions and strategic maneuvering in a nondeterministic environment are critical for path formation.[5]

It is not a great leap to accept that technology and the institutions in which it is embedded coevolve from path dependency (For the definitive discussion of evolutionary economics, see Nelson & Winter, 1982; on

---

[4]Critiques from the economic mainstream include Liebowitz and Margolis (1990, 1995). Sabel (1998) argued that in theoretical terms, path dependency removes human choice and collective action from the evolutionary process. One curious oversight in the QWERTY versus Dvorak debate is the fact that Dvorak was introduced in the 1930s. Dvorak had a much more difficult task, that is, dislodging an already established technology, QWERTY. A more comprehensive discussion and critique is provided by Ruttan (chap. 4, this volume) and Bassanini and Dosi (chap. 2, this volume).

[5]See Pinch (chap. 14, this volume) for an exposition of the social construction of technology perspective. We sympathize with the social constructionist perspective, but find the implicit deliberative and planned meaning of "construction" quite dangerous. Here, Cook and Seely-Brown's metaphor of a "generative dance" (chap. 1, this volume) is closer to our perspective on how technologies and institutions evolve. There are no blueprints and no certainty.

embeddedness, see Granovetter, 1985). Hirsch and Gillespie (chap. 3, this volume) point out the nested, intertwined nature of many path dependent phenomena, that is, a particular technological choice is often an outcome of the interaction of a number of path dependent processes. Implicit, but not well developed, is the recognition that any path is, in fact, built by actors creating, using, and reshaping the infrastructure of institutions, routines, and organizations in which the technology is embedded. This activity includes suppliers, but ranges further to include capital equipment makers, specialized financial institutions, marketing and distribution organizations, educational institutions, and a myriad of other organizations, many of which are specialized in the needs of a particular industry. Often, perhaps more important, is creation on the demand side where occupying a market space and developing customers can be critical for the adoption of an innovation. Christensen (1992) showed in the hard disk drive (HDD) industry that the emergence of new customers was critical for the survival and growth of new entrants. At times, this also extends to creating new distribution, marketing, and retail networks. In other cases customers have to be mobilized; for example, what Lampel (chap. 11, this volume) terms *technological spectacles* or *races* establish the characteristics of particular brands or even technological solutions (see also Rao & Singh, chap. 9, this volume).

Definitive explanations of industrial emergence and firm clustering in specific regions remain elusive. In a parallel to the path dependence and dominant design theories (Arthur, 1994; David, 1986; Henderson & Clark, 1990), industrial geographers such as Storper and Walker (1988) found that there are periods of locational opportunity before an industry clusters and locks-in into specific locations. Both economists dealing with innovations and industrial geographers studying regional industrial growth find that often there is an initial period of openness with a number of contenders prior to the selection of a dominant design or dominant location. It is at such moments that the small events can result in the long-term differences.

Arthur (1994) explicitly argued that positive feedbacks led to the clustering of the electronics industry in Silicon Valley. In an industry in which spin-offs are frequent, the industry will tend to cluster in a certain region when there are agglomeration economies. In the abstract model, the actual location selected is random and the spin-offs or agglomeration economies reinforce clustering in a particular region. This formal model is powerful, but even at the earliest stage, not all regions have equal competencies.

SILICON VALLEY—AN INTRODUCTION

Silicon Valley, as we shall see, had many antecedents and is the outcome of a pastiche of forces, accidents of history, planning, and human foresight. Although integrated circuitry was the triggering technology for the development of the Silicon Valley agglomeration, it was not the only electronics technology in which it became a leader. For example, in the 1940s, Silicon Valley already had entrepreneurs experimenting with magnetic recording techniques at Ampex. But it was IBM's decision in 1952 to open a laboratory in San Jose to develop magnetic recording techniques for data storage that created the intellectual capital, which evolved into the merchant disk drive industry (Gan 1991). This magnetic recording expertise led not only to the HDD industry, but provided a basic technology for the telephone call processing equipment industry. Another important industry, computer networking, had its antecedents in the Xerox Palo Alto Research Center (PARC) laboratory and an Exxon-founded semiconductor firm, Zilog. Much of the knowledge that led to the relational database software industry was developed in the computer science department of IBM's San Jose Laboratory. What all these fields shared was extremely rapid technical change and quickly growing markets. The proliferation of new industry segments meant Silicon Valley could grow even faster than it would have had it been entirely dependent on the semiconductor industry.

For heuristic purposes, Silicon Valley can be conceptualized as two interrelated economies.[6] *Economy One* includes existing firms producing integrated circuits, software, computer networking equipment, computers, and a myriad of other electronics products, that is, the existing high-technology firms and other institutions, such as universities. *Economy Two* is a loosely structured network of venture capitalists, lawyers specializing in high-technology, accountants, and consultants. Their intention is to facilitate the creation and growth of firms that can later be sold to larger firms or listed on the stock exchange not to ship products. The introduction of this division is not so much driven by theoretical concerns, but rather for the optic it provides for understanding the dynamics of the region. The ability to develop new firms to exploit technological opportunities is dependent on different institutions than those necessary to operate already established firms.

The two economies are interrelated because Economy Two depends on Economy One. Conversely, firms successfully nurtured by the institutions of Economy Two become members of Economy One. However, these economies are not identical. In Economy One, the firms create products and services to be sold. In Economy Two, the product *is* firms, which embody a set of technologies and routines that another firm will purchase or that capital markets are willing to invest in by purchasing equity. In both cases, the pur-

---

[6]This distinction is not theoretically driven. The purpose is to separate two quite different activities: the operation of existing firms and institutions and the operation of institutions dedicated to creating firms *de novo*. The point being that the dynamics of Silicon Valley are best understood through this analytic distinction, which is ungeneralizable to most other regions.

chases are justified by the belief that the firm will grow sufficiently to increase its value. In and of itself, innovatory activity in an already established firm does not directly benefit Economy Two of Silicon Valley. However, indirectly inventions and innovations in existing firms can be extraordinarily beneficial because they are the raw materials for new opportunities for entrepreneurship. Not surprisingly, Economy One firms are the single largest source of entrepreneurs for Economy Two.

When these two Economies are conflated or one is ignored, it is difficult to understand Silicon Valley. Regional growth in Silicon Valley is predicated on Economy Two. A suggestive study by Almeida and Kogut (1997; see also, Kogut, Walker, & Kim, 1991) using patent data and semiconductor firm location showed that the presence of the large semiconductor firms is highly correlated with the incidence of small firm establishments. They found that the large number of semiconductor firms (density) created the conditions for the establishment of still more firms. And, not surprisingly, the greatest density was in Silicon Valley. In effect, since the creation of Fairchild, the region developed a set of extrafirm institutions and routines that fueled growth and continuing reproduction. Of critical importance here is the intervening variable of an environment making large numbers of new entrants possible. This environment is composed of the institutions of Economy Two.

The dynamism of new firm creation and the wealth in Silicon Valley has drawn great interest from academics and policymakers. A number of explanations of the dynamics and operation of the Silicon Valley regional economy have been advanced. Saxenian (1994) explained Silicon Valley's success by comparing it to the relative stagnation of Route 128 in Boston during the 1980s. The heart of her argument is the proposition that Silicon Valley's firms remained flexible and interactive, whereas those established along Route 128 became hierarchical and rigid.[7] She held that Route 128 firms were vertically integrated whereas Silicon Valley firms either remained, or wisely, decided to become specialists. There are difficulties with this comparison and explanation, as it ignores the fact that Route 128 was built on minicomputer systems, whereas Silicon Valley was built on the semiconductor component—a far more general or basic electronic part (Robertson, 1995). Silicon Valley cannot be reduced to existing firms or their interactions, rather any explanation of the dynamism must also explain the regional routines and institutions that nurture new firm formation.

There are also a number of cultural explanations of Silicon Valley (Weiss & Delbecq, 1990). Here, Economy Two is equated with a culture of entrepreneurship. Yet this provides little explanation of the economic and tech-

---

[7]I do not examine this argument here. For an analysis of this position, see Kenney (1998).

nological foundations for the culture and the concrete conditions that sustain it. In this particular formulation, *culture* refers to economic activity, that is, purposeful activity directed toward financial gain. Few of these cultural explanations connect individual's actions to their pursuit of financial gains. In effect, economic acts are explained by cultural attributes and potential economic explanations are downplayed (Weiss & Delbecq, 1990). As an example, the propensity to establish new firms is attributed to a culture of startups. Curiously, these explanations make no reference to the potential for the entrepreneur to secure large capital gains and the fact that the entire infrastructure of Economy Two has evolved to facilitate and share in those capital gains. The infrastructure does not encourage startups that have no potential for realizing large capital gains; in fact, venture capitalists even have a word for firms that do not go bankrupt, but cannot be sold—these are known as "zombies." This is a pejorative term for companies that are too small and have insufficient growth potential, but already have the venture capitalists' investments.

A less prominent stream of study has designated Economy Two as the critical feature of Silicon Valley. For example, Schoonhoven and Eisenhardt (1989) argued that Silicon Valley is an "incubator region," in which there are numerous institutions whose *raison d'être* is to nurture the establishment and growth of small startup firms aiming to exploit a market opportunity. They empirically tested the incubator region concept by examining the creation and survival of new semiconductor firms in the United States from 1978 to 1986. Their findings show that the Silicon Valley firms had greater survival rates. In other words, the social institutions existing in the region provided environmental resources that incubated these new firms. Florida and Kenney (1988a, 1988b) advanced the concept of a "social structure of innovation," by which we meant an interactive set of institutions dedicated to encouraging technological innovation.[8] More recently, Bahrami and Evans (1995) conceptualized Silicon Valley as an "ecosystem" consisting of various institutions, skill sets embodied in individuals, and an entrepreneurial spirit. These three perspectives identify the ingredients that have made Silicon Valley so successful in creating new highly successful firms; however, they do not explicitly weave the trajectories of the technologies being exploited into their explanations.

The differences between firm organization in different regional industrial agglomerations can also be explained on another dimension, namely the technological dynamics the particular industry faces. For Robertson and Langlois (1995) the innovatory situation a region's industry faces in terms of

---

[8]Lynn, Reddy, & Aram (1996) advanced yet another somewhat similar concept of an "innovation community," however their concept is more general and seems to fit established industries better than it fits environments such as Silicon Valley or Route 128. Also, it is not quite as explicitly spatial.

the product cycle conditions the organization of the region's networks, interfirm interaction patterns, and firm structures. For example, the high-fashion garment district firms of Northern Italy face constant change in fashion designs, but these changes occur only along very limited dimensions, that is, in the designs, colors, fabrics, and shapes of the particular item; but the product, such as, jackets, pants, etcetera, does not change. In this situation, the change in production equipment and worker skills is gradual.

Silicon Valley faces a far more complicated set of changes including technologies, products, processes, and entire industrial categories. Not only are product generations rapid, but new product categories can emerge, even as other entire categories disappear (Kenney, 1998). This means turbulence is ongoing and interfirm and intrafirm relations are under continual stress. No firm or set of firms can be certain that its particular technology or product will survive. Requisite skills change rapidly, as established products evolve or are discontinued to be made in other regions (e.g., floppy disk drives and DRAMs), entirely new products are introduced constantly, customers and suppliers change, and new customers or suppliers emerge (for discussions of hard disk drives see Christensen, 1992; for RISC microprocessors, see Garud & Kumaraswamy, 1995; for LAN systems, see von Burg, 1998). New firms with superior technology can emerge rapidly and displace older firms committed to obsolete technologies.

The Langlois and Robertson (1995) thesis—that the industrial organization of regions and firms is correlated to the region's position on the product cycle—is an important contribution to understanding the linkage between particular types of networks and industrial regions. Their model explicitly recognizes the importance of the types of innovation or market changes facing firms, thereby incorporating technical change as a critical variable—a factor surprisingly underplayed in many explanations. For example, in the semiconductor industry, even while particular industrial segments such as DRAMs and microprocessors developed predictable trajectories and/or strong incumbent firms, new segments emerged, igniting a new cycle by lowering entry barriers and allowing new startups. Therefore, the semiconductor industry as a whole did not mature, rather the locus for new firm entry constantly shifted. In computer local area networking, there was a similar process of new firm formation at each discontinuity in the expansion of the network (von Burg, 1998). Economy Two is based on these technical and market discontinuities.

The economic dynamism of Silicon Valley is partly from the fast-growing established firms, such as Intel, Sun Microsystems, or Hewlett Packard, which have graduated to Economy One, but much more important are institutions of Economy Two that encourage new firm formation. Rapid firm formation is not unique to Silicon Valley; there have been periods and regions before that experienced rapid firm formation. For example, Rao and Singh (chap. 9, this volume) discuss the early phases of the automobile industry. For autos, a dominant design emerged and new firm entry became quite difficult. The electronics (and also biotechnology) industries experienced repeated new opportunities, so even as certain segments developed a dominant design and entrenched firms, new segments opened up creating new spaces for new firm formation.

However, displacing explanation to a set of entities such as regional institutions does not really help us understand the development of Silicon Valley. Most observers treat economic institutions as natural phenomena that exist *sui generis*, when, in fact, they are created. Economic institutions and routines such as the venture capital investment process are the outcomes of complicated evolutionary paths, reinforced by success or diminished by failure. Obviously, the greater the success of such routines the more they were reinforced; in this way, they could eventually become an attribute of the "culture."

This section argued that there are two economies in Silicon Valley and both exhibit path dependent characteristics. The actual evolutionary process is quite complicated because both Economy One and Economy Two are moving along trajectories made possible by Moore's and Metcalfe's Laws, which postulate a world where value and capabilities are increasing so dramatically that new commercial opportunities are constantly being uncovered.[9] The next section illustrates the path dependent nature of the semiconductor industry in Silicon Valley.

SEMICONDUCTORS

In 1947, the first operating semiconductor transistor was developed at Bell Laboratories in New Jersey.[10] At that time, few foresaw the vast technological possibilities that the semiconductor's evolution would make possible. Semiconductors would permit the digitalization of many analog functions and quickly displayed an improvement curve that allowed a doubling of capacity approximately every 18 months driving the cost per transistor down dramatically. This meant that problems relating to insufficient or too expensive calculating capacity were constantly being solved, permitting a constant flow of new applications (i.e., watches, calculators, mobile phones, communications computers, ever smaller computers, and various

---

[9]Moore's Law states that the number of circuits that can be placed on a given area of silicon doubles roughly every 18 months (Moore, 1965). Metcalfe's Law states that for any number of n machines linked by a network you get n squared potential value (Gilder, 1993).

[10]For excellent discussions of the development of the semiconductor, see Braun and Macdonald (1982), Riordan and Hoddeson (1997).

other artifacts that contained embedded computing power). Analog functions and signals could be replaced by the increasingly sophisticated integrated circuitry, so records were replaced with compact disks, watch gears with a chip, typewriter gears and levers with word processing programs, or human hands and brains with computer controlled machine tools. The semiconductor eventually would allow physical phenomenon to be digitized that no one could have foreseen.

The pace of change confirmed by Moore's Law meant incessant change and the continuous emergence of new business opportunities to produce either inputs to integrated circuit production, integrated circuits optimized for various functions, or artifacts using integrated circuits. In this environment, obsolescence of artifacts, technologies, and capabilities was incessant, thereby opening space for new entrants (for a discussion of this, see Kenney, 1998; Kenney & Curry, 1998). In other words, market-dislocating technological advances occurred frequently. Yet, even more important, were innovations defining new markets and repeatedly semiconductors were critical enabling components for industries ranging from computer networking equipment to personal computers and mobile phones. For semiconductor companies, so many new business opportunities emerged that management had to decide which ones to pursue, recognizing that if a project was blocked internally, the engineers developing the technology might decide to use the knowledge to build a new firm (Intel, 1984).

Semiconductors were at the heart of a massive technological revolution (Gilder, 1989). The exponentially increasing processing power of integrated circuits permitted the creation of new products and the transformation of old products and industries. The per-unit price of information processing power embodied in integrated circuitry dropped at a 40% annual rate for more than 20 years. Braun and Macdonald (1982) provided the example of a Fairchild transistor sold in 1959 for $19.75 that in 1962 was sold for $1.80. Most significant, invariably the transistor was profitable at both prices as learning curve and mass production economies lowered costs and the design costs were amortized during the initial part of the learning curve. This incessant fall in prices meant that by 1997 the price per transistor on an integrated circuit chip dropped to less than $.000001. These extraordinary price-learning curves created fantastic opportunities for increased profitability and constantly opened new business opportunities that could be exploited by startups (Gilder, 1989).

THE GENESIS OF SILICON VALLEY

Silicon Valley is a postwar phenomenon, however there were precursor firms in the prewar period. Sturgeon (2000) maintained that the Bay Area's pre-

World War II successes in developing vacuum tubes and a number of other devices formed the base on which the postwar growth was built. For example, in 1906, Lee de Forest invented the triode in the Bay Area and later in the 1920s, Philo Farnsworth relocated from Utah to the Bay Area with the intent of developing a working television. Farnsworth received financial support from W.W. Crocker, president of the Crocker National Bank (Fisher & Fisher, 1996). In the 1930s, Hewlett Packard was founded by two engineers at the behest of Frederick Terman, the dean of engineering and later provost at Stanford University (Leslie, 1993; Lowen, 1992). However, these disparate activities did not coalesce into a coherent pattern or set of practices.

World War II was a watershed for the U.S. electronics industry and the Bay Area benefitted greatly from massive electronics-related armaments spending. In Silicon Valley, a number of startups were established to take advantage of this spending. Also, defense contractors built a number of factories and research facilities in the area. During World War II, other smaller electronics firms were also established in the area (Sturgeon, 2000).

During World War II, Frederick Terman had gone to Boston to manage Radio Research Laboratories at Harvard University. After the war he returned more committed than ever to establishing an electronics industry in the Stanford vicinity. For the next 20 years, he encouraged major East Coast electronics firms to establish research and development (R&D) facilities close to Stanford University (Leslie, 1993). Also, he urged Stanford students, such as the Varian Brothers, to form electronics companies.

Terman's single most important intervention was convincing William Shockley, one of the coinventors of the transistor at Bell Laboratories, to return to his hometown, Palo Alto, and establish the startup firm, Shockley Semiconductor. There was an element of good fortune at play in luring Shockley back to Palo Alto. Shockley left Bell Laboratories and wanted to launch a firm to commercialize semiconductor technology. He approached a number of institutions on the East Coast, in particular, negotiating with Raytheon, an important transistor manufacturer, about funding his proposed startup. He demanded $1 million and after a month of bargaining, Raytheon refused (Scott, 1974). He also negotiated with the Rockefeller venture capital division, but no agreement could be reached. After these failures, he began discussions with Arnold Beckman, the founder of Beckman Instruments in Los Angeles. They reached an agreement and Beckman funded Shockley to start a firm in Palo Alto (Riordan & Hoddeson, 1997).

Shockley's decision to locate in Palo Alto in itself was not significant, as Shockley Transistor never became an important firm. But, as fate would have it, a small significant event occurred. Shockley proved to be an ineffective manager, and eight of his engineers left to form Fairchild Semiconductor. These eight catalyzed the pattern of new firm formation that put the "silicon" in Silicon Valley.

When the first semiconductor startups were established in Santa Clara, there were already existing institutions in the area that could be drawn on for resources. And yet, as the path dependent model would have it, in 1955 there was no certainty that Silicon Valley would become the largest center of U.S. high-technology electronics production. Boston with MIT and Harvard developed a concentration of firms, especially in using germanium for transistors. Other regions that might have become the leader include: New Jersey with Bell Laboratories, RCA Sarnoff Laboratories, Princeton, and Rutgers University; Los Angeles with Caltech, UCLA, USC, and numerous defense contractors (see, for example, Norberg, 1976); or Dallas, Texas (the headquarters for Texas Instruments); or Chicago with its many large and small electronics companies including Zenith and Motorola. A number of smaller Chicago firms attempted to enter the industry and Motorola was very successful, although it located all of its semiconductor operations in Arizona.

The critical semiconductor startups established in the 1950s were not spinouts from either university laboratories or defense-oriented corporate laboratories. Little semiconductor technology was drawn from universities. Robert Noyce pointed out "that it was after the original success of Fairchild that the two schools [Stanford and UC Berkeley] became important supporters of the technology" (Braun & Macdonald, 1982; Moore, personal communication, May 30, 1997). Corporate success fed the improvement of the engineering departments at the Bay Area universities, as they were able to attract better students, who found employment in the growing high-technology industry igniting a virtuous circle.

It is not necessary to deny the role of research universities in Silicon Valley industrial growth to place their contribution into proper perspective. Given the large number of MIT graduates becoming important entrepreneurs, it might be possible to argue, at least, in the earliest days, MIT contributed as much to Silicon Valley's growth as did Stanford. Nonetheless, the greatest source of entrepreneurs was other companies, and corporate laboratories dedicated to civilian technology development, such as IBM's San Jose Laboratory or Xerox PARC. These were the critical sources of entrepreneurs and, as important, the source of inventions that were transformed into commercial products by startups.[11]

---

[11]In the 1950s Admiral, GE, ITT, Philco-Ford, Sylvania, and Westinghouse located laboratories in Silicon Valley (Saxenian, 1980). However, many of these would be aimed at military research, especially in the microwave field (Leslie, 1993). There were a few startups but they never created an important industry. The reason is hard to ascertain, but perhaps it is because these technologies never found a civilian market. The exception, of course, was the microwave oven developed in Silicon Valley by Charles Litton. Litton Industries was later sold to Teledyne.

For the most part, Silicon Valley firms produce either systems products, such as computers, routers, switches, and hubs, or components such as semiconductors and hard disk drives. Their customers are generally other companies, distributors, or large institutions such as universities and governments. There are exceptions, for example, some segments of the personal computer and games software industry sell (through distributors and retailers) to consumers as final users.[12] The most important of these is Apple Computer, the sole survivor of the large number of Silicon Valley personal computer startups (Freiberger & Swaine, 1984). Quite often, forays by Silicon Valley firms into consumer sales failed, such as the debacles in digital watches and calculators. The exception has been Hewlett Packard (HP), which established a strong position in scientific and engineering calculators.[13]

Emphasis on institutional markets allows the startups to enter niche markets that do not require large initial investments in manufacturing and marketing. Of course, becoming a major company means expanding the niche until it is the mainstream of a market or developing follow-on products allowing the firm to expand its offerings. For example, in the 1960s and 1970s, the entry barriers to manufacturing semiconductors were fairly low. In the 1980s and 1990s, manufacturing had become a fundamental barrier to a startup. However, even as the barrier to entering manufacturing increased, a number of firms, many of them in Asia began to offer contract manufacturing services. This, once again, lowered entry barriers and provided the complementary resources enabling the creation of a generation of "fabless" semiconductor firms headquartered and doing their design in Silicon Valley, while contracting production to other companies.[14] Essentially, the Silicon Valley's semiconductor industry was able to evolve with major changes in the structure of its value chain.

Semiconductor Firms Structure the Valley

The semiconductor industry forms the core of its namesake, Silicon Valley. However, as Braun and Macdonald (1982) indicated, this was not a foregone conclusion in 1950. In the 1950s, both large diversified electronics firms and various startups around the country began transistor production. East Coast and midwestern firms were leaders and the most important customer was the Department of Defense. By the early 1960s, it was possible to clearly define

---

[12]It is necesary to be careful here. Some major personal computer retailers, Business Land and Computer Land, were established in Silicon Valley.

[13]HP is a unique firm that should not be confused with other less diversified Silicon Valley firms. Often, HP is the exception to the point we make about Silicon Valley firms.

[14]Fabless semiconductor firms do not remain fabless. When their annual revenue surpasses $500 million, they often build or purchase a fabrication facility or, at least, dedicated capacity.

three important small companies: Transitron in Boston, Texas Instruments in Dallas, and Fairchild in Palo Alto. Most independent Boston firms, especially the leader, Transitron were incorporated in the early 1950s to manufacture transistors. The problem was that most were unable to make the transition to the use of silicon and the consequent integrated circuitry. By the early 1970s, most of the Boston firms had left the business (Tilton, 1971).

The new industry divided into three branches: the captive producers, including IBM and AT&T, two small but rapidly growing merchant producers, Texas Instruments and Fairchild, and Motorola, an established firm and a merchant producer. The merchants quickly became leaders as evidenced by the fact that Fairchild was responsible for more major product and design innovations than any other single firm, including Western Electric/Bell Laboratories (calculated from Dosi, 1984).[15] Fairchild would play an even more important role as the source of semiconductor spin-offs (Braun & Macdonald 1982). This is graphically illustrated in a Semiconductor Equipment and Materials Institute (1986) genealogy chart which includes 124 startups formed through 1986, almost all of which can be traced back through generations of startups to Fairchild. As early as 1971, observers noticed that with every new technological discontinuity, there were spurts in the number of semiconductor firms (Lindgren, 1971).

In the semiconductor industry there have been recurring waves of new entrants that correlate with the development of new technologies or market spaces. As indicated, the semiconductor industry in Silicon Valley has had at least four important technological waves. The first were memory related integrated circuits, then microprocessors, then RISC microprocessors, and, most recently, application specific integrated circuits. Even, if these were perhaps the most important waves, so many other lucrative niches emerged, such as BIOS chips, programmable logic devices, and specialized chips designed for graphics, communications, and audio, to name just a few.

The startup phenomenon is interesting, but what catalyzed this process and how it assisted in creating the institutions of Silicon Valley is even more important. We have already mentioned the rapid rate of technical change and Fairchild's leadership role, but there are some more significant social patterns that were established at and because of Fairchild. Perhaps, the most significant pattern was a reaction to the situation at Fairchild. When the eight engineers resigned from Shockley Semiconductor and received funding from Fairchild, what would prove to be the most significant contract provision was one that permitted Fairchild to purchase the entire firm in 3 years for $3 million. The $3 million in proceeds was to be divided among the eight

---

[15]The second greatest number of innovations were made by Intel Corporation, the Fairchild spin-off. In contrast, AT&T had by far the greatest number of major process innovations.

founders who owned 80% of the firm and the brokerage firm (Hayden, Stone) that arranged the financing and owned the remaining 20%.

This is very significant. The founders signed away the upside potential of their shares; in other words, their gains were capped. So, when Fairchild experienced enormous success, the founders were unable to participate in the capital gains, even though they provided all the management and all of the technical skills. They were bought out by Fairchild at a price that had no relationship to the value they created. The decision not to allow the engineers to share in the capital gains was a rational decision from the perspective of the East Coast Fairchild management. Little did they know it would have critical implications for the future.

As soon as the founders' stakes were acquired, the engineers were relatively wealthy and had no reason beyond salary to continue at Fairchild. This would be problematic, because in the 1960s, there were low entry barriers in the semiconductor industry and an exploding market making the environment conducive to starting new firms. Moreover, there was an as yet inchoate network of investors willing to back engineers with projects with commercial potential. Often, the early investors were, like Fairchild, seeking a position in the new field of microelectronics. As an incentive for potential entrepreneurs, these financiers were willing to permit the entrepreneurs to retain substantial equity in the fledgling firm. Fairchild's short sighted policy regarding equity participation ensured that there would be a flow of spin-offs and the extremely rapid technical change in semiconductors ensured that opportunities would emerge.

Because cash compensation often is relatively low, the reward for entrepreneurship is realized when the company goes public or is acquired. Substantial equity means that entrepreneurs can graduate from being salaried to being an owner. Another reason for liquefying their equity through a public offering or outright sale is the extremely rapid technical changes that can dramatically lower the value of a firm experiencing difficulties (e.g., the recent situation at Apple Computer).

The semiconductor firms had a central role in shaping the institutions, structures, expectations, and culture of Silicon Valley. Advances in microelectronics formed one of the critical technical foundations for further developments, because they were critical components in so many later products such as computer networking equipment and desktop computers, which form the basis for the current high-technology gold rush related to the Internet. Fairchild also established the pattern of spinning off and, as significant, was the source of one of the primary institutions of Silicon Valley's Economy Two, the venture capitalists. It was the financial success of the early spin-offs that encouraged other entrepreneurs and attracted more capital and venture capitalists. For Silicon Valley, the semiconductor industry proved to be a critical catalyst to the development of the institutions of Economy Two.

Venture Capital

Although venture capital was first developed as an proto institution by the Rockefeller and Whitney funds in the 1930s in New York, and fully formalized in the Boston area by the firm, American Research and Development (Florida & Kenney, 1999), Fairchild and its spin-offs catalyzed the establishment of venture capitalists in Silicon Valley. *Venture capitalists* are financiers that specialize in providing funds to high-risk startups in return for an equity stake, in the hopes that corporate success will dramatically increase the value of their stake. Because they actually own part of the company, the venture capitalists receive seats on the firm's board of directors, enabling them to take an active role in monitoring firm performance.

Before World War II, there were wealthy northern Californians who invested informally in promising young companies. However, these investors were not professional venture capitalists dedicated solely to investing in startups. (They much more resembled informal investors, who currently are called "angels.") As mentioned earlier, the investment by Sherman Fairchild and Fairchild Industries in Fairchild Semiconductor was brokered by Arthur Rock, then an employee of the East Coast brokerage firm, Hayden Stone. After Fairchild's success and other experiences securing financing for small technology based firms, Rock moved to California. In 1961, he established a partnership with a manager at the Kern County Land Corporation, Thomas Davis. Davis and Rock were the general partners and their investors became limited partners (Rogers & Larsen, 1984; Wilson, 1985). The partnership was designed to last 7 years. They raised $3.5 million in funds from individuals, such as Fairchild executives, and proceeded to invest in various high-technology firms. Davis and Rock were very successful and when the partnership was completed, their profit from the partnership was $20 million. Both Davis and Rock and the limited partners were pleased.

---

[16]Other methods for funding high technology startups had been tried. For example, federally funded small business investment corporations (SBICs) were another organizational form for funding. However, SBICs suffered many problems, including the fact that they made loans rather than took an equity stake. This meant that they had high risk but no method for participating in the capital gains. Many important venture capital pioneers did come from the SBICs. American Research and Development in Boston was a public company and suffered from the problems stemming from being listed on the market. Finally, there were the family funds, the best known of these started by the Rockefellers and the Whitneys. The family funds were very successful but could not hope to meet the increasing need for funds or undertake all the opportunities presented. In the end, all of these other vehicles would pale in comparison to the partnerships.

The partnership organization Davis and Rock developed became the dominant form for organizing venture capital funds.[16] The limited partners (i.e., investors) pay a management fee of approximately 2% to 3% for salaries and the operational expenses of the fund. The incentive for the venture capitalists is the fact that they receive at least 20% of the profits after they return the initial investment. As with stock options and equity for the entrepreneurs, the venture capitalists receive a share of the capital gains.[17] The success of the early investors attracted more capital and more venture capitalists.

In the 1960s, venture capital in Silicon Valley constituted loosely coupled groups of individual investors. Often, as investments matured and needed more capital, it was necessary to partner with East Coast firms. In the 1970s, the Silicon Valley venture capital community grew and matured as an institution. Also, the maturation of venture capital made the process of securing capital far more transparent to entrepreneurs and this encouraged more entrepreneurship. The success of the firms provided high rates of return, therefore attracting more capital and venture capitalists to the region.

Initially, the individual venture capitalists came from a variety of sources. For example, Fairchild alumni, such as Eugene Kleiner of Kleiner and Perkins, Pierre Lamond and Donald Valentine of Sequoia Capital, became very successful venture capitalists. By the mid 1980s, venture capital was becoming more organized and routinized, and often, newly minted MBAs were recruited as associates or junior partners in venture capital partnerships. Another important source of venture capitalists were high-level mangers at existing firms. If these recruits were successful, they could become full partners, or begin their own partnership.

As central actors in Economy Two, firm formation is critical to the survival of venture capitalists and, of course, the venture capitalists encourage this process. As the venture capitalists became more sophisticated, they were able to take even partially formed ideas for firms and recruit the missing management functions. Thus, someone with a plan to establish a firm could approach a venture capitalist and, if an investment was made, the venture capitalist would actively assist in the recruitment of other members of the management team. This constant need for personnel created the demand for what has now become another feature of the Silicon Valley environment, organizations specialized in recruiting for high-technology startups.

When a particular technology is considered hot either by the stock market or other firms seeking to acquire capabilities in the field, frenzied startup activity can be ignited. Sahlman and Stevenson (1985) described this in the early 1980s in the Winchester disk drive industry, and Kenney (1986) described a similar frenzy in the biotechnology market the late 1970s and early

[17]This is best illustrated in a genealogy of Silicon Valley venture capitalists prepared in the mid 1980s by the Silicon Valley venture capitalist, Franklin "Pitch" Johnson's wife, Cathie Johnson (1984).

1980s. More recently, in the mid- to late 1990s, there was a frenzy in the Internet related businesses. During such periods, venture capitalists become extremely aggressive in raiding existing firms for senior engineers and managers, both to establish new firms and complete management teams in already funded firms. This raiding of established firms earned venture capitalists the moniker of "vulture capitalists" from executives at existing firms (Florida & Kenney, 1990). In other words, by the 1980s, Economy Two had developed to such a powerful extent that it now was actively creating its resources (i.e., entrepreneurs) for spin-offs. The culture of spin-offs did not simply emerge, it was nurtured into existence. The actors in Economy Two needed the "startup culture" and worked to make it happen.

The evolution of Silicon Valley venture capital was path dependent, but rather than being a second-best solution, such as QWERTY or VHS, it is difficult to imagine a more efficient system for forming high-technology startups.[18] Over the last 40 years, the institution of venture capital has evolved from individuals or two-person partnerships with less than $10 million under management to partnerships consisting of five to seven general partners and numerous associates with between $300 million to $1 billion under management. These much larger funds have become increasingly well equipped to undertake more difficult financings requiring greater quantities of capital and ventures with longer term payback periods, as is the case of biotechnology (Kenney, 1986).

This outline of the Silicon Valley venture capital complex mentioned the role of executive search firms. Numerous other highly specialized professional services critical for assisting in the rapid development of fledgling firms have been created in Silicon Valley. For example, Suchman (1994) provided an excellent discussion of the role of Silicon Valley law firms that have evolved specializations in high-technology firm incorporation and intellectual property issues of various types, including separation from former employers—obviously critical issues for venture capital investors. Other significant institutions are the marketing and public relations consulting firms, the most famous of which is the McKenna Group established by the Intel alumnus, Regis McKenna. Venture capital is, perhaps, the most critical institution for Economy Two, but it certainly is not the only one.

Thoughts and Discussion

Path dependence as a concept provides an important opening for noneconomists to contribute to explaining economic institutions using

---

[18]In an earlier book (Florida & Kenney, 1990), I questioned the benefits of the venture capital-funded spin-off system for the entire U.S. economy, arguing that this was creating an economy based on breakthroughs without sufficient follow through.

history and the other social sciences. Once technology, firms, institutions, and regions are understood in more evolutionary terms, a purely economistic analysis is no longer possible. The embeddedness of economic institutions in a social context becomes apparent (although it is also equally apparent that economic institutions cannot be explained simply by the social context; profits, capital gains, and technological trajectories also exist and are significant). Sociological concepts, such as embeddedness are both reinforced and undermined by path dependent arguments and our results. As understood by most sociologists, the "bed" is often treated as static, when, in fact, the evidence from Silicon Valley suggests it is also in the process of evolution.

Product cycle-related explanations for the organizational structures in Silicon Valley are fruitful for deepening our understanding of the effects constant technical change can have on the structure and organization of a regional economy (Robertson & Langlois, 1995). However, they do not provide strong explanations for the actual evolution of the Silicon Valley economy. Explanations of Silicon Valley, as a region that facilitates innovation and new firm formation, capture Economy Two, but also are somewhat ahistorical, as they implicitly understand the evolutionary component and would be much improved if they explicitly recognized it (Bahrami & Evans, 1995; Florida & Kenney, 1988a, 1988b, 1990; Schoonhoven & Eisenhardt, 1989). Saxenian (1994) paid particular attention to the interfirm relationships in Economy One, arguing that their success is predicated on an industrial district system based on close supplier relations, but this says little about Economy Two. Each explanation provides important insight into the nature of the Silicon Valley economy, however, considering the emergence of Silicon Valley from a path dependent perspective provides a much richer picture.

We proposed a heuristic distinction between Economy One and Economy Two for the purpose of developing a clearer understanding of the new firm formation component of Silicon Valley's dynamism. We singled out the evolution of the merchant semiconductor firms as components of the industry that provided resources to fuel the establishment of Economy Two and, reflexively, was a product of Economy Two. The evolution of the semiconductor industry had a clear path dependent character. A series of small events, such as the set of decisions by various participants including Shockley led to the creation of Shockley Semiconductor in Silicon Valley, Shockley's and Fairchild's management styles, and early funding patterns. These were the seed events necessary to create the trajectory, enabling the formation of the institutions that comprise the current Silicon Valley. With the spin-offs and the continuing rapid advance of the technology, the semiconductor industry also clustered and locked in to Silicon Valley. Further, advances in semiconductor technology were made and exploited in Silicon Valley. In describing this process, path dependence serves us well.

Path dependence also has its limits. It does not provide such a strong explanation of the creation and growth of complementary institutions. Our discussion of venture capital provided insight into the creation of new institutions that also developed paths or evolutionary trajectories. The institutional creation process reinforced positive feedback loops or virtuous circles. The result was Economy Two, which has now become a rich ecosystem continually fed by the nutrient of rapid technical change and large capital gains. The raw materials for Economy Two are a constant influx of technologists, managers, and capital drawn to the region (and internally generated by the region) by the opportunities fueling the process. Path dependence, if construed as a relatively one-dimensional model, does not do full justice to the manner by which Silicon Valley was created an ecosystem.

Still, path dependence provides an important new way of thinking about economic institutions. Here, contrary to the arguments by Ruttan (chap. 4, this volume), we believe Arthur and David must be given great credit; they have broken taboos and liberated thought from the unnatural fetters imposed by neoclassical economics. The criticism that the world of economic institutions is more complex than highly abstract path dependence models is not surprising and we concur. And yet, as this chapter has contended, some of the observations derived from path dependence, and its extension to path creation, provide significant insight into the development of contemporary Silicon Valley.

ACKNOWLEDGMENTS

The authors thank David Angel, James Curry, Giovanni Dosi, Richard Florida, Richard Langlois, Frank Mayadas, Richard Nelson, and Joel West for comments.

REFERENCES

Almeida, P., & Kogut, B. (1997). The exploration of technological diversity and the geographic localization of innovation. *Small Business Economics, 9*(1), 21–31.

Arthur, W. B. (1988). Competing technologies: An overview. In G. Dosi, C. Freeman, R. Nelson, G. Silverberg, & L. Soete (Eds.), *Technological change and economic theory* (pp. 590–607). London: Frances Pinter.

Arthur, W. B. (1994). *Increasing returns and path dependence in the economy.* Ann Arbor, MI: University of Michigan Press.

Bahrami, H., & Evans, S. (1995). Flexible re-cycling and high-technology entrepreneurship. *California Management Review, 37*(3), 62–89.

Braun, E., & Macdonald, S. (1982). *Revolution in miniature: The history and impact of semiconductor devices.* Cambridge, England: Cambridge University Press.

Christensen, C. (1992). *The innovator's challenge: Understanding the influence of market environment on processes of technology development of the rigid disk.* D.B.A. dissertation, Harvard University.

David, P. (1986). Understanding the economics of QWERTY: The necessity of history. In W. Parker (Ed.), *Economic history and the modern economist* (pp. 138–158). Cambridge, MA: Basil Blackwell.

David, P. (1999, September 31). At last, a remedy for chronic *QWERTY* scepticism. Unpublished manuscript, All Souls College, Oxford Univeristy.

David, P. (1997, November). Path dependence and the quest for historical economics: One more chorus of the ballad of QWERTY. Unpublished manuscript, All Souls College, Oxford University.

Dosi, G. (1984). *Technical change and economic transformation.* London: Macmillan.

Fisher, D., & Fisher, M. (1996). *Tube: The invention of television.* Washington, DC: Counterpoint.

Florida, R., & Kenney, M. (1988a). Venture capital-financed innovation and technological change in the USA. *Research Policy, 17,* 119–137.

Florida, R., & Kenney, M. (1988b). Venture capital and high technology entrepreneurship. *Journal of Business Venturing, 3*(4), 301–319.

Florida, R., & Kenney, M. (1990). *The breakthrough illusion: Corporate America's failure to move from innovation to mass production.* New York: Basic Books.

Florida, R., & Kenney, M. (1999). *Financiers of innovation: Venture capital, technological change and industrial development.* Princeton, NJ: Princeton University Press.

Foray, D. (1991). The secrets of industry are in the air: Industrial cooperation and organizational dynamics of the innovative firm. *Research Policy, 20,* 393–405.

Freiberger, P., & Swaine, M. (1984). *Fire in the valley.* Berkeley, CA: Osborne/McGraw Hill.

Gan, J. (1991). Staking a claim in the west. *Almaden Views,* (Winter), 1–4.

Garud, R., & Kumaraswamy, A. (1995). Coupling the technical and institutional faces of Janus in network industries. In W. R. Scott & S. Christensen (Eds.), *Advances in the institutional analysis of organization: International and longitudinal studies* (pp. 226–242). Thousand Oaks, CA: Sage.

Gilder, G. (1989). *Microcosm.* New York: Basic Books.

Gilder, G. (1993, September 13). Metcalfe's Law and Legacy. *Forbes ASAP.*

Granovetter, M. (1985). Economic action and social structure: The problem of embeddedness. *American Journal of Sociology, 91,* 481–510.

Henderson, R., & Clark, K. (1990). Architectural innovation: The reconfiguration of existing production technologies and the failure of established firms. *Administrative Science Quarterly, 35*(1), 9–30.

Intel Corporation. (1984). *A revolution in progress.* Santa Clara, CA: Author.

Johnson, C. (1984). *West coast venture capital—25 years.* Palo Alto, CA: Asset Management.

Karnøe, P., & Garud, R. (1998). *Path creation and dependence in the Danish wind turbine field.* Copenhagen, Denmark: Institute of Organization and Industrial Sociology, Copenhagen Business School.

Kenney, M. (1986). *Biotechnology: The university-industrial complex* New Haven, CT: Yale University Press.

Kenney, M. (2000). The temporal dynamics of knowledge creation in the information society. In I. Nonaka & T. Nishiguchi (Eds.), *Knowledge Creation.* New York: Oxford University Press.

Kenney, M., & Curry, J. (1998). Knowledge creation and temporality in the information economy. In R. Garud & J. Porac (Eds.), *Cognition, knowledge, and organizations* (pp. 149–170). Connecticut: JAI Press.

Kogut, B., Walker, G., & and Kim, D.-J. (1991). The role of large firms in the entry of start-ups: Centrality and cooperation in the semiconductor industry (Working paper No. 91–102). Reginald H. Jones Center for Management Policy Strategy and Organization, The Wharton School, Philadelphia, PA.

Langlois, R., & Robertson, P. (1995). *Firms, markets and economic change.* London: Routledge.

Liebowitz, S., & Margolis, S. (1990). The fable of the keys. *Journal of Law and Economics, 33,* 1–24.

Liebowitz, S., & Margolis, S. (1995). Path dependence, lock-in, and history. *Journal of Law, Economics and Organization, 11*(1), 205–226.

Leslie, S. (1993). *The cold war and American science.* New York: Columbia University Press.

Lindgren, N. (1969). The splintering of the solid state electronics industry. *Innovation*, 1(8), 2–16.

Lowen, R. (1992). Exploiting a wonderful opportunity: The patronage of scientific research at Stanford University, 1937–1965. *Minerva*, 30(3), 391–421.

Lynn, L., Reddy, N., & Aram, J. (1996). Linking technology and institutions: The innovation community framework. *Research Policy*, 25, 91–106.

Moore, G. (1965). Cramming more components onto integrated circuits. *Electronics*, 38(8), 114–117.

Nelson, R., & Winter, S. (1982). *An evolutionary theory of economic change*. Cambridge, MA: Harvard University Press.

Norberg, A. (1976). The origins of the electronics industry on the Pacific Coast. *Proceedings of the IEEE*, 64(9), 1314–1322.

Riordan, M., & Hoddeson, L. (1997). *Crystal fire*. New York: Norton.

Robertson, P. (1995). Book review of regional advantage: Culture and competition in Silicon Valley and Route 128. *Journal of Economic History*, 55(1), 198–199.

Robertson, P., & Langlois, R. (1995). Innovation, networks and vertical integration. *Research Policy*, 24(4), 543–562.

Rogers, E., & Larsen, J. (1984). *Silicon Valley fever*. New York: Basic Books.

Sabel, C. (1998, February 10). Intelligible differences: On deliberate strategy and the exploration of possibility in economic life. [On-line]. *http://www.columbia.edu/~cfs11/IntelDif.html*

Sahlman, W., & Stevenson, H. (1985). Capital market myopia. *Journal of Business Venturing*, 1, 7–30.

Saxenian, A. L. (1980). Silicon chips and spatial structure. Masters thesis, University of California, Berkeley.

Saxenian, A. L. (1994). *Regional advantage*. Cambridge, MA: Harvard University Press.

Schoonhoven, C. B., & Eisenhardt, K. (1989, August). The impact of incubator regions on the creation and survival of new semiconductor ventures in the U.S., 1978–1986. Report to the Economic Development Administration, U.S. Department of Commerce.

Scott, O. (1974). *The creative ordeal: The story of Raytheon*. New York: Atheneum.

Semiconductor Equipment and Materials Institute. (1986). *Silicon Valley genealogy*. Mountain View, CA: Semiconductor Equipment and Materials Institute.

Storper, M., & Walker, R. (1988). *The capitalist imperative: Territory, technology, and industrial growth*. London: Basil Blackwell.

Sturgeon, T. J. (2000). How Silicon Valley came to be. In M. Kenney (Ed.), *Understanding Silicon Valley: Anatomy of an innovative region*, pp. 15–47. Stanford: Stanford University Press.

Suchman, D. (1994). *On advice of counsel: Law firms and venture capital funds as information intermediaries in the structuration of Silicon Valley*. Unpublished PhD dissertation, Stanford University, California.

Tilton, J. (1971). *International diffusion of technology: The case of semiconductors*. Washington, DC: The Brookings Institution.

Voltaire, F. (D. Gordon, Trans.). (1999). *Candide*. Boston: Bedford/St. Martin's.

von Burg, U. (1998). *Plumbers of the internet: The creation of the local area networking industry*. Unpublished PhD dissertation, St. Gallen University, Switzerland.

Weiss, J., & Delbecq, A. (1990). A regional culture perspective of high technology management. In M. Lawless & L. Gomez-Meija (Eds.), *Strategic management in high technology firms* (pp. 83–94). Greenwich, CT: JAI Press.

Wilson, J. (1985). *The new venturers*. New York: Basic Books.

6

Standards, Modularity, and Innovation: The Case of Medical Practice

Richard N. Langlois
The University of Connecticut

Deborah A. Savage
Southern Connecticut State University
Yale University

The economics of standards and standard setting has grown to considerable prominence in the last few years.[1] Buttressed by influential neoclassical models of network externalities (Farrell & Saloner, 1985; Katz & Shapiro, 1985), this intellectual edifice has as its keystone David's (1985) famous history of the QWERTY keyboard. In David's account, the now-familiar arrangement of keys is a paradigmatic instance of path dependency. The choice of the QWERTY design was essentially a matter of historical accident; and, once that arrangement became dominant, the spiraling benefits of its network of complementary capabilities—notably touch-typing skills—effectively "locked" users into the QWERTY standard.

This chapter takes up the issue of standard setting both in theory and in terms of a historical case study, namely, the setting of standards for the American medical profession in the early 20th century. As a contribution to the literature on standards and path dependency, however, this chapter diverges somewhat from the beaten path. First of all, our case involves behav-

---

[1]For an overview, see David and Greenstein (1990).

ioral standards, not technological standards. Second, our case is one in which the process of path creation did not obviously involve historical accident. Instead, history seems to have been shunted onto its track by underlying economic forces. Indeed, even at the most plastic early stages of the standard-setting process in medicine, seemingly promising proposals for reform and for alternative institutional design were unsuccessful when they did not fit with the underlying logic of professional production. In this respect, our case has more in common with Chandler's (1977) account of the evolution of the modern corporation than it does with QWERTY.

The structure of standards and standard making that emerged from the early century, we argue, was one well adapted both to the provision of medical care in the United States and to the generation of innovation, at least until the modern era of high-cost and high-technology medicine. Even in this modern era, however, reforms are likely to fail that do not take into account the nature of professional production and of behavioral standard setting.

STANDARDS AS INSTITUTIONS

Most analyses of path dependency and lock-in have focused on technological systems in the narrow sense. Favorite examples have included computers, telecommunications systems, and—perhaps especially—various kinds of home entertainment systems like stereos (Langlois & Robertson, 1992), VCRs (Cusumano, Mylonadis, & Rosenbloom, 1992), or high-definition television (Farrell & Shapiro, 1992). In most of these cases, the issue is one of the compatibility of physical components or electronic signals. Occasionally, human behavior is part of what is standardized or coordinated; in the QWERTY case, it was the inertia of human touch-typing skills that generated increasing returns to the dominant standard. But few in the Davidian tradition have focused directly on standards as coordinating human behavior rather than as connecting technology.[2] And this is perhaps surprising. For standards are at base a kind of social institution; and social institutions are recurrent patterns of behavior that help to coordinate human activity (Langlois, 1986b; North, 1990).

As Kindleberger (1983) pointed out, there are basically two classes of standards: those that create economies of scale and those that lower transaction costs. In the former case, economies arise from the increase in the extent of the market that results from reduced variety. For example, in the 1910s, the Society of Automotive Engineers set standards for automobile

---

[2]One exception is David (1987) himself, who distinguished between standards of technical design and standards of behavioral performance, even if his own focus seemed to be on the former.

parts that winnowed the kinds of steel tubing in use from 1,600 to 210 and the types of lock washers from 800 to 16 (Epstein, 1928). Independent parts suppliers could then take advantage of longer production runs to reduce costs, which especially helped the smaller car companies that did not have high internal demands for parts.

In the second class of standards, benefits arise because the standards help to reduce the transaction costs of coordination and monitoring. Standards assist in coordination by helping to align expectations. In the classic case, for example, the convention that we all drive on the same side of the road is a standard that reduces the "transaction" costs of ascertaining the intentions of each oncoming driver, not to mention the resource costs of failed coordination. As David (1987) pointed out, behavioral standards of this kind can be thought of as ensuring "interface compatibility" (p. 214) much as do standards of technical design, because such standards help to coordinate the way individuals "connect together (p. 214)." Standards can also reduce measurement and monitoring costs. A single standard of weights and measures, for example, makes the comparison of goods in exchange easier and increases the cost of cheating. More generally, normative standards can reduce costs of monitoring by providing a benchmark against which quality or performance can be judged.[3] This meaning of *standards* vill figure prominently in our discussion of medical practice.

The transaction-cost properties of standards are not entirely unrelated to their economies-of-scale function. By regularizing expectations, standards increase the predictability of the extent of the market, which is crucial for large-scale investments in machinery and a more elaborate division of labor. Indeed, the essential tension between flexibility and commitment is perhaps the most intriguing aspect of standard setting. To use the language of Garud and Jain (1996), standards can be at once *enabling* and *constraining*. They can be enabling because they create an orderly framework within which economies of scale can develop and technological change can progress effectively. But they can also be constraining in that their necessary rigidity makes costly and thereby inhibits even potentially beneficial change that would require altering the standard. Much of the allure of David's keyboard story comes from the contention that QWERTY is not the best of all possible configurations and that lock in has

---

[3]In a sense, standards are always normative in that they take the form: "Do it this way." This is true whether the standard is an injunction to drive on the right of the road or a technical specification constraining design choices. The difference between a coordination standard and a normative standard is that the former is always self-enforcing, whereas the latter sometimes requires a more complex enforcement mechanism. For example, the standards of cleanliness and efficiency that McDonald's sets for its franchise holders require monitoring of individual locations by company inspectors.

prevented change to a better keyboard.[4] This same logic is true more generally of social institutions. The convention that we all drive on the same side of the road is a standard that brings order out of disorder[5] and increases the efficiency of driving; but to change such a convention can be difficult, as countries like Sweden and Okinawa discovered when they switched sides of the road. We return to the problem of change (innovation) in a system of standards.

STANDARDS AND ROUTINES

The idea that predictable, inflexible, standardized behavior generates economies is quite a general one. Influenced by the work of Herbert Simon on rule-following behavior, Nelson and Winter (1982) proposed that much of economic behavior can be discussed in terms of *routines*. Routines are habitual patterns of behavior that embody useful knowledge. Much of this knowledge is in the form that Polanyi (1958) described as tacit; it is skill-like knowledge that cannot be articulated or transmitted explicitly but that must be acquired over time through a process of apprenticeship and trial-and-error learning. And, as Polanyi also suggests, the development of such tacit knowledge economizes on the scarce resource of *conscious attention*, thus generating economies.

Although routines and standards are clearly related, they are not identical. As Kindleberger (1983) pointed out, standards are *public* goods; they reflect interpersonally shared knowledge. We might even say that a standard is a certain kind of "public" routine that helps to coordinate private (individual or intraorganizational) routines.

But routines are not only about coordination. As we saw, routines embody potentially useful—we might even say productive—knowledge. In the terminology of Ryle (1949), they reflect "knowledge how." In some cases, such useful knowledge can be knowledge about how to transact, the possession of which thus reduces transaction costs. My internalized knowledge that I always ought to keep to the right (not the left) as another car approaches might be an example, at least if we construe the interaction between oncoming drivers metaphorically as a transaction. But the skillful exercise of a particular technique for suturing an incision would also be a routine, and not one obviously involving the reduction of transaction costs. Useful knowledge applied to problems of transacting is a special case

---

[4]Liebowitz and Margolis (1990, 1994) however, challenged both David's specific contention about QWERTY and his implied contention that problems of lock-in to suboptimal paths are important and ubiquitous phenomena.
[5]There is actually a technical sense in which conventions bring order out of disorder; they reduce the entropy of the behavioral environment (Langlois, 1986a; Schotter, 1981).

of a more general phenomenon. As Winter (1988) has suggested, one needs to have economic capabilities (an effective repertoire of routines) in order to be able to transact as well as to be able to produce.[6]

Like standards, *routines* can be both enabling and constraining. The possession of an effective repertoire of routines would be crucial to the successful production of product A; but possessing that repertoire might also inhibit a transition to the production of product B. Routines are generally as hard to unlearn as to learn, which may give the advantage in situations of radical innovation to those who have never learned the routines in the first place.[7] This is no doubt what Schumpeter (1934) had in mind when he wrote that "new combinations are, as a rule, embodied, as it were, in new firms which generally do not arise out of the old ones but start producing beside them; ... in general it is not the owner of stage-coaches who builds railways" (p. 66).

But perhaps the interesting aspects of enablement and constraint are those involving the interaction of standards (public routines) with the private routines of individuals and organizations. This is what Garud and Jain (1996) seemed to mean when they talked about the degree to which the technological environment is "embedded" in the institutional environment. When there are no standards, there is complete flexibility, but very little enablement, as "customers and vendors might be prone to wait for the emergence of a dominant design before they are indu ced to make significant investments" (Garud & Jain, 1996, p. 393). But when standards are too tight, they can suffocate progress, leading to a "stuck" technology with little innovation of any kind. Only when the institutional environment (the standards) "just embeds" the technological matrix do those standards most fully enable, and not constrain, technological development. In such a "just embedded" world, technology and standards coevolve, "each of these reciprocally and continually shaping the other" (Garud & Jain, 1996, p. 393).

Now we describe a system in which technology is just embedded in the institutional structure, although we broaden the idea of technological environment to include (primarily in this case) the routines of human behavior. At the level of these routines, what we tell is a story in which new paths are constantly being created and in which lock-in to particular sets of standards (that is, to standardized behavioral routines) is a "wolf" who never quite

---

[6]Indeed, there is a developing literature on economic capabilities (Langlois & Robertson, 1995; Teece & Pisano, 1994) that starts from the difficulties of acquiring production knowledge (embodied in routines) rather than from the kinds of informational problems that create transaction costs. Langlois and Robertson (1995) pointed out that an approach to economic organization that begins with the idea of routines is more consonant with the broader *new institutional economics*—which is concerned with institutions as recurrent patterns of behavior—than is an approach that takes the transaction as the unit of analysis (Williamson, 1985).

[7]Langlois and Robertson (1995, chap. 6) provide a more careful analysis of the phenomenon of economic inertia.

gets to the door. At the higher institutional level, of course, the just enabled system we describe is itself a "path," and we attempt to shed some light on how that path was staked out and on the ways in which it is now appearing to diverge in the woods. Before turning to the case, however, we need to explore the underlying economic considerations that, we argue, helped nudge the institutional structure in the direction it went.

PROFESSIONAL PRODUCTION

Mintzberg (1979) defined as a "bureaucracy" any organization in which behavior is standardized. The hierarchical organization familiar from Weber (1946, 1947) is what he calls the *machine bureaucracy*. But the professions are also bureaucracies in Mintzberg's sense, in that professions coordinate the economic activities of their members, using standardized routines inculcated through what are normally lengthy processes of training and apprenticeship. And, although we might quibble with the term *bureaucracy* in this case, it is certainly true that standards—publicly shared routines—are crucial to professional production.

In a profession, routines are largely shared in the sense that the abilities and choices of an individual practitioner are shaped by the abilities of those with similar or complementary skills. Although each practitioner produces independently, all practitioners execute their routines in an environment created by other professionals. For example, a lawyer is constrained by the cumulative precedents of previous cases, most of which were decided long before the current generation entered the profession. At the same time, the creative application of existing law generates new opportunities for future practitioners. This is true, too, for physicians, whose day-to-day decisions are affected by the previous treatments administered to patients by other physicians. In all these cases, the shared routines imply public "interfaces" among practitioners, what engineers call "next-bench design": One can rely on the fact that other engineers, or lawyers, or surgeons have made decisions in ways that one can reconstruct by virtue of one's own training and experience (Feynman & Leighton, 1988).

But what makes a profession different from a machine bureaucracy is the extent to which, along some dimensions, conduct is *not* standardized. As Stinchcombe (1990, chap. 2) so nicely puts it, professionals are information-processing systems who must wield and apply a wide repertoire of routines to fit widely varying concrete circumstances. The reasons for this reflect the landscapes of both supply and demand. If the price is low enough, a single kind of car—any color Model T you want so long as it's black—will attract buyers, because, in the end, that single kind of car can adequately accommodate a wide range of concrete circumstances. Moreover, there are large economies of scale to centralized fabrication of cars. By contrast, the product—the service—that professionals offer typically requires finer tuning to a wide variety of concrete circumstances. This is so for reasons that have to do both with preferences and with the technology of provision. (One is willing to tolerate more standardization in a suit of clothes than in a law suit.)

As a result, professionals do not standardize the application of their routines (as does a machine bureaucracy) but only the "toolkit" of routines from which they draw. The particular concrete application of the routines requires on-the-spot professional *judgment*, a capability Knight thought essential in any situation of true uncertainty (see Langlois & Cosgel, 1993). Like more specific professional routines, judgment is a tacit skill that is cultivated in professional training. Moreover, as we have hinted and as we will discuss in more detail, professionals must also employ judgment in the creation of new routines. As we will see, it is the standardization on toolkits rather than on concrete behavior that provides the necessary contextual flexibility for innovation, making the system of professional practice "just imbedded" in the standards that guide it.

Thus, although professions are not exempt from the Smithian division of labor, and although the professional bureaucracy attempts to create economies of scale through the promulgation of fairly standard toolkits of routines, both the division of labor and standardization take place in a decentralized context. As Savage (1993, 1994) argued, these considerations lead naturally to a theory of the professions as a "production organization."

The conventional neoclassical treatment of the professions focuses almost entirely on the demand side—on the interaction between the professional and the consumer. And it frames this interaction exclusively in terms of the problems of principals and agents (Shaked & Sutton, 1981). In fact, however, one cannot understand the organization of the professions without also considering the supply side (e.g., how professions are organized to employ knowledge usefully and to coordinate among specialized producers). As Jensen and Meckling (1992) point out, economic organization must solve two different kinds of problems: "the rights assignment problem (determining who should exercise a decision right), and the control or agency problem (how to ensure that self-interested decision agents exercise their rights in a way that contributes to the organizational objective)" (p. 251).

Efficiency demands that the appropriate knowledge find its way into the hands of those making decisions. There are basically two ways to ensure such a "collocation" of knowledge and decision making: "One is by moving the knowledge to those with the decision rights; the other is by moving the decision rights to those with the knowledge" (Jensen & Meckling, 1992, p. 253). Markets (in the widest sense of the term) take the latter approach.

The Coase theorem suggests that, so long as decision rights are well defined and alienable, those rights will tend to end up in the possession of those whose specialized knowledge can make the most of them. This also solves the agency problem, because the alienability of the right means that market prices can track the value of the right, which, in turn, creates an incentive for the owner to maximize value by using the right appropriately. But there are also potential costs to such extreme decentralization. These might include the familiar sorts of transaction costs arising from moral hazard and asset specificity. More interestingly, however, they may arise from "dynamic" transaction costs (Langlois, 1992), the costs of bringing otherwise decentralized knowledge together and coordinating it, especially in circumstances involving learning and the generation of new productive knowledge.

The form of organization called *the firm* is one way to surmount such dynamic transaction costs in some circumstances (Langlois & Robertson, 1995). By concentrating decision rights at the top, the firm can in principle overcome both the narrowness of knowledge of the individual participants and the vestedness of decentralized decision rights. But such centralization of authority comes at the cost of misaligned incentives to the extent that it removes decision rights from the hands of those who must actually execute the routines of production. In a mature firm of the sort Chandler (1977) described, such problems of agency are tolerable because operations are typically characterized by repeated, consistent replication of known routines. Such routines tend to be measurable at various stages of production, and so lend themselves relatively well to formal monitoring schemes, including documentation, accounting trails, and supervision of employees (Barzel, 1982). Clearly, organizations of this sort are not obviously well adapted to the problems of professional production.

But firms and markets are not the only alternatives. There is a growing literature on *hybrid* forms of organization, forms that offer distinctive solutions to the problems of rights assignment and agency. Principal among these forms are *networks* (Powell, 1990); the professional network is a particularly important example.

Through formal and informal arrangements, professionals share rent-earning competences without ceding autonomy to a central hierarchy. When professionals locate together in a network, they do not take a joint equity position or even sign a contract. Although remaining legally independent, they make a long-term commitment of their substantial human capital to a "hubless" network organization. Because networks do not integrate ownership, they have a horizontal rather than a hierarchical coordinating structure. In fact, network members remain competitors across many dimensions.

Unlike the ideal of a price-mediated market, however, a professional network is able, with the help of standards, to provide the function of knowledge integration and coordination. Without the exchange of cash payments, members exchange information and technology, and collaborate in production (that is, share routines) without authoritarian supervision, and without integrating external management functions into their day-to-day operations.

Networks are able to transform tacit knowledge into capabilities much more valuable than any individual practitioner could have acquired alone. This is so because the network provides incentives to share skills. Practitioners recognize that they are dependent on the distinctive competences of other practitioners, and that it will be in their individual best interests to share these competences. Indeed, professionals often borrow the routines of others for both exchange and production. For example, von Hippel (1989) showed that engineers routinely share technical information with rivals. Such sharing is most likely to occur when a professional is attempting to solve a new or difficult problem. To put the matter differently, the knowledge, routines, and capabilities that give economic value to professional production lie in the interface between individual practitioners and the system. In the terminology of sociologists, professionals have complex relational roles (Barley, 1990).

MODULARITY AND INNOVATION

Sanchez and Mahoney (1996) argued for a kind of duality between the structure of products and the structure of organizations. Modular organizations are conducive to modular products, that is, to products with standardized interfaces; at the same time, modularity in product designs is conducive to modular organizational design. Professional networks are very much the kind of loosely coupled system they describe as modular. In this case, however, the "product" is in the nature of behavioral routines, and these are modular in the sense that standards have created a widely shared toolkit from which professionals can draw their repertoires. At the same time, the organization (the network) is modular in the sense that, at least in principle, there are sharply defined boundaries between the subspecialties that make up the profession. Mintzberg (1979) talks of the process of "pigeonholing" (p. 352), in which clients are sorted into categories according to which subset of standardized tools best fit their needs. Professional subspecialties are one kind of pigeonhole. And all professionals know more or less the kind of skills each subspecialty represents and how and when to interact with it.

In practice, of course, the boundaries between professions are not always sharp. Boundary disputes occur frequently, and boundaries can overlap for long periods of time (Halpern, 1992; Savage, 1993, 1994). But this only means that, as in technological systems, professions reflect continual interaction among standards, boundaries, and exogenous conditions. The result is the kind of ongoing process of modularization and remodularization that Garud and Kumaraswamy (1995) liken (in the technological context) to re-building a ship plank by plank even as it sails.

The degree and character of the modularity of a system determines the pace and direction of technological change in that system. From the perspective of product design, a system in which the parts are standardized but the connections among the parts are not lends itself to *architectural innovation* (Henderson & Clark, 1990), that is, to innovative recombinations of standardized parts. By contrast, a system in which the interfaces among the parts are standardized lends itself to *modular innovation* (Langlois & Robertson, 1992), that is, to innovation in which the parts improve in performance without changing the way in which they are hooked together. A more significant distinction in this context would be that between *systemic innovation* and *autonomous innovation* (Langlois & Robertson, 1992; Teece, 1986). A systemic innovation is one that requires simultaneous change in several stages of production. In the case of professional production, this would mean change that spills across professional boundaries. By contrast, autonomous innovation is change that can take place safely within existing boundaries.

Within the boundaries of a professional subspecialty, practitioners engage in architectural innovation. As we saw, in day-to-day practice, the professional uses judgment to select and apply more or less standard routines in new combinations in response to unique circumstances. In the large, however, professions tend to lend themselves more naturally to autonomous innovation. Individual practitioners can improve their repertories of routines, and those improvements can diffuse quickly to others with similar training, so long as the boundaries between standard subspecialties change relatively slowly. As Mintzberg (1979) noted, it is "the pigeonholing process that enables the Professional Bureaucracy to decouple its various operating tasks and assign them to individual, relatively autonomous professionals. Each can, instead of giving a great deal of attention to coordinating his work with his peers, focus on perfecting his skills" (p. 353).

Another way to look at the issue of modularity and innovation is to think of the structure of an organization as defining a kind of cognitive or perceptual structure (Langlois, 1997). In a sense, this is the Sanchez and Mahoney argument in a different guise. The cognitive structure of an organization is not just a matter of information flows along an organization chart; rather, it is strongly influenced by the modular character of its products and of its structure. For example, when IBM switched from a regime of architectural innovation to one of modular innovation based around the 360/370 series of computers, the organization's perceptual system changed dramatically (Langlois, 1997). In the case of professional networks, as we see, the decentralized structure creates a perceptual system that is open to a range of new ideas and that, like a market, is capable of rapid trial and error learning[8] (Nelson & Winter, 1977). But it is also a system whose perceptual ability is circumscribed by the standards that constitute and modularize it.

In order to make these arguments more concrete, we turn to our case study: the history of standard setting and innovation in medicine (here defined broadly to include surgery as well as medicine proper). This is a story of the process by which, and the reasons for which, medical "toolkits" came to be standardized. And it illustrates one possible way in which paths can be created.

MEDICAL STANDARDS

As Savage (1993, 1994) argued, each profession possesses a core competence that defines what we might call its natural boundaries, that is, the boundaries one would observe in the absence of important institutional overlays, such as legal restrictions.[9] Surgeons and physicians, for example, were distinct professions since at least the Middle Ages. Once members of the guild of barbers, surgeons have always possessed a core competence in wound management (Wangensteen & Wangensteen, 1978), in contrast to the physician's core competence in diagnosis. By the late 19th century, a convergence of changes in technology and practice emerged to greatly amplify the competences of physicians and, especially, of surgeons.[10] The adoption of anesthetics by the late 1840s, coupled later in that century with the techniques of antiseptic and aseptic practice (which preceded a very slowly emerging consensus around the germ theory of disease) turned surgery from a painful and often deadly route of last resort into a powerful and widely applicable approach to therapy.[11] Improved diagnostic techniques, including

---

[8]Garud, Kumaraswamy, and Prabhu (1996) discussed another kind of network—what they call a generative network—in similar terms. Although professional networks are a somewhat different kind of learning system (largely because of their modular structure), they share with generative and other networks the ability to tap into a wider variety of information sources than can a more centralized organization.

[9]This is in contrast to many traditional accounts of professions, which see boundaries solely as the product of legal or other institutional restrictions.

[10]In the terminology of Tushman and Anderson (1986), these innovations were competence enhancing for physicians and surgeons. Such need not always be the case. As Savage (1993, 1994) argued, technological change has been competence destroying for pharmacists, whose traditional core competence had been the certification of the strength and purity of medicinal drugs. The increased complexity of pharmaceuticals and the higher capital intensity of testing equipment has shifted that competence upstream to the drug manufacturers.

[11]The full emergence of modern surgery had to await the development of antibiotics drugs (such as sulfa) in the 1930s. These "three As"—anesthesia, antisepsis, and antibiotics—are generally seen as the transforming forces of modern surgery. As we argued, however, they amplified rather than created the surgeon's core competence in wound management, greatly extending the boundaries of that specialty into realms of therapy that had been the province of the physician as well as into new realms of possibility.

laboratory tests and Roentgen's X-ray device, had a similar, if less dramatic, effect for physicians (Gelijns & Rosenberg, 1995).

As the 20th century began, the medical professions started to consolidate the dramatic changes of the 19th century. In a sense, these professions experienced the kind of maturation process one often finds in the life cycle of technological innovations (Abernathy & Utterback, 1978). One part of the consolidation process was the setting of standards (notably in the areas of medical education) and the keeping of medical records.

Traditionally, the education of physicians and surgeons was a matter of apprenticeship. By the 19th century in the United States, medical school training (often in conjunction with internship) had become the norm, with some 85% of physicians trained between 1811 and 1820 having taken medical degrees (Rothstein, 1987). But even late in the 19th century, many of these medical schools were small proprietary affairs that granted degrees easily and quickly. The conventional account of the transformation of medical education that occurred in the early 20th century lays great stress on the normative character of the standards that were adopted. Many point to the influential report in 1910 by Abraham Flexner, who was working for the Carnegie Foundation for the Advancement of Teaching in close collaboration with the Council on Medical Education of the American Medical Association. Flexner deplored the state of most American medical schools and argued for standards based around the example of Johns Hopkins, America's premier medical school at the time. The Flexner report, and the press coverage it generated, is often credited with a precipitous decline in proprietary medical schools and the rapid tightening of requirements by those (largely now affiliated with major universities) that remained (Kaufman, 1976).

On closer examination, however, it is clear that the standards involved were as much conventional as normative. For one thing, Flexner himself emphasized facilities, physical plant, and such measurable variables as enrollments, entrance requirements, and number of teachers, not the quality of teaching in the medical schools (Rothstein 1987). In essence, Flexner was implicitly arguing for the comparison of medical schools with the outward characteristics of a (presumed high-quality) exemplar, not for the direct assessment of quality.

Moreover, it is clear that the changes toward standardization—and away from proprietary medical schools, which had always had a high mortality rate—were essentially in place by the time Flexner wrote (Ludmerer, 1985; Rothstein, 1987). In part, American medical education in the 19th century had reflected the diverse and still largely rural character of the country. By century's end, however, may of the same forces that Chandler (1977) described in manufacturing were also overtaking the medical professions, namely, the increased scale and integration of the American economy in an era of population growth and reduced transportation and communications costs.[12] One historian of medical education puts it this way:

> Medical education could not remain immune from the processes that were so radically transforming the traditional way of living and working in America. The country had become much more closely integrated than at any previous time in history, and as a result of new technological breakthroughs in transportation and communication, Americans had become much more mobile as well. Medical education felt the effect of these changes. If a college student in Indiana wished to attend a medical school in California, or if a medical student at Oklahoma, a two-year school, desired to do his clinical training in Minnesota, or if a graduate of the University of Michigan wanted to practice in the state of Wisconsin, a certain amount of uniformity in medical education was mandatory. Absolute equivalency was not necessary, only a high degree of similarity. However, this was the direction that medical education in the early 1900s was already taking. The innumerable discussions occurring among representatives of medical schools from all parts of the country served to reconcile differences and guarantee that courses and requirements would mean the same thing everywhere. (Ludmerer, 1985, p. 89.)

As we saw, standardizing professional education is one important way to produce standardized professional "toolkits" and to help ensure "next-bench design" among cooperating but decentralized professionals. We can read the revolution in medical education of the early century as generating on a national scale a process that is typical of professional production.

One might expect that, as medical schools became more closely integrated with hospitals, many of the same forces that helped standardize medical education would also tend to standardize hospitals. To the extent that this happened, however, it did so in an indirect way. If one imagines hospitals to be hierarchical firm-like organizations, then it should clearly be the managers of hospitals at the forefront of standardizing hospital structure and practice. And, in 1913, the American Hospital Association (AHA), a group dominated by physicians who had become hospital superintendents, decided in principle on standards for hospitals. But the organization lacked the political clout and the resources to implement its recommendations (Stevens, 1989). In part, this impotence reflected the odd position of the superintendent, who was caught between donors and medical staff (especially admitting physicians and surgeons) without the authority typical of managers in corporations. As Stevens (1989) noted, the superintendents tended as a result to organize themselves "around the hospital's hotel-management and financial concerns rather than in relation to its medical function—the entire purpose for which, in theory, the hospital existed" (p. 73).

From the point of view of the theory of professions just outlined, this result may seem less surprising. As Savage and Robertson (1998) argued, a

---

[12]Indeed, it is a recognized shortcoming of Chandler's (1977) account that he ignores the role in economic growth of noncorporate sectors like the professions (Landes, 1991; Supple, 1991).

hospital is not an organization like a firm. Rather, it is an example of what they call the *coancillary institution* of a profession. For reasons that we have already suggested, decision rights in medical practice reside largely with the decentralized professionals. Those professionals also possess significant decision rights over institutions like hospitals that are complementary to professional practice but that the professionals do not own. In this light, it is not surprising that hospital superintendents should have been relegated principally those (complementary) tasks that fall outside the competences of the medical personnel.

Nor, thus, is it surprising that the effective move to standardization in hospitals was taken by the American College of Surgeons (ACS), not by the AHA. In 1913, the ACS set up a committee on the standardization of hospitals, installing as chair a Boston surgeon named Ernest Amory Codman (Howell, 1995; Stevens, 1989). Codman proposed what he called an "end-result system," that is, a system of monitoring hospital care (especially surgeries) directly by outcomes. Part of the system would require that hospitals set up detailed systems of record keeping. The policy the ACS eventually settled on in 1917 jettisoned the end-result system but retained, and, indeed, focused on, the standardization of patient records. Coupled with the original mission of the ACS—to recognize and thereby certify the competence of surgeons—the resulting system of standards fit well with the professional nature of surgical production.

Stevens (1989) saw these developments as a matter of surgeons jealously guarding their prerogatives:

> Standardizing the surgeon emphasized professional authority. Standardizing the work suggested the surgeon was a mere craftworker, or even a mechanic working in an organization. Scientific management in industry ranged from time and motion studies to sweeping organizational analyses—activities that surgeons did not want. Codman's methods would put trustees and administrators as policemen over medical work.... Surgeons, as a group, were willing to upgrade standards but not to lose their professional prerogatives to individuals they did not trust. (p. 77)

But, as Savage and Robertson (1998) argued, the issue was not one of trusting hospital administrators. The issue was one of the optimal collocation of knowledge and decision rights; and an arrangement in which professionals retain authority and autonomy is arguably a system of monitoring with characteristics superior to one in which the hospital is recast as a hierarchical organization in the model of a firm.

Moreover, it is in the network's interest to discourage the kind of monitoring Codman was proposing; this is so because end-result monitoring (or *outcomes monitoring*, as it is now called) changes the focus from the network to the individual practitioner. The potential punitive uses of external monitoring would cause individual practitioners to be less likely to share information about their practices and would therefore diminish the value of shared competences, with destructive repercussions for the network and, withal, for patient choice. It is characteristic of professional networks to provide practitioners with up-to-date information about the latest developments in their profession, as occurs at professional meetings. But external monitoring presupposes some proprietary use of information, again lessening the incentive to share.

In a sense, the model of self-governance that developed out of the ACS standardization program may be thought of as a way of building monitoring into the production process rather than imposing it from outside. With standardized and open records, the ACS hoped to disseminate information about and to evaluate the procedures and methods of their fellow surgeons. They instituted a policy of regular local meetings at which clinical experiences were reviewed and interpreted. The process enabled surgeons to educate themselves about the safety and efficacy of competing procedures, and to decide which to include in surgical training and continued education. It was also the College of Surgeons—the specialty most dependent on emerging hospital capabilities—that recommended that hospitals adopt an open, but defined, medical staff model.[13] The ACS recognized that the future lay not in keeping fellow surgeons out, but in monitoring the quality and ensuring the cooperation of those who shared hospital resources. Over time, other specialties adopted similar formats, and in 1952 the various accrediting organizations merged to form the Joint Commission of Medical Accreditation, which is now called the Joint Commission on Accreditation of Healthcare Organizations.

MEDICAL INNOVATION

We interpret the processes of standardization that took place in the early 20th century not as a fundamental break with the existing system of medical production in the United States but rather as an accommodation to that existing system of technological and demographic changes that the 19th century had left behind. By standardizing the professional (medical education) and by standardizing the interfaces among professionals (medical records), the medical profession, like a Chandlerian firm, had seized the scale economies of a national market. And, by linking these standards to the open-staff

---

[13]In an open-staff model, the physicians and surgeons affiliated with a hospital evaluate the credentials of other professionals who wish hospital privileges. But the incumbent staff consider only the technical credentials of the applicants, not the commercial feasibility or desirability of admitting them to staff. (Thus the staff is "open" to anyone who is deemed competent.) "Economic credentialling" is becoming more common today, however, as hospitals in the era of managed care move closer to a closed-staff model and to a more hierarchical organizational form (Savage & Robertson, 1998).

model of hospital governance, the profession had brought the new technologies under professional control (like operating theaters and X-ray machines) whose scale of operation was larger than the individual practitioner. In this way, the medical profession arguably retained the porous network structure it had long possessed.

It is our contention that these standard-setting events of the early 20th century provided the institutional framework that governed medical production at least until the modern era of managed care. These events created the path down which American medicine traveled. Perhaps things might have happened differently. If the AHA or ACS had been able to impose end-result standards of the sort Codman had proposed, perhaps hospitals would have become something more like Chandlerian firms, which would have affected the rate and direction of technological change (among many other things). On the other hand, our reading of the history does not suggest that the path taken resulted in an obviously inferior or undesirable system; quite the opposite. It is not inexplicable that the ACS, not the AHA, set the standards, and that Codman's ideas fell so easily by the wayside. As we suggested, there is an important logic to the location of medical decision rights in the hands of the decentralized practitioners rather than in any central location. It would have required some large countervailing consideration—as may be present today in the form of more or less exogenous changes in the system of health-care financing—to overcome the organizational economics of professional production.

Our argument is that, by permitting a continuation of the earlier professional form of organization, the standards of the early 20th century turned out to be enabling in the sense of Garud and Jain (1996). As we saw, innovation in a professional network is likely to be both architectural (involving the individual practitioner's toolkit of routines) and autonomous (focused, at least initially, within existing professional or subprofessional boundaries). In addition, innovation in such a network is likely to benefit from a wide variety of information sources. Because professionals possess only a narrow range of capabilities (in at least some contrast, perhaps, to the research laboratories of the large corporation), they must of necessity rely on knowledge developed elsewhere, often in distant fields of knowledge, and must adapt those outside discoveries to their own local needs. Moreover, because professionals possess localized expertise that is hard to transmit to others, those professional are more likely to see opportunities for innovation than are the holders of the complementary outside knowledge. As a result, we would expect practitioners to be important sources of changes in practice and in technology. This was certainly the case in the 19th century. And, despite the increasing complexity of technology, it remains the case to a large extent even today.

Most of the important surgical advances in the 19th century were the result of professions. These included anesthesia, (suggested in one of its forms by a dentist from Hartford, Connecticut), the various tentative approaches to antiseptic and aseptic technique, and even the development of rubber gloves, which required the collaboration of a rubber company (Wangensteen & Wangensteen, 1978). Many diagnostic tools, including the electrocardiogram and the electroencephalogram, were also developed by physicians.

What is more striking is that medical practitioners remain important in technological change into the modern era. The principal change is that today it is work at academic medical centers rather than by isolated practitioners that seems to be important. Gelijns and Rosenberg (1999) noted that, although academic medical centers conduct almost no basic research in medical devices, they are significant players in the development of new devices—a pattern in stark contrast to that in other academic fields. In their study of innovation in endoscopy, Gelijns and Rosenberg (1995) discovered that innovation was a network process in which the network of medical researchers allied itself with other networks, especially technologists in industrial firms who possessed complementary knowledge. In one case, however, it was the medical researchers who actually solved important technical and manufacturing problems that had eluded the industrial scientists with whom they collaborated. In some cases, more than collaboration was required. In a study of the development of medical lasers, Spetz (1995) found that advances often came from academic physicians who had significant training in other fields, notably physics.

CONCLUSIONS

We argued that standard setting is a phenomenon that goes well beyond the well-trod ground of technical standards. Behavioral and other knowledge standards are a fertile area of study. We also argued that the creation of paths in the process of standardization does not always involve the arbitrary force of historical accident. Sometimes the underlying economics of production provide the contours in which standardization must ultimately rest. In the case of medical standards, the path that was created in the early 20th century in the United States was strongly influenced by the economics of professional production, by technological change in medicine and surgery, and by the dynamic of American economic growth in the period. Once established, however, those standards proved "enabling" in that they created a decentralized network that was open to ideas from outside and was able to collaborate easily in the interdisciplinary fashion that proved crucial for the development of new devices and techniques.

ACKNOWLEDGMENT

Deborah Savage gratefully acknowledges the support of the Robert Wood Johnson Foundation. Opinions expressed in this chapter are those of the authors and not necessarily those of the Foundation.

REFERENCES

Abernathy, W., & Utterback, J. M. (1978, June, July). Patterns of industrial innovation. *Technology Review*, 41–47.

Barley, S. R. (1990). The alignment of technology and structure through roles and networks. *Administrative Science Quarterly*, 35, 61–103.

Barzel, Y. (1982). Measurement costs and the organization of markets. *Journal of Law and Economics*, 25, 27–48.

Chandler, A. D., Jr. (1977). *The visible hand: The managerial revolution in American business*. Cambridge, MA: The Belknap Press of Harvard University Press.

Coase, R. (1960). The problem of social lost. *Journal of Law and Economics*, 2, 3.

David, P. A. (1985). Clio and the economics of QWERTY. *American Economic Review*, 75(2), 332–337.

David, P. A. (1987). Some new standards for the economics of standardization in the information age. In P. Dasgupta and P. Stoneman (Eds.), *Economic policy and technological performance* (pp. 206–239). Cambridge, England: Cambridge University Press.

David, P. A., & Greenstein, S. (1990). The economics of compatibility standards: An introduction to recent research. *Economics of Innovation and New Technology*, 1(1–2), 3–41.

Epstein, R. C. (1928). *The automobile industry: Its economic and commercial development*. Chicago: A. W. Shaw.

Farrell, J., & Saloner, G. (1985). Standardization, compatibility, and innovation. *Rand Journal of Economics*, 16(1), 70–83.

Farrell, J., & Shapiro, C. (1992). Standard setting in high-definition television. *Brookings Papers on Economic Activity* (Microeconomics), 1–77.

Feynman, R. P. [as told to Leighton, R.]. (1988). *What do you care what other people think?* New York: Norton.

Garud, R., & Jain, S. (1996). The embeddedness of technological systems. *Advances in Strategic Management*, 13, 389–408.

Garud, R., & Kumaraswamy, A. (1995). Coupling the technical and institutional faces of Janus in network industries. In W. R. Scott & S. Christensen (Eds.), *The institutional construction of organizations: International and longitudinal studies*. Thousand Oaks, CA: Sage.

Garud, R., Kumaraswamy, A., & Prabhu, A. (1996). Networking for success in cyberspace. Unpublished manuscript, New York University.

Gelijns, A. C., & Rosenberg, N. (1995). From the scalpel to the scope: Endoscopic innovations in gastroenterology, gynecology, and surgery. In N. Rosenberg, A. C. Gelijns, & H. Dawkins (Eds.), *Medical Innovation at the Crossroads. Volume V: Sources of medical technology: universities and industries* (pp. 67–96). Washington, DC: National Academy Press.

Gelijns, A. C., & Rosenberg, N. (1999). Diagnostic devices: An analysis of comparative advantages. In D. C. Mowery & R. R. Nelson (Eds.), *The sources of industrial leadership* (pp. 312–358). New York: Cambridge University Press.

Grossman, S., & Hart, O. D. (1986). The costs and benefits of ownership: A theory of vertical and lateral integration. *Journal of Political Economy*, 94, 691–719.

Halpern, S. A. (1992). Dynamics of professional control: Internal coalitions and crossprofessional boundaries. *American Journal of Sociology*, 97(4), 994–1021.

Henderson, R. M., & Clark, K. B. (1990). Architectural innovation: The reconfiguration of existing product technologies and the failure of established firms. *Administrative Science Quarterly,* 35, 9–30.

Howell, J. D. (1995). *Technology in the hospital: Transforming patient care in the early twentieth century.* Baltimore. MD: Johns Hopkins University Press.

Jensen, M. & Meckling, W. (1992). Specific and general knowledge, and organizational structure. In L. Werin & H. Wijkander (Eds.), *Contract Economics,* pp. 251–274. Oxford, England: Basil Blackwell.

Katz, M., & Shapiro, C. (1985). Network externalities, competition, and compatibility. *American Economic Review,* 75(3), 424–440.

Kaufman, M. (1976). *American medical education: The formative years, 1765–1910.* Westport, CT: Greenwood Press.

Kindleberger, C. P. (1983). Standards as public, collective and private goods. *Kyklos,* 36(3), 377–396.

Landes, D. S. (1991). Introduction: On technology and growth. In P. Higonnet, D. S. Landes, & H. Rosovsky (Eds.), *Favorites of fortune: Technology, growth and economic development since the Industrial Revolution* (pp. 1–29). Cambridge, MA: Harvard University Press.

Langlois, R. N. (1986a). Coherence and flexibility: Social institutions in a world of radical uncertainty. In I. Kirzner (Ed.), *Subjectivism, intelligibility, and economic understanding: Essays in honor of the eightieth birthday of Ludwig Lachmann* (pp. 171–191). New York: New York University Press.

Langlois, R. N. (1986b). The new institutional economics: An introductory essay. In R. N. Langlois (Ed.), *Economics as a process: Essays in the new institutional economics* (pp. 1–25). New York: Cambridge University Press.

Langlois, R. N. (1992). Transaction-cost economics in real time. *Industrial and Corporate Change,* 1(1), 99–127.

Langlois, R. N. (1997). Cognition and capabilities: Opportunities seized and missed in the history of the computer industry. In R. Garud, P. Nayyar, & Z. Shapira (Eds.), *Technological learning, oversights and foresights* (pp. 71–94). New York: Cambridge University Press.

Langlois, R. N., & Cosgel, M. M. (1993). Frank Knight on risk, uncertainty, and the firm: A new interpretation. *Economic Inquiry,* 31, 456–465.

Langlois, R. N., & Robertson, P. L. (1992). Networks and innovation in a modular system: Lessons from the microcomputer and stereo component industries. *Research Policy,* 21(4), 297–313.

Langlois, R. N., & Robertson, P. L. (1995). *Firms, markets, and economic change: A dynamic theory of business institutions.* London: Routledge.

Liebowitz, S. J., & Margolis, S. E. (1990). The fable of the keys. *Journal of Law and Economics,* 33(1), 1–25.

Liebowitz, S. J., & Margolis, S. E. (1994). Network externalities: An uncommon tragedy. *Journal of Economic Perspectives,* 8(2), 133–150.

Ludmerer, K. M. (1985). *Learning to heal: The development of American medical education.* New York: Basic Books.

Mintzberg, H. (1979). *The structuring of organizations.* Englewood Cliffs: Prentice-Hall.

Nelson, R. R., & Winter, S. G.. (1977). In search of more useful theory of innovation. *Research Policy,* 5, 36–76.

Nelson, R. R., and Winter, S. G. (1982). *An evolutionary theory of economic change.* Cambridge, MA: The Belknap Press of Harvard University Press.

North, D. C. (1990). *Institutions, institutional change, and economic performance.* New York: Cambridge University Press.

Polanyi, M. (1958). *Personal knowledge.* Chicago: University of Chicago Press.

Powell, W. W. (1990). Neither market nor hierarchy: Network forms of organization. *Research in Organizational Behavior,* 12, 295–336.

Rothstein, W. G. 1987. *American medical schools and the practice of medicine : A history.* New York : Oxford University Press.

Ryle, G. (1949). Knowing how and knowing that. In *idem, The concept of mind* (pp. 25–61). London: Hutchison's Universal Library.

Sanchez, R., & Mahoney, J. T. (1996). Modularity, flexibility, and knowledge management in product and organizational design. *Strategic Management Journal, 17*, 63–76.

Savage, D. A. (1993). *Change and response: An economic theory of professions with an application to pharmacy*. Unpublished PhD dissertation, the University of Connecticut.

Savage, D. A. (1994). The professions in theory and history: The case of pharmacy. *Business and Economic History, 23*(2), 130–160.

Savage, D. A., & Robertson, P. L. (1998). The maintenance of professional authority: The case of physicians and hospitals in the United States. In P. L. Robertson (Ed.), *Authority and control in modern industry* (pp. 155–172). London: Routledge.

Schotter, A. (1981). *The economic theory of social institutions*. New York: Cambridge University Press.

Schumpeter, J. A. (1934). *The Theory of economic development*. Cambridge, MA: Harvard University Press.

Shaked, A., & Sutton, J. (1981). The self-regulating profession. *Review of Economic Studies, 48*, 217–234.

Smith, A. (1976). *An enquiry into the nature and cause of the wealth of nations*. Oxford: Clarendon Press, Glasgow edition. [First published in 1776.]

Spetz, J. (1995). Physicians and physicists: The interdisciplinary introduction of the laser to medicine. In N. Rosenberg, A. C. Gelijns, & H. Dawkins (Eds.), *Sources of medical technology: Universities and industries* (pp. 41–66). Washington: National Academy Press.

Stevens, R. (1989). *In Sickness and in wealth: American hospitals in the twentieth century*. New York: Basic Books.

Stinchcombe, A. L. (1990). *Information and organizations*. Berkeley: University of California Press.

Supple, B. (1991). Scale and scope: Alfred Chandler and the dynamics of industrial capitalism. *Economic History Review, 44*, 500–514.

Teece, D. J. (1986). Profiting from technological innovation: Implications for integration, collaboration, licensing, and public policy. *Research Policy, 15*, 285–305.

Teece, D. J., & Pisano, G. (1994). The dynamic capabilities of firms: An introduction. *Industrial and Corporate Change, 3*(3), 537–556.

Tushman, M. L., & Anderson, P. (1986). Technological discontinuities and organizational environments. *Administrative Science Quarterly, 31*, 439–465.

von Hippel, E. (1989). Cooperation between rivals: Informal know-how trading. In B. Carlsson (Ed.), *Industrial dynamics: Technological, organizational, and structural changes in industries and firms*. Dordrecht, the Netherlands: Kluwer Academic Publishers.

Wangensteen, O. H., & Wangensteen, S. D. (1978). *The rise of surgery: From empiric craft to scientific discipline*. Minneapolis: University of Minnesota Press.

Weber, M. (1946). *From Max Weber: Essays in sociology*. (H. H. Gerth & C. W. Mills, Eds. & Trans.). New York: Oxford University Press.

Weber, M. (1947). *The theory of social and economic organization* (A. M. Henderson & T. Parsons, Trans., T. Parsons, Ed.). New York: Oxford University Press.

Williamson, O. E. (1985). *The economic institutions of capitalism*. New York: The Free Press.

Winter, S. G. (1988). On coase, competence, and the corporation. *Journal of Law, Economics, and Organization, 4*(1), 163–180.

7

Complexity, Attractors, and Path Dependence and Creation in Technological Evolution

Joel A. C. Baum
University of Toronto

Brian S. Silverman
Harvard University

Many dynamic systems fail to reach equilibrium and, consequently, appear random to the casual observer. Processes that appear random may, however, actually be chaotic. Chaotic processes follow rules, but even simple rules can produce extreme complexity. In particular, the behavior of chaotic processes is very sensitive to small differences in initial conditions. Systems with very similar initial states can follow radically divergent paths over time. As a result, chaotic systems frequently exhibit highly path dependent behavior and historical accidents may "tip" outcomes strongly in a particular direction.

This can be devastating; consider an interorganizational system whose members engage in organization-level learning but do not arrive, through population-level learning, at a standard. Such a system will be outcompeted by a system that converges, permitting standardization and coordination (Miner & Haunschild, 1995). Conversely, system-level learning that is too rapid may be equally devastating. If a system arrives at a standard before its members have had the chance to discover effective routines, the standard itself will be suboptimal. Settling too early on a suboptimal standard can be disastrous for a population; lacking variation, its members may not be able respond effectively to a competing population that learns a better version of the organizational routine (Miner & Haunschild, 1995).

Recent advances in chaos theory suggest that adaptive systems tend to steer themselves to "the edge of chaos" by regulating their level of autonomy/mutual dependence, both among components and between a system as a whole and other systems in the environment with which it interacts (Kauffman, 1993; McKelvey, 1998; Thietart & Forgues, 1995). Such regulation of interdependence is hypothesized to benefit a system by admitting order and change, structure and surprise. Consider, for example, a competitive or collaborative interorganizational system. If the organizations comprising the system are too tightly coupled, there may be excessive interdependence and rigidity; if every act of one organization influences others throughout the system, then the repercussions of any given action have the potential to destabilize the entire system. Change is difficult to mount and may even lead to system collapse, exposing the system to competition from other, more adaptive interorganizational systems. Coupling that is too tight thus leaves no room for either desirable individual autonomy or interorganizational learning. If, in contrast, the organizations in the system are too loosely coupled, there is no coherence. Coordination is problematic, knowledge fails to either diffuse or accumulate, competitive and collaborative moves become random, and confusion sets in. As a result, the system faces the possibility of being outcompeted by another, more convergent system that gains advantages of standardization and coordination (Miner & Haunschild, 1995). In short, "too much structure creates gridlock … too little structure creates chaos" (Brown & Eisenhardt, 1998, p. 14). The edge of chaos lies between these extremes, where partially connected interorganizational systems never quite reach equilibrium but never quite disintegrate. It is a transitional realm in which interorganizational systems, characterized by a relatively few simple structures, enjoy a balance between interdependence and autonomy that generates complex, unpredictable, adaptive behavior.

How complex are trajectories of technological evolution produced by competitive and cooperative interorganizational systems? In this chapter, we explore empirically the degree to which patterns of change in laser technology are characterized by complexity. The evolution of laser technology has featured a wide variety of advances in a number of related disciplines and fields. Our study design involves systematically tracing out the event historical evolution (as measured by every patent granted by the U.S. Patent and Trademark Office) of seven technologies (defined according to U.S.

Patent Classification codes) related to the laser industry from January 1974 through December 1992. We analyze the historical sequence of patenting in each technology class, using techniques for detecting and describing nonlinear dynamics (successive time plots, Lyapunov exponent, correlation dimension, and surrogate testing). The result is a characterization of the multiple technology trajectories that have come to define the modern laser industry. The characterization includes an assessment of the extent to which orderly, chaotic, or random system dynamics have become apparent in this system. To the extent that this system is well characterized by ideas from chaos theory, we identify a myriad of innovative techniques for gaining new insight and understanding that may render complex organizational–technological systems (more) predictable.

BACKGROUND

The two branches of the "science of complexity," *chaos theory* and *complexity theory*, have rapidly come to dominate the study of complex systems characterized by unpredictability and uncertainty, and interrelationships and interactions between various entities (Prigogine & Stengers, 1984). Chaos theory is a scientific discipline based on "the qualitative study of unstable aperiodic behavior in deterministic nonlinear dynamical systems" (Kellert, 1993, p. 2). A system is dynamic because its equations are capable of describing changes in the values of system variables from one point in time to another. Complexity theory is concerned with adaptive (nondeterministic) systems composed of agents who change the rules of conduct as the system evolves. In other words, in a complex adaptive system, "agents interact in a manner that constitutes learning" (Stacey, 1995, p. 335). Table 7.1 summarizes the basic features of deterministic and adaptive systems. Notwithstanding their important differences, properties of deterministic systems turn out also to hold for adaptive systems.

There has been a great deal of interest in recent years in the application of complexity theory to a variety of real-world, time-dependent systems, driven by the promise of this new branch of mathematics to untangle and elicit order from seeming disorder. Many time series that have proven difficult to analyze with conventional linear models, most notably in the natural sciences and in business finance, are now proving susceptible to analysis using nonlinear and nonparametric methods (Tong, 1990). The unraveling of these adaptive systems has been aided by the discovery of mathematical expressions that exhibit aperiodic behavior. The most famous, and oldest, example of these is the logistic equation, which was originally conceived as a model of population growth (Fig. 7.1). Another classic example is the system of equations first used by Lorenz (1963) to model atmospheric interac-

Table 7.1

Basic Characteristics of Deterministic and Adaptive Systems

Deterministic Systems	Adaptive Systems
1. Large numbers of interacting agents	Large number of interacting agents
2. Stable interaction rules	Changing interaction rules
3. Few interaction rules	Many interaction rules
4. Homogeneous interaction rules	Heterogeneous interaction rules
5. Single-loop learning	Double-loop learning

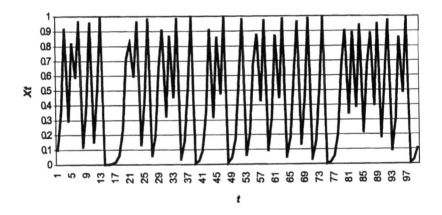

Logistic equation: $X_{t+1} = rX_t(1-X_t)$;
Initial conditions: $r = 4; X_0 = .1$

FIG. 7.1. The logistic equation ($t_0 \ldots t_{100}$).

tions (Fig. 7.2). Mathematical techniques and methods for studying nonlinear dynamics have been developed to characterize and, to some extent, explain noisy time series by analyzing these expressions. Systems are understood by looking for patterns within their complexity—patterns that describe the potential evolution of the system. Notably, many of these techniques were derived not by mathematicians, but by biologists and physicists trying to gain information from sampled time series that were short and noisy (Sprott & Rowlands, 1995).

Many adaptive systems in the natural world are now known to exhibit chaos or nonlinear behavior, the complexity of which is so great that they were previously considered random. Chaotic systems have now been recognized and studied in biology, physics, economics, and management (Anderson, Arrow, & Pines, 1988; Baum & Silverman, 1999; Brock & Malliaris, 1989; Cheng & Van de Ven, 1996; Kauffman, 1993; McKelvey, 1998; Waldrop, 1992). Complexity theory has contributed significantly to understanding heart arrhythmias and brain functioning related to epileptic attacks and to predict avalanches on the Colina volcano in Mexico (*Newsweek*, 1992). Fractal mathematics are vital to improved information compression and encryption schemes needed for computer networking and telecommunications. Genetic algorithms are being applied to economic research and stock market predictions. Engineering applications range from factory scheduling to product design, with pioneering work being done at firms such as DuPont and Deere & Co. The Simlife, SimAnt series of computer programs were developed from complexity research.

In the contemporary management literature, however, complexity theory tends to be invoked primarily at a metaphorical level (e.g., Morgan, 1996; Nonaka, 1988; Peters, 1988). Increased turbulence in the business world and accelerating pace of change appear sufficient to motivate use of the label *chaotic*. Despite limited theoretical attention (Baum, 1999; Brown & Eisenhardt, 1998; McKelvey, 1998; Stacey, 1995, 1996; Thietart &

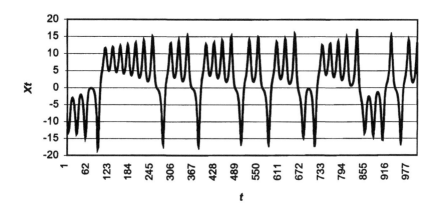

Lorenz equations
$$x_{t+1} = x_t + a(y_t - x_t)\delta$$
$$y_{t+1} = \delta(bx_t - y_t - x_t z_t) + y_t$$
$$z_{t+1} = \delta(x_t y_t - c z_t) + z_t$$
Initial conditions $b = 28; a = 10; c = 8/3;$ and $\delta = 0.01$

FIG. 7.2. Lorenz equations ($t_{4000} \ldots t_{5000}$).

Forgues, 1995) and scant empirical evidence (Baum & Silverman, 1999; Cheng & Van de Ven, 1996; Ginsberg, Larsen, & Lomi, 1996, Gresov, Haveman, & Oliva, 1993, Levinthal, 1997; Levy, 1994; Sorenson, 1996), unabashed claims are being made about complexity theory being the "next major breakthrough in management" (Vinten, 1992, p. 22). We agree that application of ideas from chaos theory may prove fundamental to the study and management of technological evolution, which itself is generated by complex, adaptive, interorganizational systems characterized by a wide array of dynamic behaviors. But we also believe that systematic empirical research is needed to verify and better comprehend the behaviors of these dynamic systems.

A TYPOLOGY OF DYNAMIC SYSTEMS BEHAVIOR

How can differences in the behavior of dynamic systems be characterized? Research in physics and biology reveals complex systems as progressing through stages of order and chaos from highly ordered, regular, predictable and controllable state through conditions of differing mixes of order and disorder. Importantly, research reveals that systems do not tend to gravitate toward random behavior but, rather, toward an area of complexity "at the edge of chaos" that lies between randomness and order. Studies of phenomena as disparate as sandpiles, earthquakes, and artificial life have found that systems move toward complex behavior (Bak & Chen, 1991; Kauffman, 1991, 1993; Langton, Farmer, Rasmussen, & Taylor, 1992). Complex behavior enables the system to maximize benefits of stability while retaining a capacity to change (Kauffman, 1993 Waldrop, 1992).

Dynamic systems exhibit three basic patterns of behavior—*orderly, chaotic* and *random*—whose characteristics and predictability are summarized in Table 7.2. Each pattern is characterized by the dominance of specific *attractors*, which differ among themselves in terms of the mix of order and disorder. Each attractor has a different form and a different causal pattern.

Trajectories and Attractors.   A crucial first step in determining the attractor of a dynamic system is the choice of embedding dimension (Brush, 1996). The *choice of embedding dimension* can be conceived as choosing the number of previous points that must be used to accurately estimate the next point. Ruelle (1981) and Takens (1981) first described a simple method for analyzing chaotic series called *time-series embedding*. It can be illustrated simply by observing the plots of pairs of points x_t and x_{t+n} for a time series. Figure 7.3 shows how plotting pairs of successive pairs of points x_t versus x_{t+1} and x_t versus x_{t+2} from the logistic time series given in Fig. 7.1 produces a recognizable pattern. Figure 7.4 shows the result for the Lorenz series in Fig. 7.2 for

TABLE 7.2

Patterns of Behavior in Dynamic Systems

Pattern	Attractor	Behavior	Microtrajectory Predictable?	Macroattractor Predictable?
Orderly	Point, limit cycle, and torus	Behavior that repeats itself exactly	Yes	Yes
Chaotic	Strange	Behavior that repeats itself a little differently each time and bifurcates into different behaviors	No	Yes
Random	Random	Behavior out of which new order emerges	No	No

successive pairs of points x_t versus x_{t+5} and x_t versus x_{t+10}. Note how, once embedded, the seemingly random *trajectories* of the logistic and Lorenz time series in Figs. 7.1 and 7.2 form complex geometric shapes beyond which the series never strays. Thus, the output of chaotic systems is point by point unpredictable, but forms a recognizable "macro" pattern over time if properly observed. The shapes in Figs. 7.3 and 7.4, which reveal the complex (but predictable) order underlying the trajectories, are known as *attractors*. An attractor is a set of points that represents the possible states, or *phase space*, which a time series generated by a dynamical system tends to take over time. Implicit in the notion is that a particular state in phase space specifies the system completely; it is all we need to know about the system to have complete knowledge of the immediate future. Thus, given x_t we can obtain a very good estimate of x_{t+n} by interpolation. This principle extends to multiple dimensions, and in general can be written as:

$$X_t = x_t, x_{d+t}, x_{2d+t}, x_{3d+t}, \cdots x_{nd+t}$$

where X is the embedded vector, d is the separation and n the embedding dimension. Takens (1981) showed that this principle generalizes, and that given a chaotic series correctly embedded, there exists a smooth function that would model it perfectly. He also demonstrated that, for any system whose attractor is d dimensional, n need not exceed $2d + 1$.[1] Both the cor-

---

[1] For example, the attractor for a point has $d = 0$, for a line has $d = 1$, for a plane has $d = 2$, and for a cube has $d = 3$.

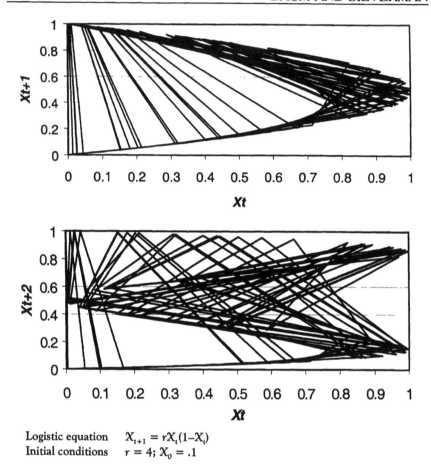

Logistic equation $X_{t+1} = rX_t(1-X_t)$
Initial conditions $r = 4; X_0 = .1$

FIG. 7.3.   Successive time plots of logistic time series.

rect embedding dimension and the smooth function can be discovered empirically using a box-assisted correlation (Brush, 1996).

Orderly Behavior: Point, Limit Cycle and Torus Attractors.   The orderly pattern coincides with *Point* and the *Limit Cycle* attractors in which behavior repeats itself exactly; the behavior of the system is both stable and predictable at both microtrajectory and macroattractor levels. In the Point attractor, behavior repeats itself like a free-falling pendulum that always comes to rest at the same point. Innovative behaviors always revert to a particular state. In the Limit Cycle attractor, behavior is drawn to a cyclically

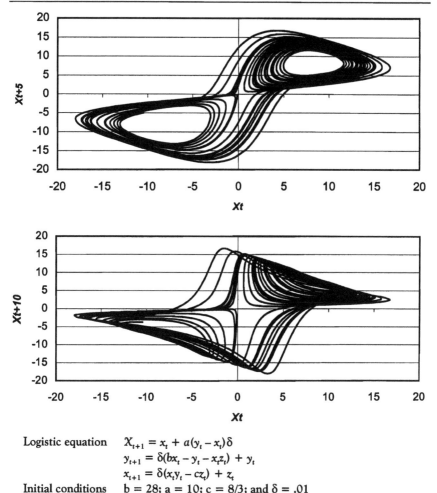

Logistic equation $X_{t+1} = x_t + a(y_t - x_t)\delta$
                         $y_{t+1} = \delta(bx_t - y_t - x_tz_t) + y_t$
                         $x_{t+1} = \delta(x_ty_t - cz_t) + z_t$

Initial conditions $b = 28$; $a = 10$; $c = 8/3$; and $\delta = .01$

Fig. 7.4   Successive time plots of Lorenz time series

changing set of 2, 4, 8, or 16 states. Behavior repeats itself like a thermostat
that maintains the temperature between two points, or a street light that
goes on and off according to the amount of daylight. Limit cycle behavior
can be found in examples of populations of animals that oscillate in a cyclic
manner. These are simple, linear, close to equilibrium systems. These attractors are typical of physical and mechanical systems. In this pattern, the system continually repeats a limited number of behaviors. When disturbed by
changes in the environment, the system always returns to the same behav-

iors. The system does not change; it appears to be overwhelmingly rigid. In *Torus* attractors, behaviors rotate vertically and cyclically around a cylinder like the circumference of the circle. Each behavior is similar to the behaviors that preceded it but slightly different. Instead of identical repetitive behavior, as in point and limit cycle attractors, there is *similar* behavior, which introduces a slight measure of disorder and nonlinearity. Behavior can be predicted with great certainty, but not exactly.

In Fig. 7.5 we show four time series based on the logistic equation along with one random time series, plotted both as X_t versus t trajectories and X_t versus X_{t+1} successive time plots to reveal underlying attractors. For $r = 1.8$, the figure shows the trajectory of the behavior moving to a fixed point—the equilibrium value of .444. The behavior illustrated in the related successive time plot conforms to a Point attractor. For $r = 3.2$, in the limit, the value moves back and forth between the two points, .513 and .799. For $r = 3.5$, the value moves back and forth between the four points .383, .827, .501 and .875. The behavior of the logistic equation for $r = 3.2$ and $r = 3.5$ depicts Limit Cycle attractors.

Chaotic Behavior: Strange Attractors. The chaotic pattern is reflected in *strange* attractors. In this pattern, the system applies a variety of behaviors that develop and change as need arises. Systems characterized by the chaotic pattern show the greatest adaptability to changing environmental conditions (Kauffman, 1993; Waldrop, 1992). Change in chaotic systems tends to be small most of the time, but occasionally large-scale change occurs (Bak & Chen, 1991; Phelan, 1995). In strange attractors, behavior is characterized by irregularity, uncertainty, and difficulty in predicting and planing ahead. Behavior bifurcates from similar, uniform behavior, to an array of possible behaviors. Strange attractors can bifurcate between 2, 4, 8, or 16 patterns.

Chaotic systems display a sensitivity dependence on initial conditions that makes accurate predictions of future conditions virtually impossible. Nevertheless, the overall behavior does have a discernable attractor that can be identified throughout time and that bounds possible behavior. Strange attractors bound systems to a limited space by defining behavioral extremes rather than by attraction to a central point. Not surprisingly, since micro-, point-by-point trajectories are predictable for both point and limit-cycle attractors, successive time plots for $r = 1.8$, 3.2, and 3.5 in Fig. 7.5 do not provide new insight beyond the plots of time-series trajectories versus time. However, for $r = 4.0$, the microtrajectory of values against time is apparently random and unpredictable. Yet, as the successive time plot for these data reveals, the values are actually moving back and forth, predictably, along a parabola; an example of a *Strange* attractor.

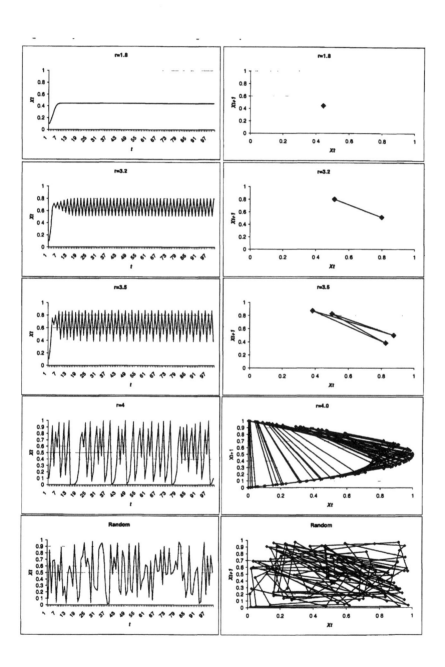

FIG. 7.5.   Trajectories and attractors for logistic and pure random processes ($t_0 \ldots t_{100}$)

The ordered and chaotic behavior of the logistic equation are summarized in Fig. 7.6 using a bifurcation diagram, which plots the limiting iterated
values of the logistic equation for the range of $r = 2$ to $r = 4$. The figure
clearly shows the *point* solutions for X when $r < 3$, the *limit-cycle* solutions for
X for $3 < r < 3.6$, and the *chaotic* region characterized by values of X that
span a wide range for $r > 3.6$. Each doubling is called a *bifurcation* because a
single point solution splits into a pair of solutions. Successive doublings occur with ever increasing rapidity as one moves from left to right in the figure.

Random Behavior: Random Attractor.   In the random pattern, unpredictability characterizes the dynamics of events. Orderly behavior does not
develop. Novelty is at its extreme and the system appears to be in a constant
agitation. The pattern that borders and constrains the behavior of chaotic
processes disappears and behavior faces no limitations of imposed order and
regularity. The behavior of Strange attractors is called chaotic because prediction at the microscale is limited but, nevertheless, there is a macrolevel pattern behind the disorder (i.e., the attractor is predictable). Chaotic processes
are not random; they follow rules, although the rules (even simple ones like
the logistic equation) can produce extreme complexity.

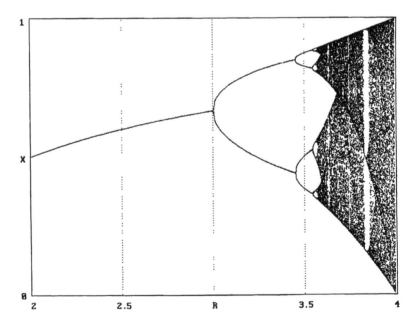

Logistic equation: $X_{t+1} = rX_t(1-X_t)$

FIG. 7.6.   Logistic bifurcation diagram.

In the random attractor, disorder reigns and allows no place for order. As the successive time plot for the random X versus t trajectory in Fig. 7.5 shows, a purely random process (i.e., Brownian motion) fills the phase space without any apparent order. Thus, neither the microtrajectory nor macroattractor is predictable for random systems. In adaptive systems, *random behavior* is a transition period where the old order has broken down and a new order has not yet emerged to replace the old. Complex systems reach this state when internal cleavages and external perturbations move the system beyond a point from which the only alternatives are transformation or disintegration.

Attractors, Path Dependence and Path Creation

The enabling and constraining effects of attractors on activities taking place in an evolving technological system can be represented by the framework in Fig. 7.7. Garud and Jain (1996) proposed this semiotic square framework originally to aid in understanding the effects of standards on enabling or constraining technological evolution. Here, we extend its application to the effects of attractors on the behavior of complex, dynamic, interorganizational systems underlying technological evolution.

Position 4. In position 4, innovation trajectories are neither enabled nor constrained, are characterized by random attractors. Here a lack of standardization creates an "open space" reigned by ambiguity and incompatibility in which an order to constrain and enable technological evolution has yet to emerge (Garud & Jain, 1996). In this "dis-embedded" position, history does not matter.

Position 2. In contrast, position 2 is path independent. It represents a highly ordered situation in which standards constrain technological evolution completely, reducing degrees of freedom to zero and making it impossible to shift to a new technological trajectory (Garud & Jain, 1996). Such "constrained–not enabled" situations are consistent with point attractors, in which behavior repeats itself, or limit cycle attractors, in which behavior is attracted to a cyclically changing set of states. In either case, when disturbed by innovative behavior or by the environment, the system always returns to the same behaviors regardless of initial conditions; technologies are "stuck" (Arthur, 1989).

Position 3. Position 3 also represents an ordered situation, but less so than position 2. In this case, standards established to reduce ambiguity and permit interconnectivity do not reduce degrees of freedom to zero. As a re-

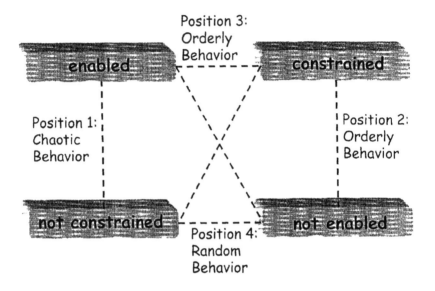

FIG. 7.7.   Enabling and constraining effects of attractors. Adapted from: Garud & Jain (1996, p. 392). Adapted with permission.

sult, technical evolution in simultaneously enabled and constrained, permitting new paths to be created within boundaries defined by the past (Garud & Jain, 1996). Enabled–constrained situations are likely to be characterized by *torus* attractors, in which current behavior is similar to behaviors that preceded it but slightly different, which introduces a measure of disorder. Behavior can be predicted with great certainty, but not exactly.

Position 1.   Finally, Position 1, in which technologies are enabled but *not* constrained, is "just embedded," permitting connectivity as well as the possibility to migrate to new trajectories—processes of path dependence and creation operate in the evolution of the technology (Garud & Jain, 1996). Such conditions are most compatible with strange attractors, which possess the greatest adaptability to changing environmental conditions (Kauffman, 1993). Behavior is characterized by irregularity, uncertainty, and sensitivity to initial conditions, which increases difficulty in predicting and planing ahead. Behavior is quasi- or boundedly independent, stretching and folding within a low-dimensional space; instead of being history independent, as in pure randomness where data points completely fill the phase space (see Fig. 7.5). Nevertheless, the overall behavior does have a pattern, defined by its attractor, that can be identified and there are boundaries that set limits to behavior. Thus, "just embedded" conditions enable the technology to maximize benefits of stability while retaining a capacity to change (Garud & Jain, 1996; Kauffman, 1993 Waldrop, 1992).

Implications for Technological Evolution

Evidence appears to support the idea that competitive and cooperative interorganizational systems are complex systems poised on the edge of chaos. In interorganizational situations, organizations must continuously coevolve, adjusting to the adaptations of opponents (partners) (Barnett & Hansen, 1996; Brown & Eisenhardt, 1998; Cheng & Van de Ven, 1996; D'Aveni, 1994; Ginsberg, Larsen, & Lomi , 1996; Levinthal, 1997; Lomi & Larsen, 1999; McKelvey, 1998; Moore, 1993; Sorenson, 1996). An organization may reach local optima in its fitness landscape, but it must be able to adapt rapidly if competitors or the environment change in such a way as to make its behavior suboptimal. Because environmental change is unpredictable, internal disorder provokes multiple responses, one (or several) of which can become a new organizational equilibrium whose characteristics cannot be predicted a priori (Thietart & Forgues, 1995). Although individual organizations appear often to stray far from chaos's edge into rigid configurations of strategy, structure and process (Miller, 1986; Miller & Friesen, 1984), creating a problem of too much continuity and order and not enough exploration of new alternatives (Levinthal & March, 1993; March, 1991), competitive and cooperative interorganizational systems are far less constrained by systems of goals and incentives and authority structures, permitting the proliferation of new ideas and technologies (Miner & Haunschild, 1995; Powell & Brantley, 1992; Powell, Koput & Smith-Doerr, 1996). At the same time, they are not so loosely coupled that they exhibit no coherence (Baum & Silverman, 1999; McKelvey, 1998; Powell et al., 1996).

Consistent with this characterization, industries and their underlying technologies typically evolve through long periods of incremental change with the occasional discontinuous change or punctuated equilibrium (Dosi, 1988; Tushman & Anderson, 1986, Utterback, 1994). This suggests that interorganizational–technological systems follow a power-law distribution or punctuated equilibrium indicative of a complex system (Bak & Chen, 1991). Organizations and leaders in a wide range of industries tend to hold their position for relatively long periods of time (often for decades), suggesting that their source of competitive advantage may be somewhat sustainable, or at least renewable (Chandler, 1990; Hamel & Prahalad, 1994). Research on technological discontinuities similarly reveals change ranging from minor process improvements to completely new product classes; smaller improvements are much more frequent than large ones (Tushman & Anderson, 1986). And, these shocks are just as likely to make the system less turbulent (competence enhancing) as more turbulent (competence destroying).

Understanding technological evolution, therefore, requires consideration of path dependence and path creation processes operating at any moment in the evolution of a technology (Arthur, 1988, 1989; Karnøe & Garud, 1997). Path dependence refers to historical forces that shape the *future* evolution of a technology. Path creation refers to the enactment of *novel* approaches that break from the past. Complexity theory provides a rich theoretical framework to characterize as well as techniques for detecting nonlinear dynamics that can potentially help account for and describe the dynamic interorganizational systems that give rise to path dependent evolution of technologies.

The ability of competitive and cooperative interorganizational systems to accumulate knowledge is constrained in both orderly and random systems. The system is either insensitive to inputs of new information, the schemes and models it uses to process information distort it, or it has problems turning meaningful information into action. An adaptive interorganizational system must absorb new information, process it so that it becomes meaningful, learn from the experience of using it, and build experienced-based models that guide its functioning in a changing environment. The random pattern, without organized models and frameworks, is not conducive to absorbing information, to making it meaningful, and to action based on it. Organizational learning, which is a prerequisite for innovation, requires a certain level of order and continuity to enable an organization to preserve its cumulative historical experience. On the other hand, a too orderly pattern prevents the interorganizational system from taking action to create new directions as a result of new meaningful information. The organizations in the system stick to their old models and schemes and processes all new information through their filters. They do not embark in new directions when they receive novel information that necessitates changes in innovative behavior. Both too much order and too much noise are unsuitable conditions for innovation. The edge of chaos may be the place for assimilating information and maximal organizational learning and innovation, and where path dependence and creation are in balance.

In sum, technological evolution, the product of competitive and cooperative interorganizational systems (e.g., Powell & Brantley, 1992; Powell et al., 1996; Schumpeter, 1934), often appears to display complex dynamics, at times seeming totally unpredictable (Cheng & Van de Ven, 1996). Although the trajectories of evolving technologies generated by competing firms may look random they may in fact be characterized by chaotic behavior, with self-reinforcing trends. Casual inspection of these trajectories may not, however, be very helpful in distinguishing between complex system dynamics and behavior that is essentially random (Cheng & Van de Ven, 1996). We now turn to an exploration of technology trajectories in lasers to examine these issues further.

The Evolution of Laser Technology

A laser (acronym for Light Amplification by Stimulated Emission of Radiation) is a device that produces an unconventional type of light. Compared to conventional light sources, laser beams remain narrowly focused over exceptionally long distances. Laser light is also "coherent," which means that all light waves coming out of a laser are lined up with each other (Hecht & Teresi, 1982). As a result, laser light is extremely intense. These attributes, among others, make modern lasers particularly useful for an immense range of commercial, medical, scientific, and military applications.

The concept behind the laser rests on the use of chain reactions to amplify light. Quantum theory proposes that the electrons in an atom move in distinct orbits, each of which exhibits a distinct energy level, and can jump from one orbit to another. If, say, a hydrogen atom is exposed to a photon (a unit of light energy), then its electron will absorb the photon and jump to a higher energy level ("excited" state). However, because atoms tend toward lower energy ("ground" state), the atom will emit a photon and its electron will return to its ground state. If an atom that is already excited encounters a photon, its electron is stimulated to emit a photon without absorbing the incoming photon. The laser is based on the principle that bombarding atoms with photons and catching sufficient electrons in excited states—by using mirrors and an enclosed space to reflect photons back onto a group of atoms—can stimulate a chain reaction of photon emission and generate "a burst of light energy all at one wavelength" (Hecht & Teresi, 1982, p. 17).

The history of laser innovation displays the punctuated equilibrium, and characteristic power law distribution, that typifies much technological evolution (Nelson, 1994). The laser was first described theoretically in the late 1950s and first built in 1960. Yet most if not all of the scientific knowledge necessary to develop the laser was widely available for as many as 30 years before its development (Rather, 1980; Bertolotti, 1983). Why did the development of the laser lag so long? The idea of stimulating emission requires the existence of a "population inversion," which occurs when there exist more excited than unexcited atoms. The idea of a population inversion existing in the real world was sufficiently radical that "for many years physicists found the prospect unthinkable" (laser inventor Arthur Schawlow, paraphrased in Hecht & Teresi, 1982, p. 18).

Once Charles Townes of Columbia University demonstrated a working population inversion for stimulating microwave emission in the early 1950s, this "mental block" was broken, and working lasers followed shortly thereafter. Laser development has continued to proceed through a series of cycles in which each successive major innovation (e.g., development of ion-based lasing; demonstration of X-ray lasing) sparked a sustained flurry of subsequent activity to refine, extend, and expand applications for the resulting laser technology (Bromberg, 1991; Hecht, 1992).

The history of laser innovation also underscores the importance of both interdependent and autonomous research efforts. Several laser historians attribute the rapid advances in laser research to a combination of (a) diffusion of ideas within the laser research community via conferences, formal and informal interorganizational links, published papers, and patents; and (b) willingness of individual organizations and researchers to pursue avenues that most of the laser community had dismissed as fruitless. For example, the first published theoretical description of a laser (Schawlow & Townes, 1958) resulted from a consultancy relationship between a Columbia University scientist and Bell Labs, thus linking basic research ideas to industry interests. Immediately upon publication, a host of organizations launched a research "race" to build a working laser (Hecht, 1992). These organizations and researchers used direct collaboration and conferences such as the annual International Quantum Electronics Conference to gain insight into developments at other organizations, and to alter research direction and research funding accordingly (Bromberg, 1991).

As it happens, the Schawlow and Townes (1958) paper explicitly stated that artificial rubies would not be a feasible medium for lasers (rubies were a common medium for earlier masers). Most organizations followed this conventional wisdom and focused on running electric current through gas. However, researchers at Hughes Aircraft persisted in efforts to construct a rubybased laser, and in 1960 "shocked the world" by winning the race to build a working laser (Hecht, 1992).[2]

Although laser historians have tended to focus on the inception of the industry due to its drama and controversy (a 30 year-long dispute between Schawlow/Townes and a former Columbia graduate student, Gordon Gould, over ownership of the original concept of the laser), these patterns continue through the present. The number of interorganizational collaborations and venues for interorganizational communication on laser technology have increased, and specialized trade journals now diffuse information

---

[2]"[Theodore] Maiman jokes that right after scientists heard that he had fashioned a ruby rod into the first laser, nearly everybody with a crystal in his or her lab tried putting mirrors on the end of it to see if it would lase. The amazing thing is that many of these scientists succeeded (Hecht & Teresi, 1982, p. 30)."

among organizations as well. For example, the 1975 Conference on Laser Equipment and Applications included presenters from at least 15 organizations, including six universities, two U.S. government labs, six private firms, and one public utility. The July 1975 issue of *Laser Focus* included a detailed summary of each of these presentations (Ion-laser work, 1975). This conference also provided an opportunity for organizations to develop informal technological roadmaps, as when Lawrence Livermore National Laboratory chief Edward Teller laid out a vision for the timing of medium-term laser technology developments. Firms, universities, national laboratories, and independent research labs continue to build both formal linkages (alliances) and informal linkages (personnel flows; personal contacts) to access information and shape technological developments and funding priorities (U.S. Congress, 1980)—reflecting what some have called the "synergism … both among the technologies and among the organizations [in the laser research community]" (Begley, 1980, p. 253).

Thus, the innovation system that spawned the modern laser appears to have been characterized by factors common to chaotic processes. The punctuated equilibrium pattern of innovation is consistent with a power law distribution characteristic of complex behavior. Moreover, the competitive and cooperative interorganizational systems involved in laser research appear to exhibit the type of partial connections among organizations (and among individual researchers) that exemplify systems inhabiting "the edge of chaos."

Measuring Laser Technology Evolution: Patent Trajectories

Our study of laser technology trajectories relies on analysis of patent data. Patent data have been compared favorably to research and development (R&D) statistics for the study of corporate or national technological competence (Pavitt, 1982; Patel & Pavitt, 1994; see Silverman, 1996 for a more extensive comparison of various measures of technology). Among other advantages, patents provide a paper trail of individual innovations and follow a consistent policy of information disclosure. Indeed, patents have been used by scholars to track crude innovation trajectories (Schmookler, 1962) as well as to inform policy debates concerning trends in national innovative capabilities (Griliches, Pakes, & Hall, 1987; Mansfield, Schwartz, & Wagner, 1981).

Along with these advantages of patent data come some limitations. Two limitations are especially relevant: (a) patents do not (and indeed can not) encompass all technological innovation, and (b) individual patents vary widely in technological and economic importance. In answer to the first criticism, Patel and Pavitt (1994) argued that much of the literature on "converging indicators" of technological skill (e.g., Irvine & Martin, 1983) suggested that patented and unpatented knowledge are highly correlated. Regarding the second criticism, Trajtenberg (1990) and others demonstrated a correlation between citations to a patent and its technological and economic importance. Thus, citation weighting of patents may offer a method for ameliorating the variance in individual patent worth. It is worth noting that the varying importance of patents could best be addressed through indepth archival or survey research, which suggests a complementarity between patent data and traditional methods of tracing trajectories.[3]

Our patent data were drawn from the Micropatent database, which contains information concerning every patent issued in the United States since 1975. This information includes: date of application; company to whom the patent is assigned (if any); and U.S. patent classification (USPC). The U.S. Patent and Trademark Office identifies patent class 372—coherent light generation—as the technology class most centrally associated with lasers. However, as Patel and Pavitt (1994) among others have argued, most technological systems rely on multiple subtechnologies; we expect that laser systems are no exception.

We used two methods to identify patent classes associated with laser technologies. First, we identified 50 laser companies by selecting randomly from firms listed under the heading *Lasers* in *Thomas' Register of American Manufactures* for the years 1983 and 1990. We then identified every patent assigned to these companies between 1975 and 1996. Next, we identified the most frequently assigned USPCs in this sample of patents. Although this sample included 26 classes, only three accounted for 5% or more of the total: 235 (registers); 372 (coherent light generators); 356 (optics—measuring and testing). These three classes comprised 60% of the patent sample.

Second, we searched the Micropatent database for all patents assigned to any organization with the keyword "laser" in its name. We then identified the most frequently assigned USPCs in the resulting sample. Although this sample included assignments to 54 classes, six classes accounted for 5% or more of all patents: 372 (coherent light generators); 606 (surgery); 369 (dynamic information storage and retrieval); 356 (optics—measuring and testing); 359 (optics systems); 219 (heating, electric). These classes comprised 64% of this sample. Note that two of three classes identified by the first

---

[3]Most studies of technology trajectories rely on case analysis of a series of commercialized or failed innovations (e.g., Arthur, 1988; Dosi, 1982; Garud & Jain, 1996; Karnøe & Garud, 1997; Sahal, 1985). Such analyses provide insight into historical accidents that enable some innovations to gain acceptance whereas others languish and provide rich detail on the significance of various patents. In turn, patent-based trajectories complement case analysis by offering a macrocontext against which to interpret and evaluate qualitative and other archival research.

method also appear among the top six according to the second method, so that the combination of approaches identified seven distinct technology classes heavily related to laser patenting (see Appendix A for sample patents from each class).

Finally, we identified every patent granted in the United States between 1975 and 1996 that was assigned to any of these seven classes. We aggregated these to generate patent counts by date of application (categorized in 3-month periods).[4] We used the resulting counts for the years 1974 through 1992 for our study.[5]

Detecting Chaos and Complexity in Laser Technology Trajectories

Recently developed techniques facilitate the differentiation among chaotic (Position 1 in Fig. 7.7), random (Position 4), and orderly (Positions 2 and 3) behavior. Two features of chaotic processes provide a basis for rendering such a differentiation. One is the extreme sensitivity of chaotic processes to initial conditions, in contrast to orderly processes, where different starting points converge to the same sequence of points on a simple attractor.[6] The second feature is that chaotic processes display predictable attractors that bound possible system behavior whereas purely random processes do not. We apply several diagnostic tests, described briefly, and in more detail in Appendix B, to measure these characteristics in laser technology trajectories.

Lyapunov Exponent. The standard method of identifying extreme sensitivity to initial conditions involves the calculation of the Lyapunov exponent. This exponent measures the rate at which neighboring points on an attractor diverge as they move forward in time. Positive Lyapunov exponents represent an increasing divergence over time and, consequently, indicate sensitivity to initial conditions. Positive exponents may therefore indicate chaos or randomness. In contrast, zero and negative exponents rep-

---

[4]It is customary in patent studies to use the date of application, rather than the date of granting, as the time at which an innovator has access to the to-be-patented technology.

[5]Griliches (1990) pointed out that only 67% of all patents that will eventually be granted are granted within 3 years of application, whereas more than 90% of such patents are granted within 4 years of application. As a result, to avoid truncation problems, we used patent counts through 1992, even though we have data on patents granted through the end of 1996. Analysis of the patents in our sample indicates that less than 1% of granted patents are granted within the same calendar year of application, but roughly 33% are granted by the end of the next calendar year. We therefore begin our study in 1974, the year before our first year of granted patent data.

[6]Although random processes generally exhibit no impact of initial conditions on subsequent behavior, initial conditions for a given random process may influence later outcomes.

resent cyclic and mean reverting behavior, respectively, both of which are characteristic of orderly systems.

Correlation Dimension. The standard method of identifying bounded system behavior involves the calculation of the correlation dimension. This measure indicates an attractor's dimension d. A major difference between chaotic and random processes is that, although both appear random to the eye and to standard linear time series methods, chaotic behavior lies within a low-dimensional space defined by its attractors, whereas random behavior is unbounded. Thus, a random system will typically have d slightly less than or equal to the embedding dimension, whereas a chaotic system will typically have a low d (Brock, Hsieh & LeBaron, 1991; Dooley & Van de Ven, 1997).

Surrogate Testing. The validity of the conclusion that a time series is chaotic based on its Lyapunov exponent and correlation dimension can be evaluated empirically using surrogate testing (Theiler, Eubanks, Longtin, Galdrikian & Farmer, 1992). In surrogate testing, to test the null hypothesis that the time series is linear (periodic), phase randomization is used to generate surrogate time series that have similar properties to the time series in question (Kaplan & Glass, 1995). A phase-randomized surrogate time series has the same length, distribution and linear dynamical structure as the original. The only aspect altered is the non-linear structure of the data. If a time series has low correlation dimension (d) because it is periodic in nature, then surrogates will also have a low d. If, however, the time series has low d because of nonlinearity, then surrogates will have d slightly less than or equal to the embedding dimension because phase randomization will have removed the low dimensional nonlinear structure.

An Example. Table 7.3 shows values of the Lyapunov exponent (L), correlation dimension (d), and the results of surrogate testing for the logistic and random time series from Fig. 7.5. L is negative across embedding dimensions 3–6, indicating the time series are periodic for the logistic equation with unrestricted growth rate r set ≤ 3.5, which are characterized by point and limit-cycle attractors. In contrast, L is positive across embedding dimensions 3–6, indicating either chaotic or random behavior for the logistic equation $r = 4.0$, which is characterized by a strange attractor, and the random time series. For all logistic time series, d is stable and low across embedding dimensions 3–7 indicating periodic or chaotic behavior. We examine embedding dimensions 3–7 because real time series rarely have $d < 3$ and time series with $d > 7$ are generally considered random. In contrast, for the random time series d increases approximately linearly with the embedding

TABLE 7.3
Detecting Chaos in Logistic and Random Time Series

Embedding Dimension	3	4	5	6	3	4	5	6	7
Coefficient	$L=$	$L=$	$L=$	$L=$	$d=$	$d=$	$d=$	$d=$	$d=$
Logistic ($r = 1.8$)	-1.178	-1.112	-1.069	-1.033	1	1	1	1	1
Logistic ($r = 3.2$)	-1.046	-1.018	-.994	-.987	1	1	1	1	1
Surrogates					1	.354	.477	1.026	1.070
					(0)	(.102)	(.157)	(.256)	(.295)
Logistic ($r = 3.5$)	-.901	-.937	-.921	-.903	1	1	1	1	1
Surrogates					.754	1.639	3.284	2.813	2.549
					(.071)	(.156)	(.943)	(.340)	(.442)
Logistic ($r = 4.0$)	.962	.984	.759	.711	1.408	1.396	1.084	1.122	1.589
Surrogates					2.605	3.437	4.204	4.532	5.709
					(.268)	(.878)	(.712)	(.484)	(.808)
Random	.587	.235	.058	.146	2.825	3.861	5.040	5.354	6.658
Surrogates					3.519	4.134	5.191	5.553	6.925
					(.375)	(.380)	(.576)	(.576)	(.648)

Note. L is the Lyapunov exponent; d is the correlation dimension. Means (standard deviations) for surrogate ds for logistic and random time series are based on five phase-randomized series. Surrogate ds for logistic ($r = 1.8$) equal 1, with standard deviation 0 for all embedding dimensions.

dimension. Surrogate testing confirms the chaotic nature of logistic behavior for $r = 4.0$. Although surrogate ds for logistic time series based on $r \leq 3.5$ are stable across embedding dimensions 3–7 at values well below the embedding dimension, surrogate ds for the logistic equation with $r = 4.0$ increases approximately linearly with the embedding dimension. This occurs because phase randomization removes the low dimensional nonlinear structure from the series.

Laser Technology Patent Trajectories

As the foregoing discussion indicates, looking at time series trajectories may not be very helpful in distinguishing complex yet predictable dynamics from those that are essentially random. As can be seen in the panels of Fig. 7.8, this is true of the laser technology innovation trajectories, defined by the event histories of patenting in seven USPCs related to the laser industry from January 1974 (shortly after the laser industry's inception) to December 1992. This figure gives detrended, 3-month moving average event frequencies for all laser-related patents combined, and separately for each of the seven laser-related USPCs.[7]

We therefore employ the methods we just outlined for detecting complex dynamics to analyze these patent trajectories. For each patent event sequence we:

1.   Compute Lyapunov Exponents for embedding dimensions 3–6.
2.   Compute Correlation Dimensions for embedding dimensions 3–7.
3.   Compute Surrogate ds.
4.   Examine successive time plots.

The end result is a dynamic characterization of the multiple technological trajectories that have come to define the modern laser industry. The characterization includes an assessment of the extent to which complex dynamics have become apparent in this system. Table 7.4 shows values of the Lyapunov exponent (L), correlation dimension (d), and surrogate testing results for the laser patent trajectories, overall and for each USPC.

L, d, and Surrogate d.   As Table 7.4 shows, L is positive across embedding dimensions 3–6 for each patent class trajectory. This indicates that

---

[7]Often chaotic behavior is superimposed on some slowly varying nonchaotic function. Effective study of chaos requires that the smooth function, which represents the slow trend, be subtracted from the data before a detailed analysis is performed. A polynomial fit is one way to accomplish this (Sprott & Rowlands, 1995). We used a fourth-order polynomial fit to accomplish this detrending for each patent trajectory.

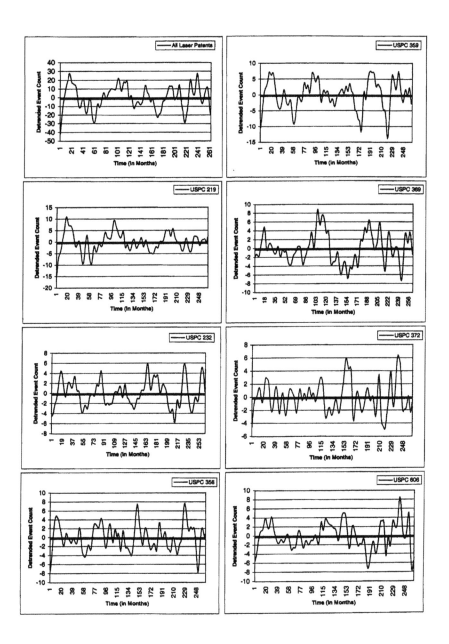

FIG. 7.8. Laser patent trajectories.

Table 7.4
Detecting Chaos in Laser Patent Trajectories

Embedding Dimension	3	4	5	6	3	4	5	6	7
Coefficient	L =	L =	L =	L =	d =	d =	d =	d =	d =
All Laser Patents	.440	.292	.279	.264	2.731	3.337	3.958	3.833	3.813
				Surrogates	2.837	3.578	3.961	4.261	4.691
					(.065)	(.346)	(.333)	(.379)	(.706)
USPC 219	.474	.394	.351	.288	2.482	2.844	3.604	3.582	3.519
				Surrogates	3.065	3.956	4.579	4.879	5.529
					(.295)	(.453)	(.390)	(.511)	(.387)
USPC 232	.481	.389	.334	.256	2.423	2.701	2.929	3.152	3.730
				Surrogates	2.908	3.705	4.359	4.920	5.442
					(.214)	(.188)	(.306)	(.238)	(.292)
USPC 356	.480	.402	.313	.250	2.919	2.984	3.216	3.754	3.881
				Surrogates	3.076	4.214	4.715	4.845	5.370
					(.292)	(.436)	(.344)	(.531)	(.332)

Embedding Dimension	3	4	5	6	3	4	5	6	7
Coefficient	L =	L =	L =	L =	d =	d =	d =	d =	d =
USPC 359	.525	.430	.299	.294	2.620	2.954	3.292	3.537	3.833
				Surrogates	3.046	3.606	4.161	5.085	5.591
					(.237)	(.329)	(.526)	(.657)	(.563)
USPC 369	.515	.369	.369	.326	2.378	2.655	2.939	3.496	3.382
				Surrogates	2.984	3.782	4.056	4.476	4.244
					(.311)	(.431)	(.326)	(.531)	(.258)
USPC 372	.524	.384	.286	.301	2.723	2.850	3.225	3.546	3.713
				Surrogates	2.850	3.396	3.863	4.429	4.267
					(.165)	(.159)	(.207)	(.186)	(.217)
USPC 606	.448	.331	.280	.230	2.667	3.107	3.541	3.808	3.122
				Surrogates	2.907	3.459	3.859	4.363	4.233
					(.124)	(.108)	(.342)	(.327)	(.507)

Note. L is the Lyapunov exponent; d is the correlation dimension. Means (standard deviations) for surrogate ds are based on five phase-randomized series.

each trajectory is sensitive to initial conditions, and consequently that each trajectory is characterized by either chaotic or random—but not orderly—behavior. Values for d stabilize for each trajectory in the range of 3.5 across embedding dimensions 3–7, thus indicating a relatively low-dimension attractor characterized by either periodic or chaotic—but not random—behavior.

Taken together, these results suggest that the patent trajectories for laser technologies are chaotic. As Table 7.4 also shows, surrogate testing corroborated this conclusion for four of the patent classes (219, 235, 356, and 359). For these classes, surrogate series ds increase steadily with the embedding dimension, suggesting that the low ds for the actual series resulted from nonlinear dynamics that were eliminated by phase randomization. For the remaining three (369, 372, and 606), however, surrogate series ds stabilize between 4.0 and 4.5. This suggests that the low ds for the original series resulted from linear (periodic) dynamics.

Successive Time Plots. Finally, in Fig. 7.9, we show the patent data from the panels of Fig. 7.8, but plotted as successive time plots. The figures plot consecutive pairs of values, X_{t+2} versus X_t, for a given patent trajectory.

In sum, our diagnostic tests indicate that technological evolution in lasers between 1974 and 1992 has been broadly characterized by chaotic processes, and consequently fall within position 4 of Garud and Jain's (1996) "just embedded" framework.[8] We are leery of generalizing from laser technology during a particular time period to other time periods in the history of the evolution of the laser, or to other technological fields at similar or different stages of development. In addition to the usual concerns about generalization, numerous scholars (e.g., Levin, Klevorick, Nelson, & Winter, 1987) noted that propensity to patent varies by industry and time. Nevertheless, we hope that this study indicates a first step toward empirical evaluation of the degree to which path dependence and path creation in innovation are a manifestation of complex systems.

DISCUSSION AND CONCLUSION

According to Arthur (1988, 1989), for path dependent systems, such as competing technologies with increasing returns to adoption, the

---

[8]As discussed earlier, Garud and Jain (1996) developed their framework to describe the dynamics of technological systems faced with (relatively) formal and explicit standards. Our findings suggest that even in the absence of such formal standards, laser technology evolution approximated a "just embedded" process. These results suggest a broader applicability of this framework than Garud and Jain's original vision. It also raises questions about the need for explicit standards to generate high levels of adaptiveness in a technological system.

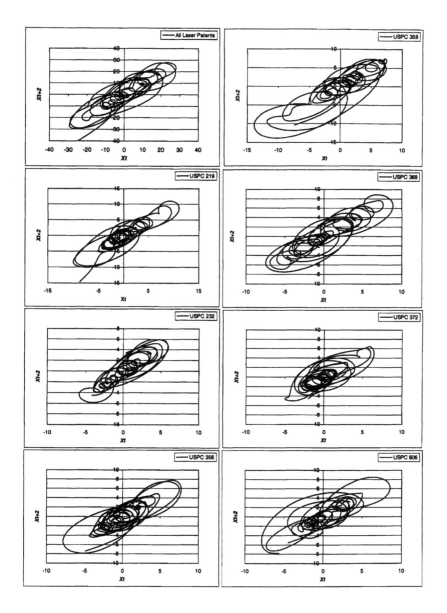

FIG. 7.9.   Laser patent trajectories, successive time plots.

macrostructure that emerges (i.e., long-run market concentrations or market shares) is typically, even inevitably, characterized by nonpredictability, potential nonsuperiority, and structural rigidity (i.e., technology lock-in). Here, rather than studying competitive outcomes (i.e., content) of technological evolution, we focus on the technological trajectories (i.e., dynamics) behind the outcomes. Complexity theory provides a model of three different basic behavioral patterns of dynamic systems: ordered, chaotic, and random. These patterns describe systems from a highly ordered condition that can be predicted and controlled through varying mixes of order and disorder, culminating in a state of randomness. The states differ from each other in terms of their dominant attractors. The first and third patterns are problematic for adaptive functioning of competitive and cooperative interorganizational systems. Within the chaotic mode, however, sustainable adaptive functioning is achieved when the system is at the edge of chaos. In this mode, the system balances itself between order and disorder, and between path dependence and path creation, so as to maximize organizational learning, flexibility, and adaptation. Complexity theory, thus, highlights the tension between exploitation of knowledge gained (path dependence) and exploration of novel actions (path creation) and illustrates how innovation processes characterized by "edge-of-chaos" behavior balance these tensions, permitting adaptive functioning of technological–organizational systems.

Using the modern laser as an example, we show how the technological trajectories produced by an interorganizational system can be related to concepts from complexity theory, and illustrate the range of possible trajectories (or macrostructures) such a system can produce. The evidence we presented suggests that, between 1974 and 1992, patent trajectories underlying technological evolution of the modern laser may be the product of a complex interorganizational system poised at the edge of chaos. This possibility opens new theoretical opportunities for understanding technological evolution and interorganizational systems.

Future Directions.   Although we have emphasized technological trajectory dynamics here, future research linking content and dynamics seems vital. In general, we wonder how technological trajectories relate to outcomes. For example, how do the features of technological trajectories relate to lock-in? The shape of technological fitness landscapes is central to predicting the occurrence of lock-in; fitness landscapes where fitness peaks proliferate and become less differentiated from the general fitness landscape are especially prone to technology lock-in. Under such conditions, payoffs to technological choices are difficult to distinguish, exploitation of good (chance) initial choices is reinforced, and lock-in to potentially inferior technologies likely. More generally, as the number and steepness of adaptive peaks on a technological fitness landscape increases, the level of fitness at any given peak and the probability of finding a better than average fitness peak falls, and the likelihood that systems will become trapped on suboptimal fitness peaks rises. Kauffman (1993) labeled this type of lock-in a "complexity catastrophe," which occurs as the falling height of accessible fitness peaks thwarts the search and selection process.

Can technological trajectories with particular features thwart such lock-in or provoke de-locking on such technological fitness landscapes? It should be possible to "tune" the interorganizational system generating the trajectory to avoid the danger of becoming trapped on poor local optima (McKelvey, 1998). For example, lock-in may be avoided by structuring the interorganizational system to partition the technological problem into subsystems, each of which optimizes its own task while ignoring the effects of its actions on the problems facing other subsystems. Subsystem boundaries permit constraints from other subsystems to be ignored, helping to avoid becoming trapped on poor local optima. Overall adaptiveness arises as collective, emergent behavior of the interacting, coevolving subsystems. Such "coevolutionary problem solving" is not likely to be useful for simple technological problems, but increases in value as technological fitness landscapes become less differentially rugged. This idea is equivalent to recommending that organizations facing difficult problems divide into departments, profit centers, and other quasiindependent suborganizations to improve their performance. In sum, we think research on "tuning" interorganizational systems to fit technological problem domains will provide basic new insights on management of technological evolution.

Future Challenges. If, as complexity theorists assert, much of the interesting and adaptive behavior in human social systems does exist at the edge of chaos, then this poses a serious research challenge because modeling complexity increases with data complexity (Dooley, 1994; Morrison, 1991). One way to represent the complexity of time-series data is algorithmic information content (AIC) (Dooley, 1994). AIC measures the compressibility of a time series (e.g., the series 123123123 could be compressed to repeat 123 three times). An orderly system with a point attractor has minimal AIC—a complete description of the system can be given with a single number. A periodic system following an orderly limit cycle has a somewhat greater AIC defined by the set of points along its deterministic path. A chaotic system has a still larger AIC related to the fractal dimension of its strange attractor. As random noise is added to the system, the data become less compressible. Random behavior has maximum AIC; the shortest description of a random time series is the series itself. Table 7.5 shows the kinds of models that are appropriate depending on AIC (Dooley, 1994).

Systems that yield time series with a high degree of order (linear deterministic) or disorder (random noise) are simple to model. Systems having a moderate degree of disorder can be modeled with linear differential or difference models. Systems with intermediate values of (dis)order—at the edge of chaos—are the most difficult to model. The conclusion: Appropriate modeling of time series generated by complex adaptive systems depends on correctly ascertaining AIC levels. If competitive interorganizational systems exist at the edge of chaos (Kauffman, 1993) or in a low-dimensional state of chaos, then assumptions of standard analytical techniques typically employed in research on such just-embedded systems may be inappropriate to modeling complex nonlinear dynamical systems. Once we know this, however, we have a better idea of what models to apply to understand the dynamics: (a) use stochastic models to explain random processes, (b) use linear deterministic models to explain orderly behavior, (c) use nonlinear dynamic modeling to explain chaotic processes (Morrison, 1991).

Adaptive behavior in human social systems at the edge of chaos also poses major theoretical and conceptual challenges. A diagnosis of chaos imposes the challenge of theorizing about phenomena that

> are neither stable and predictable nor stochastic and random; that unpredictability of behavior does not imply randomness; that [behavior] may be extremely sensitive to

TABLE 7.5

AIC and Time Series Model Types

AIC (Attractor)	Model	Example
Low (point)	Linear deterministic	Linear regression Exponential, sine/cosine Moment matching
Moderate (limit cycle)	Linear differential Linear difference	Fourier analysis Spectral analysis ARIMA linear time series
Intermediate (strange)	Nonlinear differential Nonlinear difference	Lorenz Equations Exponential time series
High (random)	Random noise	Normal distribution Probability density function

different initial conditions ... and that ... learning processes may be much more complex than simple cybernetic mechanisms imply. (Dooley & Van de Ven, 1997, p. 42)

Moreover, by definition, a chaotic system is a deterministic system: Behavior that appears random, in fact, is determined by precise laws. Finding chaos in human social systems is thus going to be challenging to defend. Such findings will have to overcome the common tendency to mistakenly contrast determinism with voluntarism rather than probabilism (Baum, 1996).

ACKNOWLEDGMENTS

We are grateful to Kevin Dooley, Raghu Garud, Peter Karnøe, Andy Van de Ven, and participants at the Multidisciplinary International Workshop on Path Creation and Dependence, Copenhagen Business School, August 1997, for comments on an earlier version of this paper.

Appendix A

USPC 219—Heating, Electric

Sample patent: 4,587,396   Assignee: Laser Industries, Ltd.
Title: Control Apparatus Particularly Useful for Controlling a Laser
This patent covers equipment that facilitates the tracing of a desired pattern with a laser beam. The equipment consists of a device for inputting the desired pattern; memory for storing the pattern; display for displaying the pattern; and a control system for controlling the laser beam.

USPC 235—Registers

Sample patent: 5,420,411   Assignee: Symbol Technologies, Inc.
Title: Combined Range Laser Scanner
This patent covers a laser scanner innovation that allows the reading of bar codes at multiple distances. The scanner uses two laser diode optical illumination systems, each of which is optimally focused for a different distance—one for contact with the bar code and one for longer range.

USPC 356—Optics, Measuring and Testing

Sample patent: 4,591,266   Assignee: Laser Precision Corporation
Title: Parabolic Focusing Apparatus for Optical Spectroscopy
This patent covers an accessory for use in conjunction with a spectrometer. The accessory uses paraboloid reflectors and collimated optical beams to provide accurate measures of focal length. The device facilitates accurate measuring and adjustment of reflectors.

USPC 359—Optics, Systems

Sample patent: 5,455,707   Assignee: Ion Laser Technology
Title: Apparatus and Method for Light Beam Multiplexing
This patent covers method and apparatus for multiplexing a laser beam with optical surfaces such as mirrors. The typical use is to enable the laser beam to switch on or off an optical device through the use of electromagnetism.

USPC 369—Dynamic Information Storage and Retrieval

Sample patent: 4,945,526   Assignee: Laser Magnetic Storage
                                                    International Company
Title: Actuator Assembly for Optical Disk Systems
This patent covers a new construction that enhances the reception of optical beams and the processing of information in optical disk systems.

USPC 372—Coherent Light Generators

Sample patent: 4,504,951   Assignee: American Laser Corporation
Title: High Speed Switching Power Supply for a Light Controlled Laser System
A high-speed switching power supply for a laser system is disclosed. A novel way of exploiting the feedback of the emitted light permits increased loop bandwidth.

USPC 606—Surgery

Sample patent: 4,895,144   Assignee: Surgical Laser
                                                    Technologies, Inc.
Title: Supply System for Sterile Fluids and Gases in Laser Surgery
This patent covers a fluid supply system for gas and sterile liquids in which a disposable cartridge is connected to a pump so that the pump never contacts the liquid.

Appendix B

METHODS FOR DETECTING CHAOS AND COMPLEXITY

Lyapunov Exponent: Sensitivity to Initial Conditions

Complex systems exhibit a sensitive dependence on initial conditions, as first pointed out by Lorenz (1963). While running computer simulations on the set of nonlinear equations previously introduced, Lorenz took a short-cut on data entry while attempting to replicate a previous simulation by rounding off initial parameters from four decimal places to two. He was surprised to find that the time paths of the equations diverged exponentially over time. He concluded that the sensitivity to initial conditions

> ... implies that two states differing by imperceptible amounts may eventually evolve into two considerably different states. If, then, there is any error whatever in observing the present state—and in any real system such errors seem inevitable—an acceptable prediction of an instantaneous state in the distant future may well be impossible. (Lorenz, 1963, p. 133)

He termed the generator of sensitivity to initial conditions *the butterfly effect.*

For a dynamic system to be chaotic, it must have a large set of initial conditions that are highly unstable. No matter how precisely initial conditions are measured, predictions of subsequent behavior become radically wrong after a short time. It is thus considered proof of the presence of chaos if sensitivity of the system under analysis to initial conditions can be demonstrated (Wolf, Swift, Swinney, & Vastano, 1985). The standard method of identifying this attribute is by the calculation of the *Lyapunov exponent.* This exponent measures the rate at which neighboring points on the attractor diverge as they are moved forward in time. Roughly speaking the (maximal) Lyapunov exponent is the time average logarithmic growth rate of the distance between two neighboring points, that is, the distance between two points grows as exponent $(t*1)$, where 1 is the exponent. The trajectories on the attractor are embedded in a multidimensional space, and so the divergence is represented as the difference between 2 n-tuples. There are as many Lyapunov exponents as there are dimensions in the state space of the system, but the largest is usually the most important. The dominant average Lyapunov exponent, L, is defined as:

$$L = \log_2 \{[\Sigma^l_{n\text{-}1} (n + 1/l_n)]/n - 1\}$$

where n indexes the samples and l is the Euclidean distance between a trajectory and its nearest neighbor. With very large amounts of data, the density of points in a region of the attractor permit calculation of local Lyapunov exponents—since the rate of divergence is not necessarily constant over the whole attractor, localized measurements can be made. In practice, data limitations typically only permit measurement of the first exponent. Positive Lyapunov exponents, which indicate sensitivity to initial conditions, are considered evidence either of chaos, because the distance grows (on average in time and locally in phase space) exponentially over time, or randomness. Negative exponents indicate mean reverting behavior, and the value zero is characteristic of cyclic behavior.

Correlation Dimension: Boundedness

Given a positive Lyapunov exponent, a second measure of chaos, the correlation dimension, can be used to distinguish chaotic from random time series (Dooley & Van de Ven, 1997). It is a measure of spatial correlation of scatter points in d-dimensional space. The correlation dimension indicates the attractor's dimension. For example, a point has $d = 0$; a line has $d = 1$, a plane has $d = 2$, and a cube has $d = 3$. The point is to determine whether the attractor for a seemingly random time series has low or high d. A major difference between chaotic and random processes is that, although both appear random to the eye and to standard linear time series methods, chaotic behavior stretches and folds within a low-dimensional space defined by its attractor, whereas random behavior is unbounded and completely fills the phase space. Thus, for a random time series, d will be slightly less than or equal to the embedding dimension.

For periodic or chaotic time series, however, the attractor has a finite dimension. If such a series is embedded in a dimension less than its true dimension, it will appear random, and thus have a d near the embedding dimension. If the series is embedded in a dimension equal to or greater than its true dimension, it will have d near its true dimensionality. For example, a time series for which the true dimension is 4.5 would exhibit d equal to 3 and 4 when plotted in embedding dimensions 3 and 4, and 4.5 for embedding dimensions 5 and higher (Dooley & Van de Ven, 1997).

Thus, in general, random processes will have high d and chaotic processes will have low d, "substantially lower than 10, perhaps 5 or 6" (Brock, Hsieh & LeBaron, 1991, p. 17). Hence identification of chaotic behavior involves testing whether or not a seemingly random time series trajectory exhibits low d. Grassberger and Procaccia (1983) provided an effective method for determining the correlation dimension of a time series.

Surrogate Testing: Empirical evaluation of the presence of chaos

The validity of the conclusion that a time series is chaotic based on its Lyapunov exponent and correlation dimension can be evaluated empirically using surrogate testing (Theiler et al., 1992). In surrogate testing, additional time series are created that have similar properties to the time series in question. To test the null hypothesis that the time series is linear (periodic), phase randomization is used to generate surrogate time series (Kaplan & Glass, 1995). A phase-randomized surrogate time series has the same length, distribution, and linear dynamical structure as the original. The only aspect altered is the nonlinear structure of the data. If a time series has low d because it is periodic in nature, then surrogates will also have low d. If however, the time series has low d because of nonlinearity, then surrogate series will have d slightly less than or equal to the embedding dimension because phase randomization will have removed the low dimensional nonlinear structure.

REFERENCES

Arthur, B. (1988). Self-reinforcing mechanisms in economics. In P. W. Anderson, K. J. Arrow, & D. Pines (Eds.), *The economy as an evolving complex system* (pp. 9–31). Redwood City, CA: Addison-Wesley.

Arthur, B. (1989). Competing technologies, increasing returns, and lock-in by historical events. *Economic Journal*, 99, 116–131.

Anderson, P. W., Arrow, K. J., & Pines, D. (Eds.). (1988). *The economy as an evolving complex system*. Redwood City, CA: Addison-Wesley.

Bak, P., & Chen, K. (1991). Self-organized criticality. *Scientific American*, 264(1), 26–33.

Baum, J. A. C. (1996). Organizational ecology. In S. Clegg, C. Hardy, & W. Nord (Eds.), *Handbook of organization studies* (pp. 77–114). London: Sage.

Baum, J. A. C. (1999). Whole-part coevolutionary competition in organizations. In J. A. C. Baum & B. McKelvey (Eds.), *Variations in organization science: In honor of Donald T. Campbell* (pp. 113–135). Thousand Oaks CA: Sage.

Baum, J. A. C., & Silverman, B. S. (1999). Complexity in the dynamics of organizational founding and failure. In M. Lissack & H. Gunz (Eds.), *Managing complexity in organizations* (pp. 292–312). New York: Quorum Books.

Begley, R. F. (1980). (Statement to U.S. Senate, in U.S. Congress). *Laser technology—Development and applications: Hearings before the subcommittee on science, technology, and space* [Serial No. 96–106]. Washington DC: U.S. Government Printing Office.

Bertolotti, M. (1983). *Masers and lasers: An historical approach*. Bristol, England: Adam Hilger.

Brock, W. A., Hsieh, D. A., LeBargh, B. (1991). *Nonlinear dynamics, chaos, and instability: Statistical theory and economic evidence*. Cambridge, MA: MIT Press.

Brock, W. A., & Malliaris, A. G. (1989). *Differential equations, stability and chaos in dynamic economics*. Amsterdam: North Holland.

Bromberg, J. L. (1991). *The laser in America, 1950–1970*. Cambridge, MA: MIT Press.

Brown, S. L., & Eisenhardt, K. M. (1998). *Competing on the edge: Strategy as structured chaos.* Boston: Harvard Business School Press.

Brush, J. (1996). *NLD toolbox user's guide (Version 2.0).*

Chandler, A. D. (1990). The enduring logic of industrial success. *Harvard Business Review, (March–April), 68,* 130–140.

Cheng, Y.-T., & Van de Ven, A. H. (1996). Learning the innovation journey: Order out of chaos? *Organization Science, 7,* 593–614.

D'Aveni, R. (1994). *Hypercompetition.* New York: Free Press.

Dooley, K. (1994). Complexity in time series modeling. *Society for Chaos Theory in Psychology and the Life Sciences, 2,* 1–3.

Dooley, K., & Van de Ven, A. H. (1997). A primer on diagnosing dynamic organizational processes (Working paper). Arizona State University, Phoenix, AZ and University of Minnesota, Minneapolis, MN.

Dosi, G. (1982). Technological paradigms and technological trajectories. *Research Policy, 11,* 147–162.

Dosi, G. (1988). The nature of the innovation process. In G. Dosi, C. Freeman, R. Nelson, G. Silverberg, & L Soete (Eds.), *Technical change and economic theory* (pp. 221–238). London: Pinter.

Garud, R., & Jain, S. (1996). The embeddedness of technological systems. In J. Baum & J. Dutton (Eds.), *Advances in strategic management, 13* (pp. 389–408). Greenwich CT: JAI Press.

Ginsberg, A., Larsen, E., & Lomi, A. (1996). Generating strategy from individual behavior: A dynamic model of structural embeddedness. In J. Baum & J. Dutton (Eds.), *Advances in Strategic Management, 13* (pp. 121–148). Greenwich, CT: JAI Press.

Gresov, C., Haveman, H. A., & Oliva, T. A. (1993). Organizational design, inertia, and the dynamics of competitive response. *Organization Science, 4,* 181–208.

Griliches, Z. (1990). Patent statistics as economic indicators: A survey. *Journal of Economic Literature, 28*(4), 1661–1707.

Griliches, Z., Pakes, A., & Hall, B. (1987). The value of patents as indicators of inventive activity. In P. Dasgupta & P. Stoneman (Eds.), *Economic policy and technological performance.* New York: Cambridge University Press.

Hamel, G., & Prahalad, C. K. (1994). *Competing for the future.* Boston, MA: Harvard Business School Press.

Hecht, J. (1992). *Laser Pioneers (Rev. ed.).* Boston, MA: Academic Press.

Hecht, J., & Teresi, D. (1982). *Laser: Supertool of the 1980s.* New York: Ticknor & Fields.

Ion-laser work featured at CLEA. (1975, July). *Laser focus,* p. 10.

Irvine, J., & Martin, B. R. (1983). Assessing basic research: Some partial indicators of scientific progress in radio astronomy. *Research Policy, 12*(2), 61–90.

Karnøe, P., & Garud, R. (1997). Path creation and dependence in the Danish wind turbine field. In J. Porac & M. Ventresca (Eds.), *The social construction of industries and markets.* New York: Permagon Press.

Kauffman, S. A. (1991). Antichaos and adaptation. *Scientific American, 265*(2), 64–70.

Kauffman, S. A. (1993). *Origins of order: Self-organization and selection in evolution.* Oxford: Oxford University Press.

Kaplan, D., & Glass, L. (1995). *Understanding nonlinear dynamics.* New York: Springer-Verlag.

Kellert, S. H. (1993). *In the wake of chaos: Unpredictable order in dynamical systems.* Chicago: University of Chicago Press.

Langton, C. G., Farmer, J. D., Rasmussen, S., & Taylor, C. (1992). *Artificial life II: A proceedings volume in the Sant Fe Institute,* Vol. 10. Reading, MA: Addison-Wesley.

Levin, R. C., Klevorick, A. K., Nelson, R. R., & Winter, S. G. (1987). Appropriating the returns from industrial research and development. *Brookings Papers on Economic Activity, 3,* 783–833.

Levinthal, D. A. (1997). Adaptation on rugged landscapes. *Management Science, 43,* 934–950.

Levinthal, D. A., & March, J. G. (1993). The myopia of learning. *Strategic Management Journal, 14,* 94–112.

Levy, D. (1994). Chaos theory and strategy: Theory, application and managerial implications [Summer special issue]. *Strategic Management Journal, 15,* 167–178.

Lomi, A., & Larsen, E. (1999). Evolutionary models of local interaction. In J. A. C. Baum & B. McKelvey (Eds.), *Variations in organization science: In honor of Donald T. Campbell*, pp. 255–278. Thousand Oaks, CA: Sage.

Lorenz, E. (1963). Deterministic nonperiodic flow. *Journal of the Atmospheric Sciences, 20*, 130–141.

Mansfield, E., Schwartz, M., & Wagner, S. (1981). Imitation costs and patents: An empirical study. *Economic Journal, 91*, 907–918.

March, J. G. (1991). Exploration and exploitation in organizational learning. *Organization Science, 2*, 71–87.

McKelvey, B. (1998). Complexity vs. selection among coevolutionary microstates in firms: Complexity effects on strategic organizing. *Comportamento Organizacional E Gestão, 4*, 17–59.

Miner, A. S., & Haunschild, P. R. (1995). Population-level learning. In B. Staw & L. Cummings (Eds.), *Research in organizational behavior, 17* (pp. 115–166). Greenwich CT: JAI Press.

Miller, D. (1986). Configurations of strategy and structure: Towards a synthesis. *Strategic Management Journal, 7*, 233–249.

Miller, D., & Friesen, P. H. (1984). *Organizations: A quantum view*. Englewood Cliffs, NJ: Prentice-Hall.

Moore, J. F. (1993). Predators and prey: A new ecology of competition. *Harvard Business Review (May–June), 71*, 75–86.

Morgan, G. (1986). *Images of organization*. Thousand Oaks, CA: Sage.

Morrison, F. (1991). *The art of modeling dynamic systems*. New York: Wiley.

Nelson, R.R. (1994). The co-evolution of technology, industrial structure, and supporting institutions. *Industrial and Corporate Change, 3*, 65–110.

Newsweek. (1992). *119*(21), New York. May 25.

Nonaka, I. (1988). Creating order out of chaos: Self-renewal in Japanese firms. *California Management Review, 30*, 57–73.

Patel, P., & Pavitt, K. (1994). Technological competencies in the world's largest firms: Characteristics, constraints and scope for managerial choice (Working paper). Science Policy Research Unit, University of Sussex, England.

Pavitt, K. (1982). R&D, patenting, and innovative activities. *Research Policy, 11*, 33–51.

Peters, T. (1988). *Thriving on chaos*. New York: Macmillan.

Phelan, S. E. (1995, August). From chaos to complexity in strategic planning. Paper presented at the 55th Annual Meeting of the Academy of Management, Vancouver, British Columbia, Canada.

Powell, W. W., & Brantley, P. (1992). Competitive cooperation in biotechnology: Learning through networks? In N. Nohria & R.G. Eccles (Eds.), *Networks and organizations: Structure, form and action* (pp. 366–394). Boston: HBS Press.

Powell, W. W., Koput, K. W., & Smith-Doerr, L. (1996). Interorganizational collaboration and the locus of innovation: Networks of learning in biotechnology. *Administrative Science Quarterly, 41*, 116–145.

Prigogine I., & Stengers, I. (1984). *Order out of chaos: Man's new dialogue with nature*. New York: Bantam.

Rather, J. (1980). (Statement to U.S. Senate, in U.S. Congress). *Laser technology—Development and applications: Hearings before the subcommittee on science, technology, and space* (Serial No. 96–106). Washington DC: U.S. Government Printing Office.

Ruelle D. (1981). Small random perturbations of dynamical systems and the definition of attractors. *Communication Mathematical Physics, 137*, 82.

Sahal, D. (1985). Technological guideposts and innovation avenues. *Research Policy, 14*, 61–82.

Schawlow, A. L. & Townes, C. H. (1958). Infrared and optical masers. *Physical Review, 112*, 1940–1949.

Schmookler, J. (1962). Changes in industry and in the state of knowledge as determinants of industrial invention. In R.R. Nelson (Ed.), *The rate and direction of inventive activity*. Princeton: Princeton University Press.

Schumpeter, J. (1934). *The theory of economic development*. Cambridge MA: Harvard University Press.

Silverman, B. S. (1996). *Technological assets and the logic of corporate diversification*. Unpublished doctoral dissertation, Haas School of Business, University of California, Berkeley.

Sprott, J. C., & Rowlands, G. (1995). *Chaos data analyzer: The professional version*. New York: American Institute of Physics.

Sorenson, O. (1996, November). The complexity catastrophe and the evolution of the computer workstation industry. Paper presented at the INFORMS College on Organization Science, Atlanta GA.

Stacey, R. D. (1995). *Strategic management and organizational dynamics*. Marshfield, MA: Pitman Publishing.

Stacey, R. D. (1996). *Complexity and creativity in organizations*. San Francisco, CA : Berrett-Koehler.

Takens, F. (1981). Detecting strange attractors in fluid turbulence. In D.A. Rand & L.S. Young (Eds.), *Dynamical systems and turbulence* (pp. 368–381). Berlin: Springer-Verlag.

Theiler, J., Eubank, S., Longtin, A., Galdrikian, B., & Farmer, J. D. (1992). Testing for nonlinearity in time series: The method of surrogate data. *Physica D, 58,* 77–94.

Thietart, R. A., & Forgues, B. (1995). Chaos and organization theory. *Organization Science, 5,* 19–31.

Tong, H. (1990). *Non-linear time series—A dynamical system approach*. New York: Oxford University Press.

Trajtenberg, M. (1990). A penny for your quotes: Patent citations and the value of innovations. *Rand Journal of Economics, 21*(1), 172–187.

Tushman, M. L., & Anderson, P. (1986). Technological discontinuities and organizational environments. *Administrative Science Quarterly, 31,* 439–465.

U.S. Congress. (1980). *Laser technology—Development and applications: Hearings before the subcommittee on science, technology, and space* (Serial No. 96–106). Washington DC: U.S. Government Printing Office.

Utterback, J. (1994). *Mastering the dynamics of innovation*. Boston: HBS Press.

Vinten, G. (1992). Thriving on chaos: The route to management survival. *Management Decision, 30,* 22–28.

Waldrop, M. M. (1992). *Complexity: The emerging science at the edge of order and chaos*. London: Penguin Books.

Wolf, A., Swift, J. B., Swinney, H. L., & Vastano, J. A. (1985). Determining Lyapunov exponents from a time series. *Physica D, 16,* 285–317.

III

Path Creation
as Coevolution

8

America's Family Vehicle: Path Creation in the U.S. Minivan Market

Joseph F. Porac
Emory University

José Antonio Rosa
Case Western Reserve University

Jelena Spanjol
University of Illinois at Urbana–Champaign

Michael Scott Saxon
Jupiter Communications, New York City

Path creation often involves the development of new product markets and thus the concept of a *market* is fundamental to the study of evolutionary processes in organizational communities. When a new market is established by novel and previously unknown products, social, cognitive, and economic processes are set in place that often culminate in buyers and sellers interacting around a dominant product design that establishes an ideal point of reference for products in the newly created market space (e.g., Utterback, 1996). When looked at in this light, path creation and market creation are often synonyms for the same underlying dynamic processes that are involved in institutionalizing a transactional space within which product attributes, standards of evaluation, quality orderings, product identities, and terms of trade are widely known and taken for granted. But, what is the nature of these processes, and how do they combine to create markets in real time?

The notion of product market is so fundamental to management theory that it is easy to overlook the fact that this notion is nothing more than a theoretical construct that has been developed to make sense of the flow of activity in competitive business environments. As such, the definition of a product market is subject to theoretical interpretation and debate. By far, the most frequently used definition of product market goes back to Robinson (1933), who saw *product markets* as arenas for the exchange of goods that are close substitutes for one another. By *substitute* it is meant that the good can be used for the same purposes. To the extent that goods are close substitutes, they are in the same product market. Market boundaries are established by gaps in demand such that products within the same market have higher crosselasticities of demand than goods that are in different markets (e.g., Auerbach, 1988).

The *gap in substitutes* definition of markets, however, is incomplete because it largely begs the question of substitution in use—that product substitutability is highly responsive to the usage contexts in which purchase decisions are made (e.g., Day, Shocker, & Srivastava, 1979). Consider the *luxury car* product category. Members of this category are substitutes only because they share attributes that are considered important to the usage definition of a luxury car. To consumers, luxury cars must have luxury appointments, comfortable seats, and powerful engines for the car to perform as luxury cars should. Products that do not meet these criteria are not luxury cars; but much is taken for granted in this definition.

The first and most obvious assumption is that we know the relationship between usage conditions and the key attributes that classify a car as a luxury vehicle, as well as the attributes that are unimportant. How do we know this? Where does this implicit knowledge come from? The second, less obvious, assumption is that we know the aspects of the car that are attributes. A "car" is a stream of information that confronts our senses with varying levels of equivocality. We parse the stream into bits that we can classify and compare. We agree on what it means to say comfortable seat and powerful engine because we know, based on our interactions with others who buy, sell, use, and write about cars, the combination of information that constitutes a "comfortable seat," and the combination that implies a "powerful engine." Abstracting these attributes from our experiences, however, is subject to each person's idiosyncratic interpretations. How is it that these very private experiences collate into taken-for-granted and widely shared attribute nomenclatures?

To be a complete approach to product markets, the claim that a market exists around close substitutes must also explain how the cognitive orderings that support substitutability arise. Such a theory of markets, unfortunately, does not yet exist. Because we don't have a good theory of markets, we cannot answer some fundamental questions: (a) How and when is a market created? (b) What determines how an infinite array of stimulus information is summarized into the constellation of attributes that define structure, and give meaning to, a product market? (c) How do specific artifacts and behaviors come to be associated with consensually understood market categories? (d) How do product categories evolve and change? and (e) How and why do product categories die?

It is our contention that these questions are answerable only when one accepts that markets are fundamentally sociocognitive in nature (Garud and Rappa, 1994). By this we mean that markets have their roots in consensual knowledge structures that define the artifacts being exchanged and coordinate transactional relationships within market networks. A market exists when a transactional network of producers and buyers develops around a set of artifacts that are subsumed under a particular identity class and that are deemed similar by virtue of their understood membership in this category. We suggest that this sociocognitive conceptualization of markets provides a robust set of constructs and frames for answering theoretical questions that have not been easily directed. *How and when is a market created?* Markets are created when potential buyers and sellers begin to connect an artifact with shared conceptual systems that define the attributes, uses, and value of the artifact in question. *What determines how an artifact is summarized and abstracted in a set of attributes?* Attribute abstraction occurs through conversations and narratives across pro'ucers and consumers. *How do specific artifacts and behaviors become associated with consensually understood market categories?* Such associations occur when knowledge structures create equivalence classes that form the basis for similarity judgments on both the demand and supply side of market transactions. *How do market categories evolve and change?* Change occurs in the knowledge structures around which markets cohere such that new attributes become associated with existing artifacts or new artifacts become assimilated into existing structures. *How and why do product categories die?* Categories die when a market's underlying knowledge structures no longer cohere in a meaningful and profitable way.

The purpose of our chapter is to "flesh out" the rudiments of a sociocognitive approach to market creation and evolution by giving explicit attention to the cognitive dynamics of markets over time. Our goal is not to provide a complete sociocognitive theory of markets. Rather, it is to sketch the broad domain of a sociocognitive approach and to show how viewing markets as conceptual systems can address heretofore intractable questions about market dynamics.

To concretize our arguments, we examine the details of an automotive product category that has been labeled as revolutionary by many experts over the course of the last decade: minivans. These vehicles are designed and built for the express purpose of transporting several people, especially

families, in comfort and have a long history in the global automotive industry. In the United States, these "people movers" did not become popular,
however, until Chrysler introduced its line of small front-wheel drive vans in
1984. Chrysler's success crystallized what had been a nascent and uncoordi-
nated market for such vehicles by stimulating a new nomenclature for describing the "minivan's" place in the world. This crystallization fueled the
introduction of imitators bent on capturing customers in this emerging
product category, and the evolution of the minivan market over the last 10
years stands as an excellent example of product competition in a differentiated characteristics space. In our analysis, we delve below the surface level
of this competition by examining the conceptual system that defined the
characteristics space itself.

In doing so, we borrow from and extend previous scholarly work in a
number of different disciplines. First and foremost, our sociocognitive analysis of markets draws from research on the social construction of technological systems (Bijker, Hughes, & Pinch, 1987; Garud & Rappa, 1994)
Constructionists have been instrumental in problematizing the materiality
of artifacts by showing how such artifacts are inextricably intertwined with
cultural and technological meaning systems. Our analysis of the minivan
market is fundamentally constructionist in nature, because we conceptualize the evolution of minivan artifacts in sociocognitive terms. We go beyond
previous constructionist research, however, by being explicit about the
knowledge structures that underlie markets, and by linking such structures
to transactional relationships between buyers and sellers. Research on the
social construction of technological systems has not been well integrated
into business strategy and management research. Our attempt to extend
constructionist principles to market transactions will hopefully be one step
toward this integration.

Secondly, our chapter builds on White's (1981, 1992) seminal work on
market networks. In extending network analysis to market contexts, White
has cogently argued that the network interface between buyers and sellers is
equivocal. This equivocality contrasts markedly with assumptions in economics concerning the transparency of supply and demand schedules, and
immediately raises the question of how a market interface can be sustained
over time when buyers and sellers have to work so hard to understand each
other. In this way, White's analysis of the market interface opens the door for
our investigation of product conceptual systems because, we will argue, it is
through such systems that a market interface is sustained over time.

Third, our analysis recognizes and derives inspiration from the work of
Abernathy and Utterback (1978), Abernathy and Clark (1985), Tushman
and Anderson (1986) and others who have articulated a theory of dominant
technological designs. Dominant design theorists have suggested that tech-

nologies evolve through a period of initial ferment, in which basic design parameters are contested via marketplace competition, followed by a subsequent period of standardization around a winning dominant design. Our analysis of the minivan product category tends to substantiate this sequence, and fleshes out some of the microprocesses that explain how design dominance was achieved through the evolution of cognitive structures linking buyers and sellers in this market space. Moreover, by revealing these microprocesses, we advance certain conjectures about how standardization in the minivan category has led to a destabilization of designs across the entire automotive industry.

Fourth, our chapter is informed by the growing recognition in management research that organizations and markets are fundamentally knowledge-based, and that economic competition is a contest over ideas, not tangible resources (e.g., Spender, 1996; Hamel & Prahalad, 1994; Nonaka & Takeuchi, 1994). Although there has been excellent research on knowledge creation and change within organizations (e.g., Nonaka & Takeuchi, 1994), empirical research on knowledge creation and change in market contexts has been scarce. Part of the problem has been the lack of a framework and approach for describing the knowledge structures that underlie market behavior. The ultimate goal of our research is to fill this gap in the organizations and strategy literatures.

Finally, our work draws inspiration from the literature on path dependence that has emerged in the social sciences within the past 10 years (e.g., Arthur, 1989; David, 1985; Liebowitz & Margolis, 1995). As originally conceived by Arthur (1989), *path dependence* is a concept that describes how small random variations in starting conditions can lead to market inefficiencies through a process of self-reinforcing lock-in around an emerging product standard. As Liebowitz & Margolis (1995) noted, however, the recent popularity of the concept has driven its meaning far beyond Arthur's strictly technical definition to make it a general claim that "history matters" in the dynamics of markets. Our story of the minivan is a case about market learning, about a battle between opposing conceptions of what a minivan should be. Within this general market dynamic, producers and consumers were materially and cognitively constrained by historical trajectories that influenced both the costs and benefits that accrued from their various market choices as well as their interpretations of market events and opportunities. In the case of minivans, we are hard pressed to find the sorts of market suboptimalities suggested by Arthur's (1989) technical definition of path dependence. Instead, the case seems more an example of what Liebowitz and Margolis (1995) termed *first degree path dependence*, the general sensitivity of present choice parameters on choices that have been made in the past. In this very general sense, market learning in the minivan case was guided by what producers and consumers already knew about motor vehicle design and usage conditions.

However, path dependence is not the most important story to tell about the evolution of the minivan market in the United States. Rather, this market reveals important combinatorial dynamics that demonstrate how producers and consumers together can create a vehicle design that goes beyond what is already known and taken for granted. The important lesson to be learned from the minivan is one of path creation rather than path dependence. This is suggested not only in the inherent revolutionary aspects of the minivan itself, but in its destablizing and creative effect on the automobile industry as a whole. It is this story that we wish to tell in this chapter.

IDENTITIES AND SENSEMAKING IN MARKET NETWORKS

Key Terms

Several economic sociologists have conceptualized *markets* as networks of individual actors, bound together in equivocal transactions that are stabilized by shared assumptions and frames-of-reference (e.g., Fligstein, 1996; White, 1981, 1992). Although still incomplete, this work is a useful starting point for a theory of how meanings become attached to artifacts and get intertwined with market transactions. A completely disaggregated view of a market as a network of transactions suggests that a market system is a lattice of dyadic ties. To begin our analysis of how this transactional network becomes part of a conceptual system, assume the extreme case where this grid of ties is all that exists. Information only flows through word of mouth and direct contacts across the market grid. Employees or customers of a firm can obtain information about possible exchange partners only through others with whom they have direct contact. There is no medium of information exchange and transfer other than direct face-to-face conversation.

To the extent that this network grid can be subdivided on the basis of two identity constructs—a "producer" community and a "buyer" community—we can label the network a *market system*, which evolves around a physical stimulus configuration called an *artifact* (see Fig. 8.1). In this case, *artifact* refers to the tangible manifestation of a set of informational cues attached to a definition that is commonly understood and taken for granted. Artifact does not refer to some stable "thing out there" but to "here and now" flows of stimulus information that impact the senses of market actors.

Artifacts of one type can be distinguished from artifacts of another type to the extent that there are "gaps of attributes" between them—that is, differences between the conceptual clustering of attributes that are commonly understood to represent the artifacts. In our view, attributes don't exist on

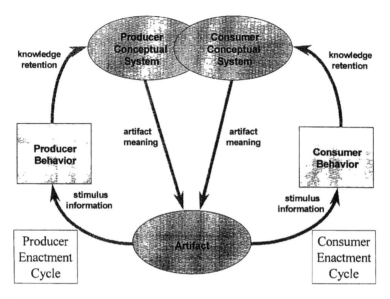

FIG. 8.1.   A sociocognitive market system.

their own "out there" either. They are inductively derived via observation and interaction with artifacts. To be sure, the use and observation of artifacts are idiosyncratic and dependent on one's vantage point and observational goals. Over time, though, as social interactions between producers and consumers take place, an explicit attribute nomenclature evolves to capture consensually understood aspects of the stimulus array. When this happens, we say that the market has stabilized, and that the nomenclatures and their associated cognitive structures become the *product conceptual system* that is at least partially shared by consumers and producers (see Fig 8.1).

Product conceptual systems can be thought of as conceptual frames composed of concept nodes and linkages between them. Figure 8.2, for example, depicts a portion of a conceptual system for the concept of *automobile*. As Fig. 8.2 suggests, automobiles are often described in terms of their uses, components, performance characteristics, and brand names. When artifacts become embedded in their own product conceptual systems, they can be called *members of the product category*. For example, minivans have come to be defined as an array of such attributes as "car-like handling," "front-wheel drive," "low step in height," "passenger side air bag," and "cargo hauling space large enough for a 4" x 8" piece of plywood," And, whereas the Dodge Caravan is a member of this product category, the Volkswagon Vanagon is not. Likewise, wines have come to be associated with "buttery textures,"

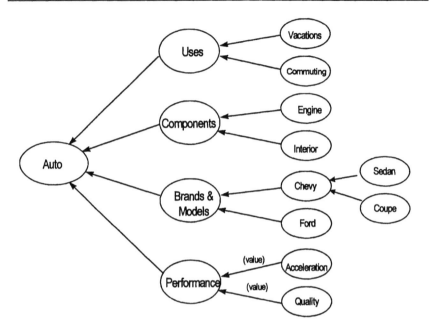

FIG. 8.2.   A partial conceptual system for the automobile.

"pear notes," and "legs," and although a Rothchild Cabernet is a member of the category, a Seagrams Passion Fruit Cooler is not. Product conceptual systems have significant cultural effects. As Bourdieu (1984) noted, understanding a product nomenclature and being skilled at it is an important factor in being considered educated and having expertise (on the producer side) and taste (on the buyer side) in the product domain.

A *producer community* consists of all market actors who are engaged in producing artifacts that qualify as members of a product category. Producer communities are often called *industries*. A *buyer community* consists of all those actors who acquire artifacts that are members of a product category for use in some aspect of their lives. Buyer communities are often subdivided into "segments" based on various tangible and intangible characteristics and how they are reflected in their choices of artifacts within the categories, or in their choices across categories. Producer and buyer communities come together when their transactions are embedded within shared knowledge about the artifact that is being exchanged.

Artifacts vary in the effort and investment their design, assembly, and distribution require. This effort is usually distributed across multiple actors, because one actor is seldom sufficient to accomplish the complete task of artifact construction. The activities that are required to design and assemble a plausible artifact may be performed within the boundaries of a firm or they may be distributed across firms. At some point, though, a transaction from the producing to the consuming side of the network structure must take place. This transfer is a market transaction and occurs across what White (1992) termed a *market interface*. In White's view, the basic characteristic of a market interface, as opposed to other forms of network organization, is an asymmetry of flow. Producers are handing off the results of their activities to buyers. That is, artifacts move in one direction, while fungible currency moves in the other direction. Both sides of the interface, however, are involved in defining the nature and quality of the product that is being exchanged. Thus, both are making comparability judgments and creating status orderings, although on the producer side the status orderings are driven by business demands for revenues and profits, whereas on the consumer side they are driven by demands for more personally defined benefits.

Markets, Equivocality, and Meaning

Market interfaces are defined as interstices between producing and consuming communities of some equivocality. The equivocality exists because producers and consumers bring distinct frames of references to the transaction, given that their behavioral backgrounds are quite different. Of course, the definition of a market transaction is relative to the artifact involved. We can discuss a market transaction involving a minivan, for example, but this transaction itself is the endpoint of a long series of other market transactions involving various components of the artifact that we call a minivan (e.g., seat assemblies, brake systems, lights, engines, etc.). Each component is embedded within its own transactional network and conceptual system, and is thus a product in its own right. White (1992) called the interconnectedness of transactions and conceptual systems the *self-similarity of social networks*. Single nodes at one level of aggregation are composed of network nodes at more basic levels. Market interfaces are combinations of self-similar transactional networks. Because of the equivocality and complexity involved in self-similar networks, a great deal of conceptual abstraction must take place in order for both producers and consumers to summarize and make sense of product and market dynamics. These abstractions eventually evolve into product conceptual systems by becoming taken for granted and commonly understood.

Conceptual systems play a constitutive role in markets by stabilizing the definition of the products and services that are being exchanged. In this way, product conceptual systems allow for the transactional network to be extended across time and space. Stabilization increases temporal extent by making it possible for both producer and buyer communities to anticipate and regulate product related decisions. Stable product definitions allow producers to plan investments in plant and equipment, as well as any number of product upgrades and extensions. Stable product definitions allow buyers to plan their purchases by ordering and labeling the basket of consumable goods at their disposal. Unstable product definitions are costly because they make it very difficult to plan production and consumption. Thus, unstable product definitions, especially when the instability hints at manipulation and self-interest by market actors, usually trigger hostile market reactions.

Temporal stabilization in turn increases spatial extent by making it possible for product information to remain valid as it moves through the transactional network. If one consumer enthusiastically recommends the purchase of a minivan to another, the validity and reliability of this recommendation depend upon the existence of a stable minivan concept that is at least somewhat overlapping with the definition held by the producers responsible for delivering the next instantiation of this concept. Without this validity and reliability, product recommendations from others would have a high probability of being fallacious in the sense that the recommended stimulus configuration would be only stochastically available. Definitional reliability also permits the formation of buyer communities whose actors are bound together by their understanding and appreciation of the nuances of stable product definitions.

Increased spatial extent also helps producers to extend themselves beyond their immediate local network to build a greater volume of business. Word of their product can spread throughout the broader buyer and producer communities. Buyers not in direct contact with the producer across the interface can move to the interface and enter into an exchange with some confidence. In addition, other producers or suppliers can pick up product related information through network contacts and plan their businesses accordingly. They may, for example, enter alternative stimulus configurations into jurisdictional contests with existing artifacts by claiming equivalence or superiority in their base attributes. The goal of such claims would be to persuade consumers on the other side of the market interface to accept this new configuration as a more plausible and agreeable rendering of an implicit product concept. These jurisdictional contests or "market share battles" are at the root of what is commonly understood to be product market competition. It is important to realize, however, that these battles assume a coherent and definable battlefield. Temporally and spatially stable product definitions create the ground on which a well-organized battle can take place.

Stable product conceptual systems are combinations of understood attributes and usage conditions shared across the market divide. A stable product market exists to the extent that there is an equilibrium consensus in core attributes and uses for artifacts considered to be members of the market. By equilibrium, we mean Hahn's (1973) definition: an economic system "is in equilibrium when it generates messages which do not cause agents to change the theories which they hold or the policies which they pursue." (p. 25). To paraphrase Hahn in the present context, a product market exists to the extent that the activities and flows within the market network are generating messages to producers and buyers that do not cause either community to substantially alter the core attribute/usage conceptual structure that both hold consensually. There may be radical changes in other subsets of producer and consumer conceptual systems (e.g., consumers using minivans as retirement vehicles, producers using plastic composites for minivan body panels), but if the core attribute/usage conceptual models are stabilized, a stable market can be said to exist.

In short, product conceptual systems are the "glue" that keeps market interfaces and their attendant producer and buyer communities connected in temporally and spatially distributed networks. The stability of market conceptual systems gives cognitive substance to White's (1992) view of the market interface. White suggested that networks cohere around identities (e.g., persons, things, or events). Product identities are fundamental to the coherence and stability of market transactional networks. The commitments binding producers with buyers have to do with a shared conception of core attributes and usages. It is this consensus that allows producers to array themselves against one another in their efforts to convince consumers that the artifact that they have produced is better than the artifacts of other producers. Similarly, this consensual conceptual structure also allows consumers to evaluate the various entrants on similar grounds and to accept or reject them with the confidence that comes from knowing the criteria on which the entrants should be judged. Markets cohere when this confidence exists on both sides, making dual commitments reasonable and prudent. This confidence, in turn, sets the foundational conditions for other market phenomena such as product market boundaries, market share battles, market research, first-mover advantage, market entry timing decisions, and product life cycles.

Market Stories and Market Dynamics

The conceptual structures behind product markets are created and shared among market actors by means of stories. Stories are critical sensemaking tools among participants in a social system (e.g., Weick, 1993; White, 1992). In product market dialogues between producers and consumers, stories establish and explain the tie between artifacts, their understood attributes, and their usage conditions. Some product market stories begin when actors within buyer and producer networks (marketers, dealers, journalists) experience new physical artifacts and summarize these experiences through dialogue with other individuals in the network. These stories are then widely broadcast through detailed descriptions such as product brochures and product reviews. When we read a product review of a minivan, for example, we collect information not only about its physical attributes, such as "three rows of seats," and "sliding doors," but also the appropriate usage conditions of a minivan, such as "family transportation," and "towing." We also collect images attached to minivans, such as "boring," or "practical." These stories help us relate to the artifact by providing a context in which to set them.

Other stories begin when actors attempt to summarize the aggregate dynamics of the market via direct and indirect observation. Thus, for example, market analysts collect information about purchase patterns and consumer preferences, and summarize this information through stories about market segments, perceptual maps, and supply and demand curves. These stories typically diffuse throughout the producer and buyer communities and help create collective beliefs about existing product market boundaries and status orderings within these boundaries. We hear arguments, for instance, that new minivan models are taking market share from older models, and listen to explanations for this effect such as the presence or absence of a driver's side fourth door. Or we hear that sport utility vehicles are gaining on minivans by siphoning off buyers who want to drive vehicles with a more "outdoorsy" image. Journalists take stock of these stories and attempt to decipher the inner workings of buyer and producer communities in stories of their own. As stories are shared on both sides of the market divide, they build consensus about market dominant logics and help both sides of the market make sense of each other's inherently ambiguous behaviors.

The story-based nature of market sensemaking implies that product conceptual systems are fashioned, maintained, and transformed over time via public and private discourse. Their dependence on stories places constraints on the durability and stability of product conceptual systems, because individual product identities are continuously evolving and changing. Product conceptual systems are always contestable and subject to revision. Thus, they are dynamic interpretive systems that reflect the tug and pull of new contingencies and new participants trying to disrupt the existing conceptual order.

These contingencies often create equivocality and uncertainty in market identities. For example, multiple descriptors can be attached to the same artifact. Sometimes these descriptors are synonyms. Thus, "sport utility vehicle" and "off-road vehicle" evoke similar stories. It is also possible for an artifact to have descriptors that are not similar, but are complementary. An artifact described as a "sport utility vehicle" may also be described as a "luxury vehicle." In some cases, descriptors may seem to conflict. A "sporty car" artifact may also be described as an "economy car," as was the case with the Pontiac Fiero in the 1980s. This type of confusion reflects sensemaking in progress and creates an underlying tension between interpretive change and stability, between the solidity of a shared understanding of product attributes, market boundaries, and market status orderings, and the blade of conceptual revisions that is always hovering somewhere in a product conceptual system. Any approach to product conceptual systems must accept this tension as a matter of course. It must also attempt an explanation of how market identities evolve through periods of market stability and market change.

Summary Assumptions

We can summarize the sociocognitive analysis of markets outlined above with a set of fundamental assumptions that underlie our approach:

Assumption 1: Underlying the behaviors of buyers and producers in the marketplace are conceptual systems used to interpret information from the marketplace (e.g., purchases, new products, new uses, promotional messages) and to inform behavior in the marketplace (e.g., purchases, consumption, usage).

Although a variety of knowledge representation schemes can be used to describe product conceptual systems, we suggest that they be viewed as complex knowledge structures composed of concept nodes and linkages between nodes. They are kind and part hierarchies within which marketed artifacts are categorized and where core sets of attributes values are attached to such artifacts.

Assumption 2: Product markets are enacted.

Enactment means that product market structures evolve from the activities of market actors who are coupled in behavior–cognition cycles (see Fig. 8.1). Producers and buyers enact markets through their behaviors (artifacts, marketing strategies, purchases, consumption), and their interpretations of the market's response to their behaviors. Enactment changes both actors and the environment, making buyers and producers both market makers and market takers. An important implication of enactment is that there is agreement among producers and buyers on how artifacts are categorized. Agreement across the market facilitates the flow of information about products and expedites the assimilation of new artifacts and new uses for old artifacts.

Assumption 3: New product market conceptual systems emerge from the experience of
new artifacts linked to existing uses or new uses linked to existing
artifacts.

In an enacted marketplace, novel experiences generate conceptual shifts.
New artifacts or uses are compared to existing categories and relations, and
difficult matches stimulate conceptual change (Thagard, 1992). New product conceptual systems will often emerge from the combination of elements
from existing systems. The combinatorial links in emerging product markets may be between conceptual systems from experientially proximate or
distal domains, and they may involve the combination of partial or whole
preexisting knowledge structures.

Assumption 4: New product markets involve transformations in the conceptual
systems of producers and consumers.

New product markets are not sustainable unless conceptual transformations take place on <u>both</u> sides of the market divide. Transformations on the
consumer side without corresponding transformations on the producer's
side result in "unrecognized needs"—unrecognized by both consumers and
producers. Transformations on the producer side without corresponding
changes on the consumer side result in "unwanted products."
Uncorresponded transformations leave conceptual traces, but they do not
remain viable elements of product conceptual systems. Unwanted artifacts
are forgotten, and unaddressed needs cease to be needs.

THE MINIVAN MARKET

Research Objectives

We investigated the veracity of the propositions just discussed by tracing
the development of the U.S. minivan market from 1982 through 1994
(Rosa, Porac, Saxon, & Runser-Spanjol, 2000). The minivan market is ideal
for studying the evolution and change of a product conceptual system.
Minivans have been an unqualified success among U.S. consumers. Indeed,
most experts consider the minivan to be a major milestone in the history of
the U.S. automobile industry. Time and again its significance has been affirmed through comments such as:

> Privately, a top General Motors executive scrambling to produce a truly competitive
> product, calls the minivan the automotive coup of the last 20 years. (*Ward's Auto
> World*, 1988)

The Dodge Caravan/Plymouth Voyager twins represented nothing less than an automotive milestone when Chrysler introduced them in 1983; it's fair to say that they recalibrated the American public's attitude toward small vans. (*Car & Driver*, 1988)

When Chrysler's first minivan exploded into the marketplace as a new 1984 model, there was universal acknowledgment—much of it grudging—that Chrysler's minivan ... was a stroke of genius. Some even claim it was the first truly original vehicle to be produced in Detroit since the Ford Model T. (Yates, 1996).

In addition to its path-breaking consequences, the minivan market has been rich in product diversity and competitive interactions. Both U.S. and non-U.S. automobile manufacturers have battled to capture profits in the lucrative people-mover segment of the industry. Most importantly, however, the minivan market has been an arena in which significant battles have been waged over the underlying definition and meaning of this vehicle type.

Our research has been oriented around two basic questions: (a) How did the minivan market, as a product conceptual system, emerge from the cognitive clutter of the auto industry in 1984, when previous attempts to institutionalize a people-mover market had been only marginally successful? (b) How did the minivan conceptual system evolve over the ensuing years and around which attribute configurations have minivan models cohered? Our research qualitatively examines the merits of four propositions:

Proposition 1: Novel artifacts or uses that achieve relatively low sales presence are either absorbed into preexisting market conceptual systems or cease to exist. New conceptual systems are triggered by artifacts that generate numerous and repetitive sales.

Most active industries are characterized by a stream of new artifacts that encounter existing market conceptual systems on entry (Pinch & Bijker, 1987). The sense making that occurs around such artifacts is heavily primed by their salience or presence in the marketplace. Novel artifacts that are substantially different from existing products trigger new conceptual systems only when such salience is high. Otherwise, the artifacts fade out of the market conceptual system or remain as anomalies (often referred to as *niche products*).

Proposition 2: New conceptual systems are characterized by unstable and inconsistent cognitive structures that lead to initial artifact diversity. As these inconsistencies are resolved via further sensemaking, conceptual structures become more stable, coherent, and compacted.

Proposition 2 derives directly from the social constructionists notion of artifact "closure" (e.g., Pinch & Bijker, 1987). It is also consistent with theories of dominant design (e.g., Abernathy & Utterback, 1978; Tushman & Anderson, 1986). However, Hypothesis 2 calls attention to the fact that it is the conceptual system, not the artifact per se, that is characterized by interpretative stability. We suspect that product conceptual systems evolve much faster than actual artifacts, and it is conceptual system coherence and stability that drive artifactual dominance, not vice versa. As conceptual systems shift, existing artifacts remain as fixed points in conceptual space and move closer or further away from the emerging category's ideal point. Some artifacts may become category exemplars, whereas others are expelled from the category.

Proposition 3: In the early stages of a product market, the artifacts seen as good members of the category will be substantially different from one another in their physical and performance characteristics.

Proposition 4: As categories stabilize, some artifacts initially seen as good category members will decline in acceptability, whereas others will improve in their acceptability.

One of the important outcomes of product category stabilization is agreement on a set of core attributes that define the category and with which artifacts must comply to be considered members in good standing. It has been shown that artifacts or models that comply well with categorydefining attributes or prototype are perceived positively, and noncomplying artifacts receive less favorable evaluations (e.g., Myers-Levy & Tybout, 1989). Clear demarcations between good and poor members of the category, however, are seldom discernible until categories stabilize. In the early stages of product market development, there may be little agreement on category defining attributes, and artifacts with divergent attribute values may have similar membership status (Garud & Rappa, 1994; Pinch & Bijker, 1987). As categories stabilize and their prototypes achieve clear definition, the acceptability of existing artifacts is likely to change. Product evaluation is thus dynamic, with the same artifact being evaluated differently depending on a category's stability. Over time, some category members are likely to decline in acceptability, whereas others will improve their standing.

Data Sources

Our method of investigation focused on the analysis of historical documents, sales records, and product specifications over time (see Rosa, Porac, Saxon, & Runser-Spanjol, 1999). We analyzed two primary sources of data:

1. Product reviews and editorials concerning minivans published in the periodicals *Car and Driver*, *Consumer Reports*, *Ward's Auto World*, and *Automo-*

tive News. We have collected all articles referencing minivans in these periodicals from 1980 through 1988. Using a variety of computer text analysis and indexing programs, we have tracked the cognitive representation of minivans over time by analyzing the attributes and evaluations of each model as revealed in the stories published in these sources. An "acceptability score" was computed for each model for each of its years of existence by coding each mention of a model attribute for its positive, negative, or neutral evaluation. These codes were summed for each vehicle and weighted by the total amount of commentary on each model.

2. Sales records collected from *Ward's Auto World* on every minivan model produced from 1984 through 1994. Our analysis led us to believe that the revolutionary nature of the minivan market is due in part to its breaching the barrier between two opposing and more fundamental product conceptual systems: the car and truck systems. The story of the minivan is very much a story of conceptual combination around these two very ingrained cognitive constructs. In the remaining portions of this chapter, we summarize some of our observations about this unique vehicle type.

The Emergence of the Minivan Conceptual System

Any historical investigation of the motor vehicle industry confronts a long-standing and commonsensical distinction between cars and trucks. This distinction is often summarized by metaphorical contrasts such as car-like versus truck-like and two-box (truck) versus three-box (cars) designs. Interestingly, however, history also reveals a number of unique attempts to combine truck and car attributes into a single hybrid vehicle. Until the minivan, most of these attempts achieved only modest success. The models we illustrate below are by no means an exhaustive list of all attempts to bridge the car–truck conceptual gap, but they do show the diverse hybrids that have been produced in the past.

The first set of photographs in Fig. 8.3 and Fig. 8.4 are of car-like vehicles that adopt truck-like features, or what are sometimes called "trucky cars." Sedan delivery vehicles like those in Fig. 8.3 were popular in the 1950s, but ceased to exist around 1960, other than for a brief reemergence between 1971 and 1975 of models based on the Chevrolet Vega and Ford Pinto. Aggregate annual sales volumes were low for all models in the category, never exceeding 50,000 units, and the models have all been discontinued. Figure 8.4 illustrates sedan-like pick-up trucks, such as the Chevrolet El Camino and Ford Ranchero. These vehicle types emerged in the late 1950s and remained in the market until the early 1980s, their disappearance concurrent with the emergence of the minivan. They were always low-volume vehicles, however, with aggregate sales for the category never exceeding 100,000

1953 Pontiac Sedan Delivery

1958 Chevrolet Delray Sedan Delivery

FIG. 8.3.   Sedan delivery vehicles.

units, in contrast to the millions of units sold of standard cars and trucks, and now minivans.

Early vans in Figure 8.5, such as the Ford Club Wagon, General Motors Greenbriar, and Dodge A–100 Van, emerged in the 1960s. These vehicles were all based on a car chassis (e.g., the Ford Falcon, Chevrolet Corvair, and Dodge Dart), but were reengineered onto a truck chassis and absorbed into the truck conceptual system by the late 1960s. Their aggregate sales volume during the years they approximated the car-like conceptual system never exceeded 200,000 units, as compared to the +1,000,000 unit sales in the 1970s for truck-like vans. Figure 8.6 illustrates trucks that have adopted car-like features, such as Dodge Town Wagons and American Motors Jeep Wagoneers. Many of these models were initially positioned as passenger cars, but evolved into being more truck-like (e.g., increased payload, four-wheel drive) and remained firmly entrenched in the truck conceptual system until the sport utility product market emerged in the late 1980s.

Sales volume differences between these low-volume hybrids and the minivan support Proposition 1. In contrast to annual sales volumes seldom in excess of 100,000 units at the height of their popularity, minivan sales exceeded 100,000 units their first year, 250,000 units their second year, and reached almost 500,000 units their third year. There is no doubt that many attempts were made to create a new conceptual system that combined car-like and truck-like attributes, but none achieved the market presence required to undermine existing conceptual boundaries and motivate conceptual shifts until the minivan. As would be expected (e.g., Weick, 1979), preexisting car-like and truck-like conceptual systems or "retentions" resisted change and forced new artifacts into compliance with the dominant

1959 Chevrolet El Camino

1962 Ford Ranchero

FIG. 8.4.   Sedan-like pickups.

1962 Ford Club Wagon

1964 Dodge A-100 Van, Rampside Pickup, and Wagon

1966 GMC/Chevrolet Greenbriar Wagon

FIG. 8.5.   Early car-based vans.

attribute kernels, or relegated them to being low-volume niche vehicles. The minivan, in contrast, quickly established a strong market presence and motivated a conceptual revolution with wide implications for the industry.

There is an interesting aspect to the success of the 1984 Chrysler minivan that speaks to the stochastic nature of market innovations. Journalistic accounts of the development of the modern minivan suggest that consumer research in the 1970's pointed to widespread U.S. consumer sentiment in favor of small people-mover vans (e.g., Barabba, 1994; Yates, 1996). General

Motors and Ford designers had begun to work on such vehicles, and in both cases the models were scuttled by top management due to high costs and low-volume projections. Marketing experts at both companies concluded that the wide diversity of consumers expressing interest in a small van was a sign that no single population segment wanted the vehicle in sufficient numbers to make its production profitable.

Chrysler, on the other hand, was on the brink of bankruptcy, and in a position to risk the estimated $700 million development costs to commercialize a vehicle that could extend its mainstay front-wheel drive K-car platform

1957 Dodge Town Wagon

1961 Dodge Power Wagon Town Wagon

1979 AMC Jeep Wagoneer Limited Four Door Wagon

FIG. 8.6. Passenger carrying trucks.

in another, potentially profitable, direction. Historical accounts suggest that these risks were palatable to Chrysler's chief executive officer (CEO), Lee Iacocca, because he had been an executive at Ford during the time of Ford's initial design work on a small van and had been a major advocate. Harold Sperling, an Iacocca lieutenant at Chrysler, had also been at Ford during this time, and was also a small van champion. Of course, Chrysler had very little to lose and much to gain by gambling on a novel passenger vehicle—a small van with front wheel drive built on a car platform. Moreover, the K-car was Chrysler's staple product, and leveraging a base technology in innovative ways made economic sense. In many ways, Chrysler's good fortune resulted from a set of externally generated and serendipitous events that forced the company to innovate along it's base technological trajectory. By recombining aspects of its core technology, Chrysler's introduction of yet another vehicle that bridged the car-like/truck-like divide was more or less an obvious executive choice.

The front-wheel drive van was an instant hit, particularly among women trying to juggle careers and parenting and finding it necessary to transport children conveniently and safely in countless car-pooling arrangements. The minivan's low height and handling capabilities distinguished it from larger truck-based vans, and made the vehicle type acceptable as an everyday people mover that could be parked in a tight garage at night. The vehicle's immediate success, and Chrysler's own marketing savvy, created a strong presence in the marketplace. The sensemaking that was triggered to explain this success generated a new nomenclature for describing the attributes of the vehicle (e.g., step-in height, 4' x 8' plywood cargo capacity, etc.) as well as for labeling the product as a whole (i.e., *mini*van). It also crystallized a product conceptual system around small vans that had heretofore been fragmented and disjointed, and prompted Chrysler's rivals to introduce their own versions of minivans shortly thereafter.

The Battle Between Car-like and Truck-like Conceptual Systems

Ford and General Motors, however, chose to differentiate their entrants into the emerging market by building their small vans on truck, rather than car, platforms. This design choice brought with it both advantages and disadvantages. Real-wheel drive and larger engines meant better towing and hauling capabilities. These models, however, had poorer handling and bulkier aesthetics. Nevertheless, the Toyota Van, Ford Aerostar, and General Motors Astro/Safari met with initial success, and triggered ongoing industry debate about whether minivans should be more car-like or more truck-like. Evidence of this early debate comes from the views expressed by both buyers and producers For example,

That's not to say there's unanimity in Detroit about what those thousands of potential customers really want—a car, a truck or something (but what?) in between. For that matter, the Big Three can't even agree among themselves just who the customers are. But they're not waiting for the marketing fog to clear: 1985 is the year the race moves into high gear, masterminded by some intriguingly different strategies on how it should be run. (*Ward's Auto World*, 1985)

Truth is, both General Motors and Ford were caught off guard by the Voyager/Caravan's huge sales success. After playing catch-up for a year and a half, they now have their own small vans, the Chevrolet Astro/General MotorsC Safari and the Ford Aerostar. But are they really competition for Chrysler's small van? We're not so sure. The product planners at Chrysler envisioned a high, boxy version of their front-wheel-drive K-cars.... The planners at General Motors and Ford took the opposite tack, each producing a downsized rear-wheel-drive truck with passenger-car amenities. (*Consumer Reports*, 1986)

Toyota introduced its minivan in 1983, General Motors in 1984, and Ford in 1985, followed by Nissan and Mitsubishi in short order. All of these vehicles had attributes that buyers and producers considered prototypical of truck-like vans. Between 1985 and 1988, intense sales competition (see Fig. 8.7) and dense product narratives across the industry touted the respective advantages of car-like and truck-like attributes, and consumers were confronted with significant choices regarding the size, shape, handling characteristics, and prices of the minivan models in the marketplace. During this time, substantial confusion existed regarding the inherent definition of a minivan, forcing producers to defend and bolster their unique positions in the conceptual space. For example, Chrysler's minivan was voted *Motor Trend's* Truck of the Year during this period, but Chrysler's management refused the award because they viewed their vehicle as primarily a car.

The Consolidation of the Car-like Ideal

In 1987, 3 years after Chrysler first introduced its minivans, six major producers accounted for almost all of the +600,000 unit sales in the United States. Chrysler, with its car-like vehicles, accounted for approximately 50% of these sales. General Motors, Ford, Toyota, Nissan, and Mitsubishi, with their truck-like vans, accounted for the rest. By 1990, however, the landscape had changed. Of the producers in 1987, Chrysler remained committed to car-like vans. Toyota had attempted to buck the trend by introducing a second truck-like van that also failed. Mitsubishi and Nissan had withdrawn their truck-like models and had car-like models under development. General Motors had introduced car-like models and had repositioned its truck-like models as mid-sized vans. Finally, Ford had a car-like model under development although continuing to market its truck-like Aerostar.

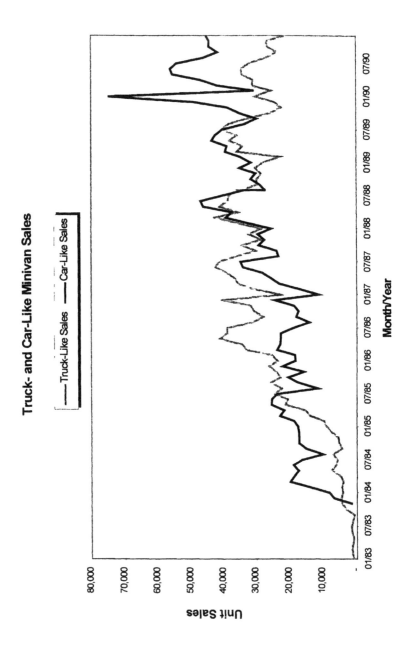

FIG. 8.7. Truck-like and car-like minivan sales from January 1983 to December 1988 (from Rosa et. al, 2000).

Over 70% of minivan sales (see Fig. 8.7) were car-like vans, and announcements by new entrants to the market all complied with the car-like conceptual system. Figure 8.7 shows sales volumes of car-like and truck-like minivans between 1984 and 1994. The figure reveals the struggle in the early years, and the dominance of the car-like conceptual system after 1989, in support of Proposition 2.

Support for Propositions 3 and 4 comes from a more detailed analysis of the text data. Proposition 3 predicted that early in emerging categories there will be considerable variation in the artifacts that were perceived as good members of the category. Proposition 4 suggests that as the category's conceptual frame stabilizes, a shift occurs in the acceptability of category members, with some improving and others declining. Figure 8.8 illustrates the growing dispersion of minivan category member acceptability scores, and the decline of some, over the 1983–1988 time period for the existing models. Note that because not all models were sold in all years, some data points are missing.

The acceptability scores for the Dodge Caravan/Plymouth Voyager and the Toyota Van make the point dramatically. These models were sold throughout the whole time period. The Toyota Van was introduced late in 1982 and received high ratings in 1983. The Caravan/Plymouth models were introduced late in 1983 to an equally positive market reception. Based on the market stories throughout 1983 and early in 1984, all these models were excellent minivans, even though they were very different in terms of their physical configurations. The Toyota was a narrow vehicle with a short wheelbase that was engineered for Japanese highways as a cargo hauler and retrofitted for passenger use. It had cab-over-engine placement, rear wheel drive, and room for seven passengers without luggage. The Dodge/Plymouth models were front wheel drive vehicles based on passenger car engineering specifications, with room for seven passengers without luggage. In addition to these basic configuration differences, car- and truck-like models differed in dozens of attributes that were mentioned in market stories (e.g., ride, handling, appointments, etc.), most of which were seen as positive in the 1983–1984 time period. It happened often that inherently incompatible attributes on different models received equally high desirability evaluations (e.g., front wheel drive and rear wheel drive were both considered excellent for minivans).

Starting in 1985, a trend begins to develop in the data series, which becomes quite pronounced by 1988. Acceptability scores for the model types start to diverge. Scores for the Caravan/Voyager models remained mostly positive and improved during the period. The acceptability of the Toyota Van and other truck like models, however, became increasingly negative. The theory suggests that as the minivan category stabilized from 1983 to 1988, a category prototype resembling the Dodge/Plymouth models developed, causing truck-like models to lose some of their membership status. The acceptability score trends illustrated in Fig. 8.8 support Propositions 3 and 4.

As additional support, starting in 1985 we find references in producer stories to upcoming models from General Motors, Ford, Toyota, and American Motors, all of which were front wheel drive vehicles based on passenger car engineering. The producers claimed that their new designs were based on consumer voice, even though we see clearly from our analyses that consumers were speaking less about minivans and that the conversations were ambiguous.

The actions of the producers in 1985 fit well into our socio-cognitive view of the market. It is evident from our data that the information emanating from the marketplace was equivocal in 1984 and 1985 as producers were making future product decisions. Not only were market evaluations of models unstable and conflicting (see Fig. 8.8), but so were sales volumes as illustrated in Fig. 8.7. Note that in the 1984–1985 time period, a time when producers initiated stories about future car-like models, the sales of truck-like units rose above the sales of car-based minivans and remained higher until after the end of 1988.

Given that market signals (e.g., sales trends, consumer stories) were so mixed in 1984–1985, it appears that producers were enacting the competitive environment, not responding to it. Into a cognitively volatile marketplace, producers introduced stories that basically said that in spite of the sales evidence, the minivan category prototype is a car-like front wheel drive vehicle. These stories circulated and entered into the conceptual system used by consumers and other producers as they developed their own representations of what a minivan should be. As the producers' stories were shared and became assimilated, they influenced the behaviors of other market actors, and in effect became a self-fulfilling prophecy. Producer actions (new product announcements) set in motion the social construction process, and may have handed the Chrysler design the dominance of the marketplace it asserted after 1988, and which it still retains today. It is possible that if producers had enacted stories in which the Toyota Van and similar models (e.g., Chevrolet Astro/General Motors Safari and Ford Aerostar) were the preferred ones, the minivan category would have stabilized around a different configuration altogether.

ADDITIONAL IMPLICATIONS OF THE MINIVAN CONCEPTUAL REVOLUTION

In addition to creating a product market with unit sales of over 1,000,000 vehicles in the United States, and involving practically all major producers

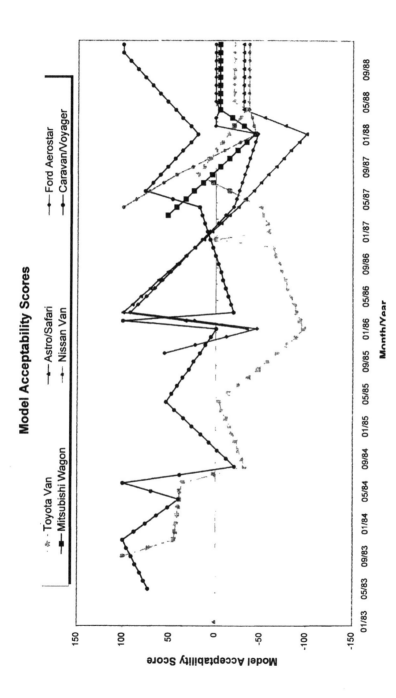

FIG. 8.8 Model acceptability scores from January 1983 to December 1988 (from Rosa et. al, 2000).

239

of motor vehicles in the world, the minivan conceptual revolution may have spurred additional conceptual shifts in the motor vehicle industry that merit investigation. One such phenomenon is the growth of the sport utility vehicle as an alternative to sedans for urban and suburban dwellers. Although the sport utility vehicle category pre-dates the minivan by several decades (Jeeps were introduced in the 1950s), its growth in sales during the late 1980s and into the 1990s has exceeded all industry projections. One possible explanation is that the conceptual instability triggered by the minivan helped redefine the conceptual market position of sport utility vehicles as well.

It is widely acknowledged that minivans gained most of their sales from traditional family sedan buyers. As the artifacts in the category converged on the car-like conceptual system in terms of attributes, many potential buyers have chosen instead to purchase the more truck-like sport-utility vehicles, although citing many of the purchase rationales that minivan buyers have used (e.g., command of the road visibility, versatility, etc.). One possible reason for this phenomenon is that as minivan models converged on a car-like set of attributes, they left a conceptual and behavioral gap for downsized, truck-like, two-box vehicles, which sport utility vehicles fit well. Analysis of the market dialogue, specifications, and attribute valuations of sport utility vehicles similar to what has been done for the minivan market may give us insight into the conceptual development of the sport utility market.

At a more aggregate level, another interesting phenomenon that may have been triggered by the minivan revolution is the increase in car-like comfort, ride, and handling characteristics of light duty trucks. The late 1980s and 1990s have also seen a dramatic surge in the level of car-like luxury, comfort, and styling applied to pick-up trucks and closely related vehicles in the U.S. market, and an equally dramatic shift in the sales distribution between cars and trucks. More and more, consumers are making trucks their vehicle of choice for urban commuting, and are equipping them with features and options historically found only in cars (e.g., high level sound systems, cruise control, cloth and leather interiors). Luxury trucks have become a standard offering by all major truck producers, and new competitors are emerging annually.

One possible explanation for this phenomenon is that the minivan conceptual revolution not only established a new conceptual system, but destabilized the preexisting ones enough for their boundaries to shift, with the truck system expanding to encompass attributes and usage conditions that had previously been linked only with cars. The destabilization of the boundaries could also force changes in the car-like conceptual system, perhaps forcing that system to retrench and become narrower in its definition of what constitutes a car. If that were the case, it may explain the increasing popularity of pure passenger vehicles (excellent performance and handling but without cargo hauling abilities) such as the Plymouth Prowler and Dodge Viper. The redefinition of the truck- and car-like conceptual systems can also be investigated using the methods described earlier, and confirmed through ethnographic research among market actors (producers, dealers, consumers, magazine publishers, etc.). Until such research is done, however, these remain interesting but speculative extensions.

CONCLUSIONS

We suggested that market networks cohere around product conceptual systems that summarize the abstracted attributes and uses of artifacts. When artifacts are embedded in a coherent and shared conceptual system, they become products that can be described, valued, and exchanged. We have used the set of artifacts that has come to be known as the *minivan market* as an example of how the dynamics of market conceptual systems link buyers and sellers within a transactional network extending over time and space. The minivan market reveals how conceptual discontinuities play out in microprocesses that lead to eventual coherence around a set of core attributes that define the artifact in question. It also reveals the potential ripple effects of a revolutionary conceptual system through the category structure of an entire industry. In the minivan's case, Chrysler followed its base technological trajectory and extended its front wheel drive platform in a novel way. In contrast to previous people moving vehicles that attempted to bridge the car/truck gap, Chrysler's vehicle was a success. This success fueled imitators who battled over the core definition of the minivan itself. These battles eventually lead to a consolidation of the definition around a car-like vehicle type, and the successful convergence seemingly led to a breach of the car/truck conceptual boundary. There is some evidence that this breach has led to a destabilization of the entire automotive product category profile. We believe this sequence of events reveals the potential of a sociocognitive analysis of market creation and evolution.

REFERENCES

Abernathy, W., & Utterback, J. (1978). Patterns of industrial innovation. *Technology Review, 80*, 40–47.

Abernathy, W., & Clark, K. (1978). Innovation: Mapping the winds of creative destruction. *Research Policy, 14*, 3–22.

Arthur, W. B. (1989). Competing technologies, increasing returns, and lock-in by historical events. *Economic Journal, 97*, 642–665.

Auerbach, A. J. (1988). *Competition: economics of industrial change*. Oxford, UK: Blackwell.

Barabba, V. (1994). *Meeting of the minds: Creating the market-based enterprise*. Cambridge, MA: Harvard Business School Press.

Bijker, W. (1987). The Social Construction of Bakelite: Toward a Theory of Invention. In W. E., Bijker, T. P. Hughes & T.J. Pinch (Eds.), *The Social Construction of Technological Systems*. Cambridge, MA: MIT Press (pp. 159-187).

Bourdieu, P. (1984). *Distinction: A social critique of the judgement of taste*. [Richard Nice, Trans.]. London: Routledge.

David, P. A. (1985). Clio and the economics of QWERTY. *American Economic Review, 75*, 194–197.

Day, G. S., Shocker, A. D., & Srivastava, R. K. (1979). Customer-oriented approaches to identifying product markets. *Journal of Marketing, 43*, 8–19.

Fligstein, N. (1996). Markets as politics: A political–cultural approach to market institutions. *American Sociological Review, 64*, 656–673.

Garud, R., & Rappa, M. (1994). A socio-cognitive model of technology evoulution. *Organization Science, 5*(3), 344–362.

Hahn, F. H. (1973). *On the notion of equilibrium in economics: An inaugural lecture*. London: Cambridge University Press.

Hamel, G., & Prahalad, C. K. (1994). *Competing for the future*. Cambridge, MA: Harvard Business School Press.

Hannan, M. T., & Freeman, J. (1989). *Organizational ecology*. Cambridge, MA: Harvard University Press.

Liebowitz, S. J., & Margolis, S. (1995). Path dependence, lock-in and history. *Journal of Law, Economics and Organization, 11*(1), 205–236..

Nonaka, I., & Takeuchi, H. (1994). *The knowledge creating company: How Japanese companies create the dynamics of innovation*. Cambridge, MA: Harvard Business School Press.

Pinch, T., & Bijker, W. (1987). The social construction of facts and artifacts In W. E. Bijker, T. P. Hughes, & T. J. Pinch (Eds.), *The Social Construction of Technological Systems. Cambridge, MA: MIT Press*. (pp. 17–50).

Porter, M. E. 1980. *Competitive strategies*. New York: Free Press.

Robinson, J. (1933). *The economics of imperfect competition*. London: Macmillan.

Rosa, J. A., Porac, J. F., Saxon, M. S., Runser-Spanjol, J. (2000). Socio-cognitive dynamics in a product market. *Journal of Marketing*.

Spender, J. C. (1996). Making knowledge the basis of a dynamic theory of the firm. *Strategic Management Journal, 17*, 45–62.

Tirole, J. (1988). *The theory of industrial organization*. Cambridge, MA: MIT Press.

Tushman, M. L., & Anderson, P. (1986). Technological discontinuities and organizational environments. *Administrative Science Quarterly 31*, 439–465.

Utterback, J. (1996). Mastering the dynamics of innovation. Cambridge, MA: Harvard Business School Press.

Weick, K. E. (1993). The collapse of sensemaking in organizations: The Mann Gulch disaster. *Administrative Science Quarterly 38*, 628–652.

White, H. C. (1981). Where do markets come from? *American Journal of Sociology, 87*, 517–547.

White, H. C. (1992). *Identity and control: A structural theory of social action*. Princeton, NJ: Princeton University Press.

Yates, B. (1996). *The critical path*. New York: Free Press.

9

The Construction of New Paths: Institution-Building Activity in the Early Automobile and Biotechnology Industries

Hayagreeva Rao
Emory University

Jitendra V. Singh
University of Pennsylvania

From time to time, new forms emerge on the organizational landscape and enhance organizational diversity. For instance, the late 1980s witnessed the spawning of the personal computer industry as computers began to be designed for personal use by individuals rather than organizations. More recently, the rise of health maintenance organizations (HMOs) has transformed the organization of health care in the United States. The rise of new forms can properly be viewed as an organizational counterpart to the biological phenomenon of speciation, earlier termed *organizational speciation* (Lumsden & Singh, 1990).

We think speciation is one important instance of path creation. A large and varied literature portrays *path dependence* as a process in which small, chance events have disproportionately large consequences and lead to a lock-in of choices (for example, Arthur, 1989; David, 1985; see Hirsch & Gillespie, chap. 3, this volume, for a detailed review). Thus, path dependent models emphasize how history constrains the evolution of technologies, products and organizational forms, and leave little room for social action. By contrast, path creation entails a break with the past and a departure from preexisting technologies, products, and forms, and implies new expectations about the future (Karnøe & Garud, 1995). Whereas new paths may arise in the economy due to various underlying causes (regulation or deregulation, for example), new technologies, and rapid changes in population demography, they can be created only if institutional entrepreneurs combine or blend existing knowledge with new knowledge (Hirsch & Gillespie, chap. 3, this volume). Thus, the emergence of new technologies (Bijker, Hughes, & Pinch, 1987; Garud & Rappa, 1994), the construction of product-markets (Porac, Rosa, & Saxon, 1997), and the establishment of new forms are all instances of path creation.

Our approach in this chapter explores how the creation of new organizational forms, organizational speciation, is deeply interconnected with the politicoinstitutional context in which it occurs. Whereas the dominant focus here is on how the politicoinstitutional context influences organizational forms, it is clearly the case that new organizational forms also influence the context. Thus, their relationship is coevolutionary (Baum & Singh, 1994; Singh & Lumsden, 1990). However, our emphasis here is to propose the relationship. The specific causal details are best viewed as a subsequent task.

We pointed out elsewhere that new organizational forms are novel recombinations of core organizational features involving goals, authority relations (including organization structure and governance arrangements), technologies, and client markets (Rao & Singh, 1999). Whereas one of the earlier uses of the concept of organizational form has limited its focus to organization structure, we believe a broader conception that includes other core features of organization is valuable. Along these lines, Lucas(1998) studied various exemplars of new forms including network, virtual, and spinout forms, among others. We believe that organizational forms occur in gestalts. Thus, a singular interest in the structure of a network form ignores many other distinctive features of the form such as the use of information technologies or authority relations, among other features. This concept of new organizational forms is distinct from the politicoinstitutional context that helps shape, and is, in turn, shaped by new forms.

As an instance of path creation in which new organizational assemblages come into existence, speciation plays a vital role in the evolution of organizational diversity (Astley, 1985). Novel social structures matter because they underpin organizational diversity. The ability of societies to respond to social problems hinges on the diversity of organizational forms, and, in the long run, a fluid environment diversity can be maintained or increased by the rise of new forms (Hannan & Freeman, 1989). Moreover, new forms are consequential motors of evolution—indeed, an important piece of organizational change, at the macrolevel, consists of the replacement of existing organizational forms by new organizational forms (Astley, 1985; Schumpeter, 1950).

Furthermore, because new forms are structural incarnations of beliefs, values, and norms, they emerge in tandem with new institutions, and foster cultural change in societies (Scott, 1995; Stinchcombe, 1965). For these reasons, how new forms are constructed and how new paths are created is one of the central questions of organizational theory (see Singh, 1995).

It is only recently that organizational theorists have begun to analyze the origins of new organizational forms from the standpoint of the random variation, constrained variation, and cultural frame institutional perspectives. The *random variation* perspective is premised on biological evolutionary models and holds that realized variations in organizational forms are random. Proponents of this view suggest that new forms arise when search routines lead to modifications of operating routines (Nelson & Winter, 1982), or when a small group of competence-sharing organizations is isolated and finds a favorable resource environment (McKelvey, 1982). Despite its appeal, the random variation perspective is of limited use because it is difficult empirically to identify competencies and routines. Another major drawback is that this perspective is silent on the specific processes that generate variations, and the content of variations. A third difficulty is that variations may be the outcome of systematic processes rather than random processes—a point of view developed by institutional theorists of organizations.

The *constrained variation* perspective asserts that environmental conditions predictably foster or diminish variations in organizational forms. Different versions of this view variously emphasize creative destruction through technological innovation (Schumpeter, 1950; Tushman & Anderson, 1986), environmental imprinting (Tucker, Singh, & Meinhard, 1990), wherein social conditions at the time of founding limit organizational inventions (Kimberly, 1975; Stinchcombe, 1965), and depict existing organizations as producers of new organizational forms (Brittain & Freeman, 1982; Lumsden & Singh, 1990). A key premise common to all versions of the constrained variation perspective is that the existence of ecological niches unoccupied by other forms is an important precondition for the birth of new organizational forms. Nevertheless, proponents of the constrained variation perspective differ with respect to the antecedents of resource spaces. If models of creative destruction hold that the demise of existing organizations frees resources for new organizations, then models of environmental imprinting stress the importance of political upheavals (see Carroll, Delacroix, & Goodstein, 1990), or entrepreneurs' access to wealth and power, labor markets, and the protective role of the state (see Aldrich, 1979). By contrast, those who portray existing organizations as producers of new organizations, hold that interrelations among existing organizations influence the branching of new resource spaces (Carroll, 1984) and the ability of existing organizations to exploit new resource spaces (Romanelli, 1989). Although the constrained variation perspective usefully emphasizes the primacy of re-source spaces, its drawback is that resources do not preexist as pools of free-floating assets, but have to be mobilized through opportunistic and collective efforts (Van de Ven & Garud, 1989). Another limitation is that the constrained variation perspective elides how formal structures become imbued with norms, values, and beliefs during the process of resource mobilization by entrepreneurs.

The *cultural-frame institutional* perspective complements the constrained variation perspective and proposes that new organizational forms arise when actors with sufficient resources see in them an opportunity to realize interests that they value highly. A core premise is that the creation of all new organizational forms requires an institutionalization project, wherein the theory and values underpinning the form are legitimated by institutional entrepreneurs (Aldrich & Fiol, 1994; DiMaggio, 1988; Scott, 1995). In this perspective, institutional projects can arise from organized politics or social movements, and in the case of the former, they resemble the latter to the extent that resources and interests are not fixed and the rules governing interaction are contested (Fligstein, 1996a; Haveman & Rao, 1997; Rao, 1998). In the cultural-frame institutional model, prevalent cultural and political frameworks constrain entrepreneurs seeking to establish new forms, but new forms can also reshape the existing politicocultural landscape (Scott, 1995). In this sense, it is compatible with research on the coevolution of technologies and institutions (Garud & Rappa, 1994; Karnøe & Garud, 1995; Van de Ven & Garud, 1994). However, the cultural-frame institutional perspective on the rise of new organizational forms is as yet emergent and needs to "direct attention ... to the establishment of field-wide environments around the forms" (DiMaggio, 1991, p. 289).

These considerations provide the motivation for us to compare the creation of the automobile and the biotechnology industries.[1] We believe a comparison of both industries is useful because it sheds light on the role of state apparatuses, professions, collective action, and consumers in the creation of new paths. To begin with, the automobile industry arose in 1895 and grew in the early part of the 20th century; the rise of the biotechnology industry occurred in the later part of the 20th century. As a result, the automobile industry evolved in a social context characterized by sparse governmental intervention and minimal professional activity. By contrast, the biotechnology industry arose in a context marked by strong governmental involvement and a central role of professionals.

---

[1]We refer loosely to the biotechnology industry here, although recognizing that biotechnology is more accurately characterized as a set of technologies that find applications in many diverse industries such as food, agriculture, bioremediation, pharmaceuticals, diagnostics, and veterinary medicine.

Thus, we compare two "extreme cases" (Eisenhardt, 1989) and hope to enlarge our understanding of how fields and forms coevolve (see Singh, 1995). We present case studies of both industries, and then derive generalizations on the institutional dynamics of path creation.

THE EARLY AMERICAN AUTOMOBILE INDUSTRY

The early American automobile industry was a new organizational form that constituted a departure from its precursor, the horse carriage industry. In contrast to the horse carriage industry, the automobile industry's goal was to develop horseless vehicles of transportation. Moreover, automobile firms used a completely different set of technologies based on steam, gasoline, and electric power to provide customers with horseless carriages. Furthermore, early automobile firms also differed from the horseless carriage industry in their structure of authority; unlike horseless carriage firms that were one man operations, early automobile firms were assemblers who put together bought-out components. Finally, early automobile firms also catered to a different class of customers; the early purchasers of automobiles were enthusiasts, and physicians who needed mobility (Flink, 1970).

Some accounts of the automobile industry emphasize and reduce the birth of the early automobile industry to the question of how gasoline-powered cars began to dominate the industry. Although the technological development of the automobile industry is interesting in its own right, it can deflect attention from the larger question of how the automobile came to be legitimated. We discuss how contending attempts at institutionalizing the industry and delegitimizing the automobile influenced the birth of the automobile industry. We focus on how a dearth of legitimacy induced institutional entrepreneurs and professionals to establish the reliability of the automobile. We also emphasize how attacks by antispeeding vigilante groups jeopardized the standing of the industry and led to efforts by automobile clubs to influence speed legislation, and how state authorities endorsed the automobile by passing laws designed to register cars and license drivers.

Some writers trace the origins of the American automobile industry to the Selden two-stroke engine design developed in 1879, or to William Morrisons's electric car of 1892, or to Ransom Olds' steam vehicle, purportedly sold to a Indian firm in Bombay. However, the first firm to make automobiles was set up by the Duryea brothers in 1895 (Flink, 1970). The appearance of the motor car spurred considerable enthusiasm. Thomas Edison (1895) proclaimed that the "horseless carriage is the coming wonder ... It is only a question of time when the carriages and trucks in every city will be run with motors" (p. 67). The first task of pioneers in the industry was to establish the reliability of the car.

Establishing the Reliability of the Automobile

As a novel technology, the automobile per se was unfamiliar to prospective consumers and putative inventors. Consumers were confused because the source of power, the number of cylinders, systems of steering and control, and the mode of stopping; these were topics of considerable controversy (Thomas, 1977). The only point of agreement about the automobile was that it could not be animal powered. Consumers were hesitant to purchase cars because they felt that they were in no position to decide, especially when the engineers themselves could not agree on the best design (Epstein, 1928). Moreover, consumers were still reeling from an earlier debacle—the explosive growth and collapse of the bicycle industry (Rae, 1959).

Consumers also could not evaluate the products offered by producers. In turn, producers suffered from a dearth of information about the strengths and weaknesses of rival designs. Hiram Maxim, a pioneer of the industry, wrote that he was "blissfully ignorant that others were working with might and main ... on road vehicles" (cited in Thomas, 1977, p. 17). Writers for automobile magazines as well as the local press complained about the proliferation of firms without track records and reputations (Flink, 1988). Many cars were unable to complete a drive successfully and had to be hauled back by teams of horses. In preparation for such eventualities, quite a few vehicles were designed with whip sockets and harness hitches (Epstein, 1928), and the misleading advertisements issued by some firms evoked complaints against the "endless amount of nonsense being published in the public press" (*Scientific American*, 1895).

It was in this context that some entrepreneurs sought to regulate quality by establishing collective trade associations. Others instituted reliability and speed competitions to demonstrate the capabilities of the horseless vehicle and to certify quality. However, the state played a minimal role in codifying quality standards and only became a customer after the industry was established. Professionals also indirectly contributed to the form by establishing standards for parts suppliers.

Regulating Quality Through Collective Action.   Trade associations and professional bodies did not exist in the first 5 to 6 years of the industry. The National Association of Automobile Manufacturers was established in 1900 in a bid to assure product quality but was superseded by the Association of Licensed Automobile Manufacturers (ALAM) which was formed in 1903. ALAM was a trade association formed to license the Selden patent and was set up ostensibly to prevent "incursion of piratical hordes who ... desire to flood the market with trashy machines" (*Motor Age*, 1903, p. 3). But the Selden patent was widely disregarded and, due to internal divisions, ALAM was unable to secure quality by enforcing its threat of litigation.

A rival association called the American Motor Car Manufacturer's Association (AMCMA) was established in 1905 and also proved to be an ineffective mechanism of collective action. Both trade associations disintegrated during the period 1909–1911, as a result of legal battles, and their secondary functions were assumed by the Society of Automotive Engineers (SAE), which was established in 1905. In turn, the SAE was a professional body of engineers, which focused on common standards for automobile components made by suppliers but had little ability to police the technical quality of the cars made by automobile producers (Flink, 1970).

Endorsement by State Authorities.   In contrast to the European experience, where national governments supervised automobile construction standards and developed national laws regulating the use of motor vehicles, the U.S. federal government was apathetic to the automobile until 1909. The governments of France, Germany and England had discerned the military potential of the automobile and offered subsidies for the development of military vehicles that inhibited the manufacture of light cars, but in the United States, the war department only began to acquire specialized automobiles in 1909. Similarly, the post office, although more alert than the war department, experimented with the use of motor vehicles to collect mail from 1896, and even started subcontracting mail collection to private entrepreneurs in a series of tests from 1901 to 1906. However, it was only in 1909 that the post office made a substantial commitment to the purchase of motor vehicles in New York City, and thereafter, encouraged the use of automobiles for free rural delivery. Thus, the federal government, despite the existence of customers with the potential to influence the market, such as the war department and the post office, exercised little influence on the rise of the automobile industry in the United States (Flink, 1988, pp. 120–125).

By contrast, municipal governments played a more active role in the emergence of the automobile industry. Initially, municipal governments were customers who used self-propelled vehicles in fire departments as early as 1903, and for police patrols as far back as 1898 in Akron, Ohio. With 25 automobiles in 1905, New York City government authorities led the nation in pioneering the use of motor vehicles. The use of motor vehicles for emergency relief in the San Francisco earthquake of 1906 soon became a model for municipal governments in the rest of the country.

Professionals and Standards.   Professionals indirectly assisted in the legitimization of the automobile by establishing standards for parts. The SAE began in 1905 with a small group of journalists and automobile engineers, and by 1910, formed a standards committee and saw a rapid rise in its membership. Many SAE members hailed from small manufacturers. Unlike large firms that had an assured supply of components because of well-established relationships or internal production, small manufacturers were worried by the failure of suppliers of components and the danger of price fluctuations in auto parts. In 1910, the SAE diagnosed the lack of intercompany standardization of components as a major cause of production problems and expenses, and initiated a program that, by 1921, resulted in the eventual formation of 224 component standards.

Professionals also contributed to the rise of the automobile by standardizing the training of skilled personnel who could make and repair motor cars. Although some makers of high-priced cars trained drivers at special schools (e.g., Packard and Locomobile), it fell to the Young Men's Christian Association (YMCA) to institute courses to educate drivers and to offer advanced training for engineers and draftsmen, which it did during the period 1903–1904. Soon, the New York School of Automobile Engineers was founded, in 1905, by a professor of engineering from Columbia, and these schools disseminated mechanical expertise that in turn enabled entrepreneurs to establish chains of service and repair garages in cities and small towns.

Races and the Emergence of Standards.   It was in this context that numerous enthusiasts and automobile clubs in the industry organized races. The first contest was the *Times-Herald* race held on Thanksgiving Day in 1895. The publisher of the *Times*, H. H. Kohlsaat, wanted to organize the competition "with the desire to promote, encourage and stimulate the invention, development and perfection and general adoption of motor vehicles" (quoted in Thomas, 1977, p. 21). Five of the 11 entrants participated, and only two vehicles were able to complete the race. The first prize of $10,000 was won by a gasoline powered Duryea car, which had a winning speed of 8 miles per hour.

Shortly thereafter, *Cosmopolitan* magazine organized a contest on May 30, 1896 and offered a prize of $3,000; this contest was also won by a Duryea automobile. Subsequently, the Rhode Island State Fair Association offered $5,000 in prize money and organized a competition that was won by an electric car. The Riker Electric car won the race but spectators found the contest to be so dull that they originated the cry "get a horse" (Flink, 1970, p. 42). The stage was set for numerous enthusiasts to sponsor reliability contests (hill-climbing runs, endurance tours, and fuel economy contests) and speed races. Contests were organized in Trenton, Detroit, Omaha, Chicago, Empire City, Brighton Beach, Florida, and at speedways such as Indianapolis and Atlanta. In 1901, the Automobile Association of America of New York City formulated a set of racing rules and assisted promoters in the organization of races.

Organizers placed few restrictions on participants in order to increase the number of entries. There were strong incentives for firms to participate in these contests because they had a chance to test technical improvements and to acquire publicity as innovators (Flink, 1988). Manufacturers sponsored cars directly in these contests, and, although a few contests (especially the Glidden tour) stipulated that cars were to be driven by their owners, firms circumvented this because any executive of an auto firm could drive the recent models himself (Flink, 1970). Some contests were one-shot exercises, and others, such as the Glidden reliability tour or the Vanderbilt Cup speed races, were organized by different groups of organizers each year. All competitions awarded prizes to the first-place contestants, whereas a few also dispensed additional prizes to second- and third-place contestants. Moreover, organizers of contests also sought to assure the viewing public of their integrity by allowing extensive press coverage and by instituting grievance procedures (Thomas, 1977). Despite the fact that these contests were free to the viewing public, there were sufficient incentives for organizers to schedule such events. Many of the organizers were enthusiastic activists committed to the development of the automobile. Thus, there was no incentive problem; organizers could exclude others from the psychological benefits of contributing to a cause or to the creation of automobile clubs (Flink, 1988; Rae, 1959).

As the public watched contests and learned about them through the media, knowledge of the automobile per se diffused across different sections of American society. Newspapers such as the *Chicago Times-Herald*, and newsmagazines such as *Cosmopolitan* sponsored contests. Specialized trade journals arose to disseminate information about the automobile, and many of them dedicated resources to the coverage of contests such as the Glidden tour, the Vanderbilt races, the Indy races, and other local contests. Another benefit of contests was that they spawned the creation of automobile clubs such as the Automobile Club of America, and, in turn, these clubs organized more contests (Flink, 1988).

As a result, the automobile began to be considered more reliable than a horse. In 1899, *Harper's Weekly* reflected the common presumption that a "good many folks to whom the horse is a wild beast feel much safer on a machine than behind a quadruped" (*Harper's Weekly*, 1899, p. 383). So widespread was this influence that the automobile no longer was a novelty purchased either by other engineers who desired to experiment with it or by physicians prizing mobility. In time, it became a necessity for the middle class. Frank Munsey noted that the "uncertain period of the automobile is now past. It is no longer a theme for jokers and rarely do we hear the derisive expression "Get a horse" (*Munsey's Magazine*, 1906, p. 404).

Apart from establishing the identity of the automobile industry, victories in contests enhanced the reputations of individual firms. The contests also provided consumers with data on the fuel economy, ruggedness, durability, speed, and dependability of the cars produced by participating firms. Winning firms reaped substantial publicity from the press coverage and proclaimed these victories in their advertising campaigns. After winning some hill-climbing contests, the Peerless Company advertised its car as "a rapid and powerful hill climber." Similarly, the St. Louis Motor Carriage Company, after faring well in some endurance contests, touted its cars as "rigs that run." The Stearns organization modestly depicted its products as "reliable gasoline motor cars," Oldsmobile grandly proclaimed that the Olds "runs everywhere," and Cadillac coined the slogan that "when you buy a Cadillac you buy a round trip" (Thomas, 1977, p. 47). Buick, after winning several contests proclaimed "Tests tell—could you ask for more convincing evidence?" Thus, the advertising campaigns planned by automobile producers were mechanisms for informing the public of their winning record (Epstein, 1928).

Attacks by Antispeed Activists; Defense by Automobile Clubs

Because the automobile threatened to displace the horse-drawn carriage, it evoked some opposition from manufacturers of horse-drawn carriages, livery stable owners, and horse-drawn vehicle driver associations (Flink, 1970). However, prominent manufacturers of horse-drawn carriages such as Studebaker, Flint Wagon, Standard Wheel Company, and Mitchell Wagon Company all entered the automobile industry and began to produce motor cars. Enterprising owners of livery stables also switched to opening garages or to starting car rental services.

By far, the most serious opposition to the automobile stemmed from vigilante antispeed organizations that were a common phenomenon at the turn of the century. Organizations such as the New York Committee of Fifty, a prominent antispeed organization, obtained data on speeds using stopwatches and convincingly showed that cars were driven at high speeds. They then lobbied for instituting speed limits. Antispeed vigilante groups posted rewards for evidence that could convict speeders; the more aggressive organizations, such as the Long Island Highway Protective Society, described speeders as "scorchers," and resorted to illegal tactics such as puncturing tires of speeding cars, and in some cases, even riddling tires with bullets.

Opposition to the motor car was prevalent in rural areas during the touring season when speeding automobiles threatened livestock and horse-drawn traffic, and raised dust that damaged crops. Some farming communities plowed roads (as in Rochester, Minnesota) to make them unusable for cars, and local businessmen in some counties were threatened with boycotts by farmers if they choose to buy motor cars. The hostility toward cars reached a peak during the period 1904–1906, when the Farmers Institute of Indiana asked for a ban on cars from using roads, and some farmers clubs adopted resolutions vowing not to support political candidates who owned cars. Influential periodicals such as the *American Agriculturist* (circulation of 85,000) and the *Farm Journal* (circulation, of 500,000) expressed concern about speeding, reckless driving, and called for legislation and enforcement of antispeed laws.

Regulation of the Automobile.   Under pressure from antispeed organizations, municipal governments took the lead in issuing ordinances mandating the registration of automobiles with distinct tags so as to identify speeders. It was only in 1903 that state-level laws superseded municipal regulations. New York was the first state government to require registrations in 1901; and by 1903, eight other states also had insisted on mandatory registration. By 1905, 26 states required registrations, and by 1915, all states had a motor vehicle registration law. Although some automobile clubs, notably in Chicago and Detroit, opposed registration requirements, they later began to view it as a blessing in disguise because it checked reckless driving, defused public opposition, and even led to greater road building, thereby promoting the diffusion of the automobile.

Municipal governments also pioneered the development of ordinances designed to regulate the licensing of drivers. After the first rules were developed in Chicago in 1899, other cities followed suit, and it was only in 1906 that state-level laws began to supersede local certification of drivers. New Jersey created a special department of Motor Vehicle Regulation and Registration in 1906, and Massachusetts followed suit shortly. Motorists saw road tests and licensing exams as essential for the development of safe and reliable motoring. By 1909, 12 states had routinized the licensing of drivers.

Automobile Clubs.   Voluntary associations of automobile users arose to defend the automobile and to propagate a favorable image of the car. Automobile clubs lobbied for favorable legislation, defended drivers against criminal suits, organized tours and races, and contributed to the development of mechanical expertise.

The American Motor League, set up in 1895, was the first attempt to organize a club, and it foundered. It was only at the turn of the century that the Automobile Club of America (ACA) was established in New York with aristocratic pretensions. By 1901, 22 local clubs had mushroomed in different cities such as Boston, Newark, and Chicago and disputed the ACA's role as arbiter of the automobile. In 1902, there were two rival attempts to form national organizations. The American Motor League (revived by its 1895 organizers), led by Charles Duryea, sought to enlist individual motorists as members and aimed to establish branches. By contrast, the American Automobile Association (AAA) began as an association of local clubs, and after some early hardships stemming from the resentment over the influence of New York clubs on the AAA, emerged as the primary representative of car owners in America.

The local clubs that were members of the AAA played an invaluable role in defusing hostility to the automobile. For example, they organized tours for underprivileged children to dissuade children from throwing stones on passing cars in places such as Chicago and New York. Similarly, automobile clubs initially opposed local ordinances on registration and licensing but quickly realized their value in creating safe roads and popularizing the motor car.

However, a key contribution of the clubs was to forestall antispeed legislation, or lobbying for laws with liberal speed limits. By 1902, only three states (New York, Connecticut, and Massachusetts) had passed antispeed laws with liberal speed limits ranging from 15 mph to 20 mph. In fact, the New York law with its speed limit of 20 mph had been drafted by the Automobile Club of America. Local automobile clubs counted prominent personalities as members and used their contacts and resources to lobby for legislation. Their lobbying included providing legislators free rides to show them that cars were safe at high speeds. By 1906, 15 states had speed limits of 20 mph, and 5 states allowed speed limits of 25 mph. Attempts to enforce speed laws by governmental authorities led to the diffusion of the "speed trap," where policemen clocking cars would quickly arrest offenders. Automobile clubs placed warning signs about speed traps, placed representatives on the spot to warn drivers, and provided defense attorneys and funds to motorists.

Automobile clubs were at the vanguard of the "good roads movement" and promoted the diffusion of the automobile in addition to diminishing public fears about speeding automobiles. Although, in the 1890s, bicycle manufacturers, dealers and owners were at the forefront of the good roads movement, by 1901, automobile clubs were more prominent. The ACA of New York spawned the signboard movement, and other clubs joined in to mark road routes, and the American Motor League inaugurated a new program for erecting danger signs.

Successful Institutionalization

By 1909, institutional entrepreneurs' attempted to establish reliability, the efforts of automobile clubs had borne fruit, and the automobile industry was firmly situated. During that year there were 735 producers, capitalized at $173,837,000, employing 85,359 workers, paying wages totaling $58,173,000, with an added value worth $117,556,000. By contrast, there were 5,492 carriage and wagon firms capitalized at 175,474,000, employing 82,944 people, paying wages totaling $45,555,000, with an added value worth $77,942,000 (Flink, 1970). By then, opposition from farmers and antispeed organizations had dimmed, as farmers' interests represented by the Granges, which were farm fraternity organizations, began to cooperate with the AAA for good roads. So strong already was the hold of the automobile that even during the 1907 financial panic, *Harper's Weekly* predicted "there is no question that the [automobile] business is going to get steadily better. There is one reason for this ... the automobile is essential to comfort and happiness" (*Harper's Weekly*, 1907, p. 1172). Charles Duryea remarked in 1909 that the "novelty of the automobile has largely worn off" (*Independent*, 1909). By 1913, reliability contests ceased, as there was little interest in competitions designed to demonstrate the capabilities of the horseless vehicle; instead, speed competitions became more popular and contributed to the rise of speed as a criterion of car design and car purchase.

THE EARLY AMERICAN BIOTECHNOLOGY INDUSTRY

The *biotechnology industry* consists of hundreds of dynamic startup firms producing a wide range of products, and research aimed at unraveling the secrets of the gene—the "control center of life" (Lee, 1993, p. 2), and attracts some of the brightest minds in the country. The establishment of Genentech in 1976 signaled the birth of an industry premised on "the application of engineering and technological principles to living organisms or their components to produce new inventions or processes" (Conte, 1994, p. 5).

Although animal breeders and farmers have been carefully selecting and breeding particular strains for centuries, they could not directly manipulate genes as molecular biologists do today. Genetic engineering technology had remarkably humble origins in a monastery in Brno (in the Czech Republic), with Gregor Mendel, who was admitted to the novitiate in 1843. He studied the common pea, reported the existence of "dominant" and "recessive" traits, and went on to outline the laws of inheritance that seemed to govern their passage across generations. By the 1880s, researchers had already documented the frenetic activity of chromosomes just before cell division. These delicate, thread-like structures, which could be observed quite easily because of their ability of absorb dyes used for staining (hence the name chromosome, which literally means "colored body"), were "passed on equally from parent to daughter cells during cell division" (Bodmer & McKie, 1994, p. 21). By 1901, the American zoologist, Walter Sutton, had shown that the patterns of inheri-

tance recorded by Mendel could be explained by the behavior of these myste-
rious chromosomes.

Over the course of the next 50 years, several important experiments fur-
thered our understanding of genes. The most important of these was the
work of Oswald Avery and his coworkers at the Rockefeller Institute in New
York, which established the critical role of DNA (deoxyribonucleic acid) in
the transmission of hereditary information. Once DNA had been identified
as the basic component of genes, researchers focused their attention toward
discovering its chemical structure. In 1953, James Watson and Francis Crick
reported that DNA had a double-helix structure and, by 1966, deciphered
how the four bases that were its essential chemical components could code
for hundreds of amino acids, the building blocks of proteins.

Yet, without a way to amplify DNA (i.e., produce it in large quantities), the
task of sequencing the genome would remain stubbornly laborious (Bodmer
& McKie, 1994, p. 54). The break came in 1973 when Stanley Cohen and
Herbert Boyer devised an elegant way to coax an *E.coli* bacterium to tirelessly
reproduce a particular DNA sequence. Developing a method to quickly replicate a piece of DNA in the laboratory was a significant achievement in itself.
What made Cohen and Boyer's work revolutionary was that the pieces of
DNA that the *E.coli* was so busy replicating were foreign DNA. Cohen and
Boyer had recombined DNA from two different types of bacteria and inserted
this into the *E.coli* (Bodmer & McKie, 1994). This unique procedure was
called *recombinant DNA technology*. By making it possible to clone a gene,
that is, to isolate and reproduce a particular gene at will in the laboratory,
rDNA technology opened up the possibility for commercial applications. This
breakthrough by Cohen and Boyer has been described as "the single pivotal
event in the transformation of the "basic" science of molecular biology into an
industry" (Kenney, 1986, p. 23).

The rDNA technique devised by Cohen and Boyer promised to revolu-
tionize genetic engineering. At the same time, it raised concerns among scientists about the safety of the procedure. In particular, biologists debated the
possible health dangers of inserting foreign DNA into bacteria.

Regulating Genetic Engineering Research: Professional Initiatives

Stanford professor Paul Berg led a group of prominent molecular biologists
who "called for a temporary moratorium on certain types of experiments"
(Kenney, 1986, p. 23). This group also proposed the formation of an advisory committee by the National Institutes of Health (NIH) to study the possible dangers of rDNA research (Kenney, 1986). In February 1975, with the
support of the National Academy of Sciences, this group organized an international meeting of scientists at Asilomar, California.

The Asilomar Conference. The conference was "for the purpose of assessing the risks of the new technology and establishing the conditions under which research could or should proceed" (Krimsky, 1982a; cited in Kenney, 1986, p. 23). It is worth noting that the organizers deliberately kept debates about the social and ethical implications of the rDNA research off the agenda (Kenney, 1986). Although its organizers originally intended to invite only scientists (Kenney, 1986), the Asilomar conference ultimately included "139 researchers, along with a number of science administrators, journalists, and experts in law and ethics" (Lee, 1993, p. 4).

The participants concluded that it was safe to continue with rDNA research "provided that biological and physical containments were followed." (Lyon & Gorner, 1995) Soon after the Asilomar conference in April 1975, a self-imposed, worldwide moratorium "involv[ing] aspects of research with tumor-inducing viruses and with genes responsible for producing potent toxins in bacteria" was put in place (Lee, 1993, p. 4).

The Recombinant Advisory Committee. The NIH Recombinant DNA Advisory Committee (RAC), which was instituted in response to recommendations by Paul Berg and his colleagues, met the day after the Asilomar Conference (Kenney, 1986, p. 24). Its task was to prepare guidelines for molecular biologists. It took 16 months of debate and dissent before these guidelines were finally announced in July 1976 (Kenney, 1986). These guidelines "were mandatory only for recipients of federal funding and exercised no control over research undertaken or funded by industry" (Kenney, 1986). Interestingly, the RAC, which was widely criticized for its tardiness by scientists who had voluntarily suspended their experiments involving genetic engineering until the issuance of these guidelines (Kenney, 1986), continues to be the target of severe criticism even today. Its critics claim that "RAC reviews constitut[e] a redundant and unnecessary regulatory hurdle that is retarding the advancement of the science" (*Cancer Researcher Weekly*, 1994).

The National Debate on rDNA Research. The Asilomar conference played an important role in raising public awareness of the development of techniques involving genetic manipulation. It had the unintended effect of "rais[ing] fears of epidemics caused by escaped organisms" (Kenney, 1986, p. 24). Concerns about the haphazard use of the technology quickly made its way into the Congress and city halls. In 1975, the same year as the Asilomar conference, the Senate's subcommittee on health, under the leadership of Senator Kennedy, discussed the role that the government ought to play in monitoring rDNA research (Krimsky, cited in Kenney, 1986). By the end of that year, articles appearing in the *New York Times*, *Washington Post* and the *Boston Globe* had begun questioning the safety of such research (Kenney, 1986).

In 1976, Harvard University's plan to construct a genetic engineering containment facility drew the attention of the Cambridge, Massachusetts City Council. "After a bitter debate the city council asked Harvard and MIT to desist from carrying out [the] experiments until a civilian review board could draw up recommendations for the city council" (Kenney, 1986, p. 24). Permission was eventually granted for the construction of the facility in January 1977, but not before the council had put its own restrictions in addition to those required by the NIH guidelines of 1976 on the scope of the research project (Kenney, 1986). The interest of local governments in overseeing genetic research carried out in laboratories within their jurisdiction and setting safety standards reflected the public concern that this esoteric discipline had aroused (Kenney, 1986).

But despite the concerns voiced by different legislators, no laws were passed to curb research activity in the field. The bills drafted by Senator Kennedy and Representative Rogers in 1977 that would require all biotechnology research to be subject to the NIH guidelines were never passed. This was in no small measure due to the intense lobbying efforts of scientists and university administrators who had organized a group, Friends of DNA, for the purpose (Kenney, 1986).

Although the safety guidelines issued by the NIH in 1976 applied only to federally funded research, they nevertheless played an important role in calming critics concerned with the haphazard use of the technology (Lee, 1993). As a result, they could be gradually relaxed and, by 1979, researchers had begun working on cancer viruses (Lee, 1993). By 1983, the original guidelines had been watered down sufficiently to allow "nearly all conceivable rDNA experiments" (Kenney, 1986, p. 27). It was the culmination of a steady process toward the relaxation of genetic research guidelines, and was evidence of the growing realization that the initial concerns about recombinant DNA technology had been exaggerated.

Patent Protection for Genetically Engineered Products

Apart from the role played by the NIH guidelines and the fact that none of the dangers of rDNA research had materialized, the lure of huge profits from the exploitation of this technology was also instrumental in silencing opposition to genetic research (Kenney, 1986). By the early 1980s, a wave of entrepreneurial activity was well under way in the field, leading to the creation of numerous biotechnology startups. Powerful multinational corporations also had begun demonstrating interest in biotechnology (Kenney, 1986) and patent protection for biotechnology products became a burning legal issue.

The Diamond vs. Chakrabarty Decision. In 1972, microbiologist Ananda Chakrabarty had filed a patent application on behalf of the General Electric Company for a bacterium of the genus *Pseudomonas*. What made this microorganism unique was that it had been genetically engineered to simultaneously break down multiple components of crude oil. No naturally occurring bacteria possessed this ability. This particular microorganism, which could prove extremely useful in the treatment of oil spills, clearly appeared to have significant commercial promise (Greenhouse, 1980). Initially, the U.S. Patent and Trademark Office (PTO) had refused to grant a patent, arguing that Congress had never intended living organisms to be patentable (Kenney, 1986). Unhappy with this decision, General Electric took its case to the U.S. Court of Customs and Patent Appeals (CCPA), which ruled in 1979 that the microorganisms were, in fact, covered by existing patent laws. This time, the federal government appealed the decision and took the case to the U.S. Supreme Court (Greenhouse, 1980).

In the landmark Diamond vs. Chakrabarty (1980) case, the U.S. Supreme Court ruled that life forms created in the laboratory were indeed patentable. As the author of the majority opinion, Chief Justice Warren Burger noted that in drawing up the patent laws, the Congress had created a distinction "not between living and inanimate things, but between products of nature, whether living or not, and human-made inventions" (p. 79). Therefore, the main issue before the court was to "determine whether [Chakrabarty's] microorganism constitute[d] a "manufacture" or "composition of matter" within the meaning of the [relevant] statute" (p. 79). Chief Justice Burger observed that although the law was sufficiently narrow in its scope to refuse patents to "laws of nature, physical phenomena and abstract ideas" (p. 136), Chakrabarty's microorganism satisfied the requirements for patenting. In particular, Justice Burger pointed out that "[Chakrabarty's] claim is not to a hitherto unknown natural phenomenon, but to a nonnaturally occurring manufacture or composition of matter—a product of human ingenuity having a distinctive name, character and use" (National Desk, 1980, p. 17). This ruling, which passed narrowly on a 5 to 4 vote, was attacked by prominent scientists concerned about the release of genetic organisms into the environment, and religious leaders concerned about the sanctity of life.

The Supreme Court ruling persuaded the PTO that life forms created in the laboratory deserved full patent protection because they could be classified as inventions and met the requirements of existing patent laws. Since the *Diamond v. Chakrabarty* ruling, patents have been awarded for virtually all kinds of nonhuman life. In April 1988, an important milestone was established when researchers at Harvard University won a patent for "OncoMouse," a mouse that contained a human gene in each of its cells, and is a useful model for studying the development of breast cancer (Lee, 1993, p. 179).

Venture Capitalists, Initial Public Offerings (IPOs) and Successful Institutionalization

"Biotechnology has emerged as an industry largely because of one economic institution: venture capital" (Kenney, 1986, p. 133). The foundation of Genentech, the first company committed exclusively to genetic engineering (Kenney, 1986) in 1976 was an early milestone in the relationship between biotechnology scientists and venture capitalists. Although this event went "largely unnoticed" at the time (Kenney, 1986, p. 25), it is an interesting story and worth recounting here.

Venture Capitalists.   Soon after he and his colleague, Stanley Cohen, had demonstrated that genes could be cloned, Herbert Boyer was approached by a young venture capitalist, Robert Swanson. Swanson wanted to know whether gene-splicing technology could be used to persuade bacteria to produce human proteins. It was a novel idea; most of Boyer's colleagues believed that it would be at least a decade before any commercial applications of rDNA technique could be found. A meeting between Boyer and Swanson to discuss this prospect lasted nearly 4 hours (although Boyer had warned earlier that he could spare only 20 minutes). At the end of that meeting, Boyer was sufficiently convinced about the promise of the emerging technology to borrow $500 from Swanson and become a partner in founding Genentech (Bodmer & McKie, 1994).

Like the meeting between Boyer and Swanson, "the formation of a genetic engineering company nearly always involved an entrepreneur soliciting various professors until he discovered one who was interested in forming a company" (Kenney, 1986, p. 137). In return for private equity placements, venture capital firms invested a substantial amount of seed money (sometimes in the range of a few hundred thousand dollars) in startups, and contributed to successive rounds of financing to support research in the fledgling enterprise (Kenney, 1986).

From 1976 to 1983, venture capital firms financed more than 100 startups in the biotech industry. By 1980, however, the concerns about safety of rDNA research had abated and the scope for commercialization of the technology became apparent to molecular biologists. Now, "scientists ... directly approached venture capitalists with business plans" (Kenney, 1986, p. 138). This subtle shift in the pattern of startup formations had two important implications for the industry.

As scientists approached venture capitalists with proposals, competition for financing became more intense, and enabled venture capitalists to exercise power and demanded that startups recruit professional managers, who by pressing for early commercialization would help them realize a tidy profit (Kenney, 1986). Increased competition for funding enabled venture capitalists to choose firms that had more than just a business plan.

In turn, venture capital firms infused capital and allowed early startups to remain private for a fairly long period of time and escape many of the regulations that constrained the activities of publicly owned companies (Kenney, 1986). Additionally, startups could tap into a "vast network of connections with business development consultants"—these included executives with experience in marketing, finance and management, and experts who could provide legal counsel on patent laws (Kenney, 1986, p. 144). Finally, venture capitalists promoted the adoption of professional management techniques by the startups. Once venture capitalists began evaluating projects on the basis of "management competence as much as on market potential" (Kenney, 1986, p. 139), biotechnology startups had to appoint experienced managers who would "allocate scientific resources optimally" and "concentrat[e] on project implementation and marketing" (Eagle & Coyman, 1981 cited in Kenney, 1986, p. 149).

Wall Street and IPO's. As competition for financing from venture capital firms had become intense by the early 1980s, and as the investment needed by an individual firm to conduct research grew quickly over time, public equity markets became important sources of capital.

Small biotechnology startups were initially reluctant to make public offerings (Kenney, 1986). Like privately held firms in other industries, they jealously guarded their relative autonomy and were keen to avoid the regulations to which public companies must adhere. These firms were also concerned that because of their small size, they would be forced to divulge events that larger firms could choose not to disclose (Kenney, 1986), and in the process undermine their competitive strategy. With negligible sources of revenue except for a few research contracts with large multinational corporations (MNCs), and large expenditures in the form of research and development (R&D), these startups were afraid that public investors would balk at their cash flow statements (Kenney, 1986). Investor enthusiasm for biotech startups and the need for money induced many entrepreneurs to go public in the 1980s. Wall Street's response to Genentech's public offering was instrumental in triggering the spurt of IPOs by biotech startups in the early 1980s.

The hugely successful public offering by Genentech on October 14, 1980, marked a turning point for the industry (Hybels, 1995). It sparked a "buying frenzy" (Kenney, 1986, p. 156), and "set a record for the fastest increase in stock price for an IPO, from $35 to $89 in twenty minutes" (Hybels, 1995, p.

12). For the first time, biotechnology companies realized that despite their huge monthly cash expenses and woeful record of commercialization, the public was willing to invest huge amounts of money in the sector. Firms rushed to seize this window of opportunity to fund their expansion, and although public offerings in the following couple of years were not as successful, they were received with considerable enthusiasm. Apparently, Wall Street was unconcerned by the fact that the biotechnology industry had produced only one marketable product by 1982 (Kenney, 1986). IPOs rebounded with the rising stock market in 1983 when there were more public offerings of biotechnology stocks "than in the previous three years combined" (Kenney, 1986, p. 157). Before 1981, there were barely 336 biotech companies in the United States. But in 1981 alone, over 100 companies were founded in the biotechnology sector (Burrill & Lee, 1993). Wall Street was sufficiently enamored of the technology to offer funding to even risky "me too" companies that had been turned away by venture capitalists (Kenney, 1986). Indeed, citing Bylinsky, Hybels (1995) noted: "By the mid-1980s some biotechnology firms, such as Nova Pharmaceutical were going public immediately after founding, when the organizations were little more than a set of agreements among venture capitalists and a few principal scientists and managers" (p. 13).

CONCLUSIONS

We began this paper with the aim of illuminating how an essential component of the construction of new forms was a politicoinstitutional process and chose two "extreme" cases to gain empirical insights. Because the early automobile and biotechnology cases pertain to two very different time periods in the evolution of American industry, some striking differences exist in the creation of both forms.

An important difference between the automobile and biotech industries relates to the role of the state. The state constrains the creation of new forms both as a collective actor and as an institutional structure (Campbell & Lindberg, 1990). In the former case, as a set of semiautonomous actors, agencies of the state grant charters, allocate finance or monopoly status, and impose taxes and regulatory controls (see Singh, Tucker, & Meinhard, 1991). In the latter, states have the capacity to define and enforce property rights (rules that determine the conditions of ownership and the control of the means of production). In the automobile industry, semiautonomous components of the government, especially municipal governments, indirectly protected the fledgling industry. The federal government played a passive role and became a consumer after the establishment of the industry. Municipal and state governments assisted the industry by routinizing registration, licensing, and speed rules. These regulations dampened the eventual viability of antispeed organizations and bolstered the standing of the automobile. By contrast, in the biotechnology industry, the capacity of the state to define and enforce property rights proved to be decisive in inducing the foundings of startups. The U.S. Supreme Court ruling in the *Diamond vs. Chakraborty* (1980) case that life forms created in the laboratory were indeed patentable spurred the growth of the biotechnology industry.

Another striking difference between the automobile and biotechnology industries concerns the significance of professions in industry emergence. Professions shape the rise of new organizational forms by providing cognitive frameworks and by spawning formal structures to create and defend jurisdictional claims (Abbott, 1990). Professional activity had minimal effects on the rise of the automobile industry because the profession of automobile engineers was not yet established. In contrast, professional intervention was decisive in the biotechnology industry. On the one hand, molecular biologists took the lead in establishing frameworks for self-regulation, and on the other hand, they were among the critical building blocks for biotechnology startups. In turn, biotechnology firms transformed the role of the molecular biologist from a pure scientist into a scientist entrepreneur (Powell & Smith-Doerr, 1994).

Nevertheless, there are two striking commonalties in the creation of the automobile and biotechnology industries. One important similarity between the early automobile industry and the early biotechnology industry is that the creation of new paths in both cases was a political process. Stinchcombe (1968) asserted that the entrepreneurial creation of new forms "is pre-eminently a political phenomenon" (p. 189) because support has to be mobilized for the goals, authority structure, technology, and clients embodied in the new form. In the case of the early automobile industry, opposition from vigilante antispeed organizations jeopardized the standing of the automobile, and had not automobile clubs played an active role in defusing opposition, the industry might have faced stringent legal constraints on the use of cars. Similarly, in the biotechnology industry, concerns about the dangers of rDNA technology and the release of reckless organisms might have led to restrictive laws had not professionals quickly devised voluntary safeguards, and forestalled governmental intervention.

Finally, the dominant actors in the creation of the early automobile and biotechnology industries were external actors rather than producers. In both instances, collective action by firms within the industry played a small role in the establishment of the new industry. In the automobile industry, collective action on the part of automobile manufacturers failed to take root in the early history of the industry, as trade associations disintegrated due to internecine warfare. By contrast, it was automobile enthusiasts and automobile clubs who established reliability contests to demonstrate the capabilities of the horseless vehicle. Similarly, business firms whose interests were threatened by the automobile such as horse carriage manufacturers, did not mount serious opposition to the automobile industry. Makers of horse carriages diversified into cars and stable owners blossomed into garage shops. Opposition to the automobile was led by vigilante antispeed organizations and farmers organizations. In turn, automobile clubs defended the industry from the assaults of antispeed activists and lobbied the government. In the biotechnology industry, incumbent pharmaceutical firms mounted little opposition to the new industry and many of them acquired biotechnology boutiques in due course. Collective action by technology startups had little do with the successful institutionalization of the industry. Instead, it was venture capitalists and investment banks promoting IPOs who made the industry viable by providing access to capital.

Taken together, both cases suggest that the construction of new paths is a politicoinstitutional process, wherein new forms have to been justified and integrated into the prevalent institutional order. Both cases also imply that there is historical variation in the salience of the state and the professions. However, both cases emphasized one side of the coevolutionary relationship between forms and culture and emphasized how forms need to be connected and linked to the prevalent cultural order. Future research needs to shed light on how new forms alter and modify the prevalent cultural order.

We would like to conclude by noting that the primary focus of this chapter has been the creation of new organizational forms. Yet, the fundamental underlying question can concern equally both the creation and demise of organizational forms. Does the creation process, followed by the growth in numbers of a specific form, ultimately culminate in the demise of the form? What factors lead to this demise? Does the politicoinstitutional context play a role in this demise process as well? We cannot do full justice to these and other important questions, but we can suggest some early ideas.

On this question, our position is distinct from organizational ecologists (Hannan & Carroll, 1992; Hannan & Freeman, 1989). They suggest that the demise of organizational forms is a consequence of competitive forces playing out.[2] Although that may sometimes be the case, we believe that institutional selection (Rao & Singh, 1998) has a crucial bearing on this question of demise. Changes in the politicoinstitutional context, whether gradual or sudden, that have little to do with competitive selection, frequently account for the demise of forms. Some of these changes may occur in the guise of regulation, yet other changes may involve more fundamental

---

[2]It is worth noting, however, that most empirical studies in organizational ecology have not focused on the demise of organizational forms. The focus, instead, has usually been on the demise of specific instances of a particular organizational form, as in the density dependence literature.

changes in the context of institutionalized myths and beliefs in society (Meyer & Rowan, 1977). The contemporary example of the U.S. tobacco industry is a case in point. Many knowledgeable observers assert that the tobacco industry will likely not survive in its current configuration. Yet, only a few years ago it was a thriving industry. The key change concerns a gradual shift in values in U.S. society about the role of smoking cigarettes, a primary product of this industry, in public health. It is likely that tobacco firms recognize the fundamental nature of this threat, considering that until recently, leading companies were willing to sign on to a $370 billion settlement with various states in order to avoid some kinds of legal action. We acknowledge that this is an oversimplification of the complex forces at play. Yet this case illustrates well the threat to a recently thriving organizational form that is now under increasing attack, and not from increased competitive forces.

ACKNOWLEDGMENTS

We thank Howard Aldrich, Terry Amburgey, Joel Baum, Paul DiMaggio, Dick Scott, Mike Tushmanan, and Sid Winter for their helpful suggestions.

REFERENCES

Abbot, A. (1990). *The System of the Professions*. Chicago: University of Chicago Press.

Aldrich, H. E. (1979). *Organizations and Environments*. Englewood Cliffs, NJ: Prentice-Hall.

Aldrich, H. E., & Fiol, M. (1994). Fools rush in? The institutional context of industry creation. *Academy of Management Review*, 19(4), 645–670.

Arthur, B. (1989). Competing technologies, increasing returns, and lock-in by small events. *Economic Journal*, 99, 116–131.

Astley, W. G. (1985). The two ecologies: Population and community perspectives on organizational evolution. *Administrative Science Quarterly* 30(20), 224–241.

Baum, J. A. C., & Singh, J. V. (1994). Organizational niches and the dynamics of organizational founding. *Organization Science*, 5, 483–501.

Bijker, W. (1987). The Social Construction of Bakelite: Toward a Theory of Invention. In W. E., Bijker, T. P. Hughes & T.J. Pinch (Eds.), *The Social Construction of Technological Systems*. Cambridge, MA: MIT Press (pp. 159-187).

Bodmer, W., & McKie, R. (1994). *The book of man—The human genome project and the quest to discover our genetic heritage*. New York: Scribner.

Brittain, W. J., & Freeman, J. (1982). Organizational proliferation and density dependent selection. In R. Kimberly & R. H. Miles (Eds.), *The Organizational Life Cycle* (pp. 291–338). San-Francisco: Jossey-Bass.

Burrill, G. S., & Lee, K. B. (1993). *Biotech 94: Long-term value, short-term hurdles*. Ernst & Young, San Francisco.

Campbell, J. L., & Lindberg, L. (1990). Property rights and the organization of economic activity by the state. *American Sociological Review*, 55, 634–647.

Cancer Researcher Weekly. (1994). C.W. Henderson Publisher, December 19.

Carroll, G. R. (1984). Organizational ecology. *Annual Review of Sociology*, 10, 71–93.

Carroll, G. R., Delacroix, J., & Goodstein, J. (1990). The political environment of organizations; An ecological view (pp. 67–100). In B. Staw & L. L. Cummings (Eds.), *Evolution and adaptation of organizations*. Greenwich, CT: JAI.

David, P. (1985). Clio and the economics of QWERTY. *American Economic Review Proceedings, 35,* 128–152.

Diamond, Commissioner of Patents and Trademarks v. Chakrabarty. (1980). No. 79–136; Supreme Court of the United States, 447, U.S. 303; 100 S. Ct. 2204.

DiMaggio, P. (1991). Constructing an organizational field as a professional project: U.S. art museums 1920–1940. In W. W. Powell & P. DiMaggio (Eds.), *The New institutionalism in organizational analysis* (pp. 267–292). Chicago: University of Chicago Press.

Edison, T. (1895). *The story of the American automobile.* Cited in R. E. Anderson, (Ed.), 1950. Washington, D.C.: Public Affairs Press, p. 96.

Eisenhardt, K. M. (1989). Building theories from case study research. *Academy of Management Review, 14,* 532–550.

Epstein, R. C. (1928). *The automobile industry: Its economic and commercial development.* Chicago: A. W. Shaw & Co.

Fligstein, N. (1996a). How to make a market: Reflections on the attempt to create a single market in the european union. *American Journal of Sociology, 102,* 1–33.

Flink, J. J. (1970). America adopts the automobile, 1895–1910. Cambridge, MA: MIT Press.

Flink, J. J. (1988). The automobile age. Cambridge, MA: MIT Press.

Greenhouse, L. (1980). Science may patent new forms of life. *The New York Times,* June 17, Sec. A, p. 1, Col. 6.

Hannan, M. T., & Carroll, G. (1992). *Dynamics of Organizational Populations,* New York: Cambridge University Press.

Hannan, M. T., & Freeman, J. (1989). *Organizational Ecology.* Cambridge, MA: Belknap.

Harper's Weekly. (1899). *The status of the horse at the end of the century, 43,* 1172; November 18.

Harper's Weekly. (1907). *The horse of the future and the future of the horse, 51,* 383; March 16.

Haveman, H., & Rao, H. (1997, August). Structuring a theory of moral sentiments: institutional–organization co-evolution in the early thrift industry. *American Journal of Sociology, 6,* 1606–1651.

Hybels, R. C. (1995). *On operationalizing the legitimation construct in organizational theory: A content analysis of press coverage of the U.S. biotechnology field, 1971–1989.* Presented a the Academy of Management national meeting in Vancouver.

Independent. (1909). *How to select an automobile, 66,* 1213; June 3.

Karnøe, P., & Raghu G. (1995). *Path creation and dependence in the danish wind turbine field.* Working paper, Institute of Organization, and Industrial Sociology. Copenhagen Business School.

Kenney, M. (1986). *Biotechnology: The university industrial complex.* New Haven, CT: Yale University Press.

Kimberly, J. R. (1975). Environmental constraints and organizational structure: A comparative analysis of rehabilitation organizations. *Administrative Science Quarterly, 20,* 1–19.

Lee, T. F. (1993). *Gene future—The promise and perils of the new biology.* New York: Plenum Press.

Lucas, H. (1996). *The T-form organization: Using technology to design organizations for the 21st century.* San Francisco: Josey-Bass.

Lumsden, C., & Singh, J. (1990). The dynamics of organizational speciation. In J. V. Singh (Ed.), *Organizational evolution: new directions* (pp. 145–163). San Francisco: Sage.

Lyon, J., & Gorner, P. (1995). *Altered fates: Gene therapy and the retooling of human life.* New York: W.W. Norton & Co.

Meyer, J. W., & Rowan, B. (1977). Institutionalized organizations: Formal structure as myth and ceremony. *American Journal of Sociology, 83,* 340–363.

McKelvey, B. (1982). *Organizational systematics: taxonomy, evolution, classification.* Berkeley: The University of California Press.

Motor Age. (1903). *Unfair public demands, 3,* 3–4; June 18.

Munsey's Magazine. (1906). *The automobile in America, 34,* 404; January.

National Desk (1980). Excerpts from Supreme Court opinions on man-made life forms, *The New York Times,* June 17, Section D, p. 17, Col. 3.

Nelson, R. R., & Winter, S. G. (1982). *An evolutionary theory of economic change.* Cambridge, MA: Belknap.

Nelson, R. R., & Winter, S. G. (1982). *An evolutionary theory of economic change.* Cambridge, MA: Belknap.

Porac. J., Rosa, J., & Saxon, M. (1997, August). *America's family vehicle: The minivan market as an enacted conceptual system.* Paper presented at the Conference on Path Creation and Dependence. Copenhagen Business School.

Powell, W. W., & Smith-Doerr, L. (1994). Networks and Economic Life, pp. 368–402. In Neil Smelser and R. A. Swedberg (Eds.), *Handbook of Economic Sociology,* New York: Russell Sage.

Rae, J. B. (1959). *American auto manufacturers: The first forty years.* Philadelphia: Chilton.

Rao, H. (1998). Caveat emptor: The construction of non-profit consumer watchdog organizations. *American Journal of Sociology,* pp. 912–961

Rao, H., & Singh, J. (1999). Types of variation in organizational populations: The speciation of new organizational forms (pp. 63–78), in Joel A. C. Baum and B. McKelvey (Eds.), *Variations in Organization Science.* Thousand Oaks, CA: Sage.

Romanelli, E. (1989). Organizational birth and population variety: A community perspective on origins (pp. 211–246). In B. Staw & L. L. Cummings (Eds.), *Research In Organizational Behavior* (Vol. 2, pp. 211–246). Greenwich, CT: JAI.

Schumpeter, J. A. (1950). *Capitalism socialism and democracy.* New York: Harper.

Scientific American. (1895). *The horseless age,* 73, 107; August 17.

Scott, W. R. (1995). *Institutions and organizations.* Thousand Oaks, CA. Sage.

Singh, J. V. (1993). Review essay: Density dependenct theory—current issues, future promose. *American Journal of Sociology,* 99, 464–473.

Singh, J. V., & Lumsden, C. J. (1990). Theory and research in organizational ecology. *Annual Review of Sociology,* 16, 161–195.

Singh, J. V., Tucker, D., & Meinhard, A. (1991). Institutional change and ecological dynamics, (pp. 390–422). In W. W. Powell & P. DiMaggio (Eds.), *The New Institutionalism in Organizational Analysis.* Chicago: University of Chicago Press.

Stinchcombe, A. L. (1965). Social structure and organizations. In G. March (Ed.), *Handbook of organizations* (pp. 142–193). Chicago: Rand McNally.

Stinchcombe, A. L. (1968). *Constructing social theories.* Chicago: University of Chicago Press.

Thomas, R. P. (1977). *An analysis of the patterns of growth of the automobile industry, 1895–1929.* New York: Arno Press.

Tucker, D. J., Singh, J. V., & Meinhard, A. G. (1990). Founding characteristics, imprinting and organizational change. In J. V. Singh (Ed.), *Organizational evolution: New directions* (pp. 181–200). Newbury Park, CA: Sage.

Tushman, M. L., & Anderson, P. (1986). Technological discontinuities and organizational environments. *Administrative Science Quarterly,* 31, 439–465.

Van de Ven, A. H., & Garud, R. (1994). The co-evolution of technical and institutional events in the development of an innovation. In J. A. C. Baum & J. Singh (Eds.), *Evolutionary Dynamics of Organizations* (pp. 425–443). New York: Oxford University Press.

10

Constructing Transition Paths Through the Management of Niches

René Kemp
Maastricht University, the Netherlands

Arie Rip
Johan Schot
University of Twente

Caminante no hay camino, se hace camino al andar[1]
(the traveler has no path, he creates one as he goes)

The phenomenon of path dependency in economic development and technological change is widely recognized. It introduces a historical element and the notion of irreversibility into economic analysis (Arthur, 1988; Hahn, 1989; Organization for Economic Co-operation and Development, 1992) and bridges the gap with sociological analysis (North, 1990).

In this chapter we are less interested in causes of path dependency than in the possibility of intentionally constructing a desirable path. There are different ways to do this. One can try to construct a path by brute force, that is, by planning a new system and working to overcome the barriers to its realization by eliminating them. This is often done in the case of physical infrastructures and complex technical systems like the grid-based electrical

---

[1]From a Spanish song by Antonio Machado.

system. One can also try to bend the development process by judiciously applying economic (or social, for that matter) incentives and disincentives, so as to make some possible paths more, and others less interesting and feasible. Thirdly, one can float with the coevolution processes and modulate them. Depending on circumstances and on the position of the actor who takes the lead (government, public–private consortium, firms, nongovernment organizations, and societal groups), one or the other approach will appear more suitable and/or more advantageous.

Elsewhere, we have argued that the third approach is important in modern societies, and that it is underdeveloped as yet (Schot & Rip, 1997). We have also articulated a method for constructing paths, called *strategic niche management* (SNM—the creation, development, and breakdown of protected spaces for promising technologies. Originally, the method was positioned as making the introduction of new technology more successful (Rip, 1992). Our exposition of the method in this chapter shows the same orientation, for example when specifying enhancing the rate of application of the new technology as the aim of SNM. But the method is now part of a broader framework: the build-up of new technological regimes and the possibility of intentionally working toward desired regime change.

The structure of the chapter is as follows. The nest section provides a discussion of patterns in coevolution processes, as a way to broaden the notion of path dependency (Rip & Kemp, 1998). We provide an analysis of such patterns and the role of niches and regimes in coevolution patterns. This helps prepare the ground for a discussion of strategies for achieving technological regime shifts. This discussion is followed by a description of the Californian and Danish wind power policies as examples of two different types of policy approaches. The case of wind energy development is interesting because it shows that an existing electric power regime can become less homogeneous and more open for change. The case is also interesting in that it allows a comparison between de facto SNM (in Denmark) and an incentive-based approach (in California). The cases do not permit us, however, to evaluate these approaches in terms of achievement of regime shifts because a regime shift has (yet) not occurred.

We then present and discuss the method of SNM as a way to create a new path through the creation and management of protected spaces for promising technologies and as a tool of transition. Of course, actual transitions are complex and contingent, and it will not be clear, at the outset, whether the path under construction will actually become a transition path. In other words, SNM may well be a necessary element to bring about a transition, but it will not be sufficient. In particular, the stabilization of a new regime as a mosaic of rules (including dominant designs and standardization) has dynamics and patterns of its own. SNM can help sway the balance there, but needs to be developed further to function as a management tool. When we reflect on the potential and the difficulties of SNM, in the final section, we come back to this issue.

TECHNOLOGICAL REGIMES AND TRANSITION PATHS

Technological change is not an autonomous force, nor is it a haphazard process; it is structured and focused, geared toward solving particular problems that have grown in the process of development, and endogenous to the structure of economic incentives, firms' capabilities, (legal) standards, and economic interests. New technologies are not created outside society, but part and parcel of social–technical transformation processes.

In this transformation process, the frontiers of knowledge shift, new problems emerge, people's preferences change, old institutions dissolve, while others are created. A recurrent theme in the social study of technology is the evolutionary nature of technology in which technical change is seen as bounded and open-ended—bounded in the short term and open-ended in the long run. Rosenberg (1994) speaks of a "soft determinism" (p. 17) that is exercised by technological trajectories. *Technological change* is a path dependent cumulative process in which the existing body of knowledge, techniques and tools determine which further steps can be taken at any time. There is an "overhang of technological inheritance" that shapes ongoing technological research.

There is also second aspect of path dependency, already discussed in an older work of Rosenberg (1976), which is that technological research is essentially a problem-solving activity and that engineering attention is drawn to devote attention to problems that present themselves. In particular, observed imperfections in existing products and processes are instrumental in focusing the attention of engineers on particular problems by throwing up signals about what technological efforts can be usefully exercised (Rosenberg, 1976). This point is developed further in the work of Nelson and Winter (1982), and Dosi (1982, 1988) who writes that technological change is cumulative, selective, and finalized into specific directions somewhat irrespective of the cost and demand structure.

Sociologists focusing on actor strategies and networks take a partially different, but overlapping view. According to one influential representative, Donald MacKenzie, technological trajectories are sustained not by an internal logic but by the interests that develop in their continuance and the belief that the trajectory will continue (MacKenzie, 1992). The very presence of trajectories is a continual and contingent achievement of actors working to stabilize technical change (see Rip, Misa, & Schot, 1995).

Implicit in these approaches is that technological change is not an autonomous, deterministic process but constrained and structured: (a) by engineering consensus about the relevant problems and how these may be solved, (b) by the available methods and techniques, (c) by the organizational and institutional context, and (d) by patterns of infrastructures and demand. The structured nature of technological change means that there is a "grammar" or rule set that shapes ongoing research and investment (Rip & Kemp, 1998). Examples of these rules are: (a) the search heuristics of engineers (about the relevant problems to be solved and how to solve them); (b) the investment selection criteria employed by private firms operating in the market place (with its rules for survival); and (c) organizational procedures, technical standards, social norms, regulatory standards, and rules of ownership and patent protection. These rules are not something that exist outside technology or the organizational and social context in which technology is developed and used but are embedded in engineering practices, process technologies, product characteristics, skills, institutions, and infrastructures of a technology. They are an integral part and structuring force of sociotechnical coevolution processes.

Elsewhere we have used the term *technological regime* to characterize this rule set (Kemp et al., 1994; Rip & Kemp, 1998). Combining both economic and sociological approaches, a *technological regime* is defined as the grammar or rule set comprised in the complex of scientific knowledges, engineering practices, production process technologies, product characteristics, skills and procedures, and institutions and infrastructures that make up the totality of a technology (for example, an internal combustion engine or gas turbine) or a mode of organization (for example, the Fordist regime of mass production). Although others (like Hughes, 1983) used the term *technological system*, we prefer to use *regime* because regime refers to rules: in particular, the rules and grammar that are implied in sociotechnical configurations. The notion of regime also allows for a better appreciation of the decentralized multiactor dynamics at play in technical change.

Technological regimes are a broader, "socially embedded" version of a technological paradigm. A technological regime combines the rules that are embedded in engineering practices and heuristics with the rules of the selection environment. Accordingly, our definition differs from that of Malerba and Orsenigo (1993) and Dosi, Malerba, and Orsenigo (1993) where a technological regime comprises a particular combination of technological opportunities, appropriability conditions, degree of cumulativeness of technological knowledge, and characteristics of the relevant knowledge base (Dosi et al., 1993). The idea of regime is used by the authors to explain differences in innovation patterns and sectional structures, to explain what Nelson and Winter (1977) called the "differential productivity growth puzzle" (p. 41). Their definition revolves around knowledge: cumulativeness, tacitness, and publicness of knowledge and how new products embodying knowledge may be protected against imitation. Our definition is less about knowledge and more about rules defining innovative activities. There is a second important difference. In Malerba and Orsenigo's (1993) definition, the demand aspects and broader social aspects (norms, lifestyles) are virtually absent, whereas in our definition, they are an important constituent part of a regime. Because of this, our definition is more in line with the way in which the concept of regime is used in studies of international relations, political science, and figuration sociology (e.g., Spier, 1995).

Technological regimes, as we define them, are configurations of science, technics, organizational routines, practices, norms and values, labeled for their core technology or mode of organization. Between the different elements, strong and weak linkages occur, creating a semicoherent structure that serves to guide engineering activities in particular directions and not in others. The prefix "semi" is important, because such structures are not perfectly coherent: there are tensions in the form of product imperfections, side effects, bottlenecks (reverse salients or system limits), and unsatisfied demands (of consumers or general public). There may also be different competing designs. Designs are not given as such but evolving. The way in which they evolve, however, is not ad random but structured. In particular, designs are structured by the accumulated knowledge, engineering practices, value of past investments, interests of firms, established product requirements and meanings, intra and interorganizational relationships, government policies, that make up a technological regime.[2] This is the key tenet of (quasi) evolutionary approaches of technological change.

The concept of a technological regime is important not only to understand the prestructured nature of technological change but also to come to terms with the problem of technological transition. Viewing it as a process of transformation of an old regime (or a set of regimes) into a new one, the question central to transition processes can now be adressed as one of how regimes change. For example are they transformed through external pressure or by internal inconsistencies and system limits? There is little theory about the issue. All we have are historical accounts of technological transitions that emphasize different aspects of what causes old systems to fall into decline and new systems to grow. One explanation emphasizes the role of system builders. In this view, new systems are created by system builders—people of imagination and persistence who perceived early on the op-

---

[2]It should be noted that it is not just a matter of mind sets of engineers who are committed to a particular technology. Also policy actors, companies, and consumers are directly or indirectly, willingly or unwillingly, committed to a particular technological regime (e.g., the fossil fuel-based electric system).

portunities offered by a new technology, who conceived the new technology as the constitutive part of a new system, and who managed the transition process toward a new system. History of technology is often written around these entrepreneurs, picturing them as people of vision and determination. The names of Edison, Insull, and Mitchell are then associated with the development of the electric power system (Hughes, 1979, 1983). There was Edison, the inventor–entrepreneur who built the first electric power system; Insull, the manager–entrepreneur, who managed the expansion of the electric system, uniting local systems to larger ones; and Mitchell, the financier–entrepreneur who introduced financial and organizational means (such as the holding company) by which the growth of the utilities could continue on a regional level (Hughes, 1979). All of them had a vision of what the new system should look like and were determined solvers of the many problems frustrating the growth of the system.

Although we do believe system builders play an important role in regime changes, in many cases the future use of the new technology is not clear at the outset nor anticipated and thus not engineered in any important way, which downplays the system building element. One example of an unanticipated use of a radical technology is the radio (Rosenberg, 1995). At the time of its invention, the radio was seen not as a means of mass communication but as an alternative to the telegraph, to communicate between two points where communication by wire was impossible. It would be used for private communication, that is, for narrow casting, not broad casting. The users that were envisaged by Marconi were steamship companies, newspaper companies, and navies (Rosenberg, 1995). A similar story can be told for the steam engine. Originally, the steam engine was viewed, and used, as a pumping device. After the 18th century, however, the pump developed into a industrial power source, a propulsion technology (first in ships, later in trains and road vehicles), and finally, an electric power technology. Computers are a recent example of a technology whose final use and dominance was not anticipated. At the time of its invention, the computer was thought to be of potential use only for rapid calculation in scientific research or data processing contexts. The prevailing view in the late 1940s was that world demand could be satisfied by just a few computers (Rosenberg, 1995).

In situations in which the new system is the long-term outcome of the planned and unplanned actions of many actors (which in our view is the common case), *niches*, limited domains in which the technology can be applied, play an important role in the transition process. Military demand often provided a niche for a new technology, but the market system sometimes provide niches as well. The steam engine was developed by Newcomen to pump up water from mines; clocks were first introduced in monasteries where life was arranged according to strict timetables; the origin of the assembly line began with the American army in Springfield, Massachusetts, where the manufacture of muskets was standardized to the extent that all components were interchangeable; and the wheel was first used for ritual and ceremonial purposes (Schot, 1998). These niches were important for the take-off of a new technological regime in several ways: (a) they helped to demonstrate the viability of the new technology, (b) they provided financial means for the further expansion, (c) they fostered support from customers, investors, suppliers and other actors, and (d) set into motion interactive learning processes, the development of complementary inventions, and institutional adaptations—in management, organization, and the overall institutional framework in which firms operate—that are all important to the wider diffusion of the new technology.

Niche developments happen in two (partly overlapping) forms: in protected spaces and in the market place. Niche development often starts in protected spaces, where regular market conditions do not prevail because of special protection in the form of research and development (R&D) programs (of companies and governments), subsidies or loosening of institutional constraints (as in skunk works*). Such protected spaces can be called technological niches to distinguish them from market niches. They often take the form of experiments. Examples are experiments with electric vehicles in various European countries and cities (Rochelle, Rugen, Gothenburg, etc.**) which would not be possible without the sponsorship and support of different actors. Technological niches often precede market niches. Both processes of niche development can occur simultaneously and reinforce each other. For example, the introduction of electric vehicles in certain market niches might stimulate the emergence of new technological niches for a new type of customer. *Niche development* is a process of niche expansion and proliferation resulting in new ecology of niches. Key processes in niche development are: the refinement and coupling of expectations, articulation of problems (like production imperfection, side effects), needs

---

*Skunk works consist of special teams of researchers, production engineers and marketeers doing dedicated work on a novel product free from normal business constraints.

**La Rochelle refers to the user experiment with electric vehicles in the French town La Rochelle. The experiment started in Dec 1993 and lasted 2 years. In this period 50 prototype Evs (25 Peugeot 106 and 25 Citroën AX) were used by private people who volunteered to use such vehicles. The experiment sprang from the existing co-operation between the 3 main actors: PSA, EDF and the Municipality of La Rochelle. The aim of the project was to investigate this market. User experiences were very promising but sales of electric vehicles were disappointing. The La Rochelle experiment showed that in the current regime of car use, electric vehicles for private use can not survive without special protection. Rügen consisted of a large-scale experiment with 60 prototypes of battery-powered electric vehicles on the Island of Rügen in Germany. The vehicles were converted internal combustion engine vehicles produced by German automobile manufacturers. The vehicles were used by 100 user organisations over a period of 4 years. At the time of the start of the experiment, in 1992, it was the biggest experiment with electric vehicles. The experiment confirmed the negative view about Battery Evs in the German car industry and their choice to focus on other options such as fuel cells.

and possibilities, and network formation (Schot, Slob, & Hoogma, 1996; see also Elzen, Hoogma, & Schot, 1996; Kemp, Schot, & Hoogma, 1998).

Niches and entrepreneurial activity are important for the take- off of a new regime, but there is a third factor: the knowledge, skills, techniques that are available at any given time plus the support from parties who stand to gain from the new development. For their development, new technologies often depend on the old technologies, in particular the accumulated knowledge, capabilities, and skills acquired in existing technological regimes. Without these skills, technologies and the support from organizations having an interest in the development of a new regime, the system builders would have gotten nowhere and niches would not have emerged. Historically, the automobile benefited from experience accumulated in bicycle and carriage production. Electrical engineering firms played an important role in modern wind turbine development, as did semiconductor firms in the development of solar cells. The development of new technology thus depends on the characteristics of the existing technological regimes and the overall sociotechnical landscape.

In our view, all three factors—system builders (entrepreneurs), niches and the institutional support and capabilities of actors in existing regimes—are necessary for a regime shift. Without a market application, the technology will remain an idea or prototype, and in order to find a market application, there must be entrepreneurs to manage the "junction between the entrepreneurial firm and multiple market places" (Star & MacMillan, 1991, p. 167). The success of entrepreneurial activities, however, will depend not just on the managerial skills of the entrepreneur but also on the knowledge, skills, and assets that are available at the time, and the prevailing cost and demand conditions.

The story of novelty creation and regime change is depicted in Fig. 10.1. A distinction is made between three levels at which processes of sociotechnical change occur: the level of individual firms and households, the level of technological regimes, and the level of sociotechnical landscapes. The technological regimes and sociotechnical landscape of roads, villages, farms, and factories provide the backdrop to novelty creation; they shape the process of novelty creation by providing the means to produce the new technology and defining economic possibilities for their use. In turn, the regimes and landscape are transformed by the novelties and practices that go with them. This is the key process: the overall sociotechnical system and landscape with its different technological regimes create opportunities for sociotechnical change. And the exploitation of these opportunities by firms and other actors opens up yet further possibilities.

Our claim is that the success of niche formation is governed by processes within the niche, and by developments at the level of the existing regime and the sociotechnical landscape. So it is the alignment of develop-

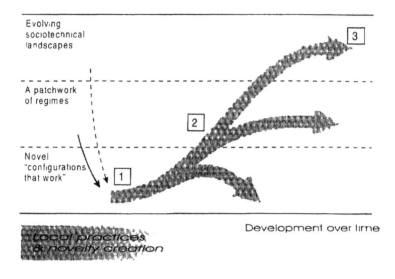

Evolving
sociotechnical
landscapes

3

A patchwork
of regimes

2

Novel
"configurations
that work"

1

*local practices
& novelty creation*

Development over time

[1] Novelty. shaped by existing regime
[2] Evolves. is taken up. may modify regime
[3] Landscape is transformed

FIG. 10.1.   The dynamics of sociotechnical change at the different levels of the technology–society relationship. Figure 10.1 shows that local practices, including those that involve novelty creations, occur in a context of regimes and the sociotechnical landscape, with exert influence on the shape and success of novel products.

ments—successful processes within the niche reinforced by changes at regime level and at the level of the sociotechnical landscape—which determine if a regime shift will occur.

This view differs from Arthur's (1988) model of path dependency. Arthur assumes two competing emerging technologies, whereas we consider the existence of a dominant, fully developed technology embedded in a technological regime that is challenged by a new emerging technology. The latter case is probably more common (see Islas, 1997).

In this scheme, technological regimes are the key element. They are both the *backdrop* and *setting*, that is, both the context and place in which novel products are conceived, developed, and introduced. Novelty originates within existing regimes, starting at the microlevel of local practices, and their success is linked in some way to structural problems or even crises of an existing system (Smith, cited in Kemp, Miles, Smith, Boden, Bruland, van der Loo, Street, Wicken, with Andersen, Freeman, & Soete, 1994). When the new technologies become more robust, benefitting from dynamic scale and learning economies and from institutional adaptations, finding new applications, and getting embedded into the social environment, irreversibilities increase and a reversal may occur. The technology now becomes a force of its own, and the context for newly emerging technologies, setting the terms of competition for other technologies. Having broken up sociotechnical relationships, it now fixes others (Rip & Kemp, 1998).

A good example of the aforementioned model is the evolution of the computer. From the turn of the century onward, a computing regime had evolved where calculations (in astronomy, in ballistics) where done by instructions to rooms full of women with calculating machines. These tasks could be substituted by the first electronic calculating machines of the 1940s. Further steps in the development of these protocomputers were geared to existing regimes, and to evolving possibilities of these new configurations (machine codes, machine languages, and then programming languages, which created a certain independence from the particular contexts of use). One can trace the developments in detail (Van den Ende, 1994). The important point is that it was only by the 1960s that the possibilities of the computer itself became the driving force of developments, also in other sectors.[3] A reversal occurred, and from the 1960s onward, one can speak of a computer regime. In other words, a transformation of regimes occurred, but not because an external input, the new computer, forced such a change; the computer could emerge and develop only within the existing regime, and an eventual transformation depended on this evolution.

HOW TO CHANGE TECHNOLOGICAL REGIMES?

Having discussed the problem of transition as a problem of regime change, we now ask the question whether it is possible to manage a transition process into a new regime (e.g., a more sustainable energy or transport system). It is clear that a shift toward an alternative technological regime presents a huge problem for public policy makers (or anyone else, for that matter). The task is no longer to control or promote a single technology but to change an integrated system of technologies and social practices in a nondisruptive way; that is, without causing social problems and welfare losses in the transition process. This is the problem that public policy makers face if they set out to promote a shift toward a different technological regime, like a transition to an alternative transport system. But how to do this? In our view there are basically three options.

---

[3]This is not to say that there had not been expectations about the potential of the computer, and attempts to realize them, often based on support from the military (Edwards, 1995). But until the 1960s, this took place in niches, and no new regime emerged.

The first strategy is to change the structure of incentives and let market forces play. This is the kind of approach favored by economists. Instead of engaging in the search for technologies to solve specific social problems, policymakers should change the structure of economic incentives: tax negative externalities and reward positive externalities. The advantage of this strategy is that decisions are decentralized, made by individual actors. Technology choices are left to the market rather than to a bureaucratic agency. Applied to environmental problems, this strategy promises that environmental benefits can be achieved at the lowest cost.

The arguments in favor are well known. There are good reasons to use incentives and even make them a central part of government policies, but there are also problems with incentive-based approaches, and these need to be pointed out. One problem is that the policy measures have to be really drastic in order to have an impact, considering the dominance of existing technologies. Even a tenfold increase in world market oil prices in the 1973–1983 period (from $3 to $30 per barrel) did not bring about a shift toward renewable energy sources; fossil fuels continued to be the main energy sources. It promoted energy efficiency improvements in energy generating and use technologies, but it did not lead to a replacement of fossil fuels by alternative renewable energy sources. The use of economic incentives can also lead to (temporary) windfall profits for manufacturers and dead weight losses for consumers. There is a third problem with the use of government incentives, which has received far less attention: The taxes and subsidies need to be accompanied by corrective measures that limit the negative effects of the wide-scale use of alternative technologies that are favored by the tax regime. No technology is perfect. All technologies have their side-effects and drawbacks. Thus, unless the use of economic incentives is accompanied by other policies aimed at limiting the harmful side effects of alternative technologies, incentive-based policies are bound to cause a transfer of problems.

The second strategy is to plan for the creation and building of a new sociotechnical system based on an alternative set of technologies, in the same fashion as decision makers have planned for large infrastructure works, like coastal defence systems or railway systems. The problem with this approach is that in modern, pluralistic, and advanced societies, a new technological system cannot be completely planned; the integrated nature of technology and complex social aspects defy a rational planning approach. In general, goals and human needs are too general and manifold to provide a precise guide to planners and technology developers. Even for firms it is often difficult to know what their own clients want. User requirements are articulated in relation to the technologies that are available; we learn what we want through experience (Truffer, Cebon, Dürrenberger, Jaeger, Rudel, & Rothen, 1999). Moreover, our wants and needs change over time.

The third and last strategy is to build on the ongoing dynamics of sociotechnical change and to exert pressures so as to modulate the dynamics of sociotechnical change into desirable directions. For this strategy, the task for policymakers is to make sure that the coevolution of supply and demand produces desirable outcomes, both in the short run and in the longer term. Rather than laying down requirements policymakers need to engage in process management to keep the process of sociotechnical change going into the right direction.[4] In contrast to the traditional policy approach, which starts from a stated goal after which a set of instruments is selected to achieve this goal, process management does not start from a quantified goal but from a set of partially conflicting goals. It aims to (a) change the rules of the game; (b) to shape the interactions and strategic games between different actors representing different interests and capabilities (for example, by empowering certain voices, making sure that the process is not dominated by certain actors); and (c) to keep the process of change going in desirable directions by counteracting undesirable effects and amplifying desirable ones. Although there are no principle problems, and the strategy may be the only feasible one in contemporary society, the practicalities are less clear. How is the modulation to be done?

Our discussion of technological transitions as a process of niche proliferation suggests one possible strategy to manage the transition process: Start by creating protected spaces, or niches, for promising new technologies. These spaces, in the form of technological niches, function as local breeding spaces for new technologies, in which they get a chance to develop and grow. Once the technology is sufficiently developed, and broader use is achieved through learning processes and adaptations in the selection environment, initial protection may be withdrawn in a controlled way. This strategy is called *strategic niche management*. It was first described in Rip (1992) and elaborated in Schot et al. (1996) and Kemp (1997). We describe strategic niche management as a way to induce and manage technological regime shifts. But before we do, it is useful to look first at some actual energy technology policies aimed at creating new paths—the Californian and Danish wind power policies—to get a better idea of the difficulties and problems in creating alternative paths of development. This case does not show a complete regime shift but it will help to clarify why strategic niche management is a potentially useful approach.

THE CALIFORNIAN AND DANISH WIND POWER POLICIES

This section describes the Californian and Danish policies toward wind power. Both policies are instructive in demonstrating some advantages of a bottom–up

---

[4]The use of process management as a means of sociopolitical governance has been advocated by various policy scientists (see for instance Glasbergen, 1994; Kooiman, 1993).

approach over a top–down approach. The policy descriptions draw on the work of van Est (1999), Righter (1994), Cox, Blumstein, & Gilbert (1991) and Grubb (1992) for the United States, and Lauritsen, Svendsen, & Sørensen (1996) and Jørgensen and Karnøe (1995) for Denmark.

The Californian Wind Power Policy

In the 1980s, California became the world center of wind energy development (Cox et al., 1991). Between 1981 and 1990, some 15,000 wind generators were erected (Righter, 1994). Installation rates in California rose from 10MW in 1981 to 60MW in 1982 and trebled in each of 1983 and 1984 to a level of 400 megawatts (MW) a year. The cumulative investment in Californian wind energy by 1986 totaled about $2 billion (Grubb, 1992). Of all the electricity produced from wind in the world during 1985, 87% was produced in California (Cox et al., 1991).

This development was largely the result of federal and California tax subsidies and buy-back regulation. Between 1930 and 1970, there was virtually no interest in wind power (Righter, 1994). There were no wind turbine manufacturers and little expertise in wind mills; wind mills had lost out against the expanding central electricity system in the 1930s. Most of the old windmills were manufactured elsewhere. Interestingly, f)r the purpose of argument, California can claim a unique windmill, in the form of a 10-ton, 70-foot, bell-shaped tube, developed by an inventor–entrepreneur, Dew Oliver, in the 1920s. The turbine looked like anything but a windmill. It consisted of a giant moveable metal tube that sat on a circular track atop a concrete foundation. Within the tube, there was a series of propellers mounted on a horizontal shaft; the propellers turned two generators, capable of producing 200 horsepower. Oliver raised $12.5 million from investors for his invention but the tunnel system generator never went into commercial operation. Oliver's business floundered and Oliver ended up in jail for breaking security laws. In a sense, this event already anticipated some later events during the "Californian wind rush" (based on Righter, 1994).

The phrase "wind rush," coined by Grubb (1992), indeed seems appropriate for much of what happened in the 1980s, when private investors, attracted by favorable tax incentives, entered the newly created market for wind power. In California, a 25% energy tax credit was offered for investment in renewable energy sources on top of the 15% federal energy tax credit for investment in renewables and a 10% federal general investment tax credit.[5] In addition, in 1978, Congress passed the Public Utility Regula-

---

[5]The California energy tax credit was eliminated on December 31, 1986, after been reduced to 15% on January 1, 1986. The federal tax credit for wind turbines, introduced in 1978 as part of the Energy Tax Act, expired on December 31, 1985.

tory Policies Act (PURPA) which required utilities to purchase electricity from independent power producers at prices corresponding to avoided costs. The policies were supplemented by a federal research and development programs for wind energy (mostly for large machines), soft loans, and accelerated depreciation schemes.

There do not exist many accounts of the California wind power policy process. The main source of information on which we draw is that of van Est (1999). According to van Est, politics played an important role in the whole development of wind power. In the United States, there was a strong soft energy movement, which propagated the use of renewables, especially solar energy. There was also a small but active solar industry. It is doubtful that the soft energy movement would have gotten very far without the 1973–1974 Arab oil embargo. When the oil embargo hit the United States in December 1973, the country was in a crisis. President Nixon unveiled Project Independence, aimed at making the United States independent of any source of energy outside its border by 1985. In 1975, the Congress launched the Energy Research and Development Administration with a division of Solar Energy. Interestingly, wind power received little attention; most of the attention and support went to solar power that, unlike wind power, was seen as a back-stop technology capable of totally replacing fossil and nuclear power plants.

Wind energy, which at the time was viewed as a limited energy source, benefited from the attention given to self-sufficiency and solar power. In California, in a 1978 revision of the 1977 solar tax-credit law, wind power was included as another renewable qualifying for support. Before 1978, there was hardly a wind power policy, and in a way the whole development of wind power in California was accidental. Wind energy was the primary beneficiary of three policies that were not really aimed at wind power, namely the (a) 1978 Public Utilities Regulatory Policies Act, which broke down the monopoly position of power utilities and established a favorable regime for independent electricity producers; (b) the National Energy Act of 1978, which provided a 15% federal energy tax credit for investment in renewable energy sources; and (c) the 25% California energy tax credit for investment in renewables, none of which was aimed particularly at wind power. As to the crucial requirement in PURPA that obliged utilities to purchase electricity from independent power producers at prices corresponding to avoided costs, this was only a minor part of a complex piece of legislation that did not receive much attention in the political process.[6] The effects it would have on the business of power production

---

[6]Some say that the utilities accepted the PURPA requirement as a goodwill gesture, without realizing how it would change the energy business. According to others, Congress mistook Section 210 of PURPA as the federal equivalent of the Private Energy Producer Act in California, designed to encourage private producers to competitively develop independent sources of natural gas and electric energy for their own use. In any case, PURPA went far beyond the California law by opening up opportunities for independent power producers that were unanticipated (van Est, 1999).

in the United States were not anticipated. The 15% tax credit under the Energy Tax Act also did not receive much attention at the time, it was passed by Congress, according to Cox et al. (1991), "almost as an afterthought".[7]

The California energy policies were more targeted toward wind power. The 1978 Mello Act established the goal that, by 1987, 1% of California energy would be produced by wind generators, rising to 10% by 2000, but these goals it seems played a minor role in public and private wind energy development decisions. A California proposal to implement a $80,000,000 state wind energy program was rejected by the Committee of Resources, Land Use and Energy and by the Senate. The Senate did approve a $800,000 plan to identify high-wind sites that, according to van Est, proved to be very influential in the startup of wind farms in California. Furthermore, the Office of Appropriate Technology (OAT), established in 1976 by Governor Brown to promote energy self-sufficiency, had a wind energy program for small wind energy conversion systems. Even though the OAT program was nothing more than a small dissemination program offering guidance to people who wanted to construct a turbine, it had an important impact in preparing the ground for wind farms (van Est, 1999). The increasing attention given to wind power led to the (in retrospect) misconceived idea (even among the utilities) that the technology was sufficiently developed and economical to use, and lured by the expectation of high profits, many companies rushed in. The profits to be captured from setting up a wind farm were indeed high. According to Cox et al. (1991) "most investors could recover about two-thirds of their investment through the reduction of their taxes in less than three years, even with no sales of electricity. If the turbine did perform as projected, the investor could expect to earn a high return (10-20%) from the investment" (p. 349).

The tax incentives and buy back rates on the basis of full avoided costs created a bonanza of opportunity for wind energy (Righter, 1994). Installation rates in California grew from 7MW before 1982 to 1,121MW in 1985. With the market base and finance of California, several companies invested heavily in wind energy technology and gained rapid experience. Major improvements in machine performance were achieved that led to a higher availability and doubled capacity rates in the 1980s. Capital costs fell from $3,100/kilowatts (kW) in 1981 to an estimated $1,250/kW average in 1986 (historic prices) (Grubb, 1992). However, the quick expansion was achieved at a considerable cost. As Grubb (1992) accounts: "Many of the early machines installed were of poor quality, and some broke in the first season of operation ... Machines were often sited carelessly and very densely,

---

[7]According to Ron White, who attended the hearings of the Joint Committee on Taxation, this was not true. The Energy Tax provision received a great deal of attention at the time. According to him, a source of the problem was the U.S. federal system, with its different levels of government, with inflexible federal policies working out differently in states.

and some were sold on fraudulent promises" (p. 170). The bad performance of the machines was not just an engineering failure; it was also caused by the perversity of the tax system in which the tax benefits were based on turbine capital costs rather than performance, and which allowed a developer to "sell inferior machines at a hefty profit" (Cox et al., 1991, p. 350). In 1985, the California wind industry produced only 45% of the energy they projected it would produce (Cox et al., 1991).

Looking back, it is clear that the Californian wind power policy was not a success. There had been too-rapid investments in an immature technology, the subsidies were too high, creating windfall gains for the wealthy, and too little had been done to prevent the construction of simply bad turbines. In an interview, Thomas Thompson of the Californian Public Utilities Commission (CPUC) admitted that the CPUC had made "two classic economic blunders" by not adjusting prices and not restricting the volume that could sign up (van Est, 1999, p. 60). The perversity of the tax credit program—where investors obtained tax benefits whether the equipment performed adequately or not—was acknowledged by the California Energy Council but they did not want to raise the issue with a new governor hostile to wind power and rising opposition to wind turbines. They felt that reopening the legislative issue was like opening Pandora's Box (van Est, 1999).

It is clear that many of these drawbacks could have been prevented by more carefully designed policies and timely revisions of the policies. The experiences did not disqualify the use of economic incentives as the basis of (energy) policy per se. As noted in Kemp (1997), there are many good reasons for using incentive-based approaches. However, it brings out some of the disadvantages of economic incentives as the sole policy instrument. By establishing a regime of tax credits and guaranteed rates, the government had little control over the number of wind farms that were established, nor did it have much influence over the choice of wind turbines in the absence of a standardized performance rating system (Cox et al., 1991). The sight of ugly, motionless machines damaged the reputation of wind energy, and was an important factor in the decision to eliminate the state tax credits on Dec 31, 1986, 1 year after the elimination of the federal tax credits, which sent many wind turbine companies into bankruptcy (in California and in Denmark) and suspended a reasonably benign energy source.

The Danish Wind Power Policy

The Danish policy toward wind power differs in several respects from that in California: It was less general, more specifically aimed at wind power, the policy was adjusted to (changing) circumstances, and, most important, combined elements of promotion and control. The Danish policy combined several elements of strategic niche management described in the next section, although it never was conceived in these terms by those involved in the creation of wind power—policy actors, politicians, wind turbine manufacturers, and grass-root entrepreneurs.

As in California, the Danish policy emerged in the second half of the 1970s, after the 1973–1974 oil crisis when oil prices were soaring and energy diversification and self-sufficiency were put high on the political agenda. An important difference with the California policy was that the Danish policymakers did not attempt to jump start the development of wind power. In Denmark, a much wider range of policy instruments were used, ranging from R&D support programs, tests, siting regulations, certification schemes, investment subsidies, to streamlining of planning procedures and longer term energy planning. Perhaps more important, besides political circumstances, the policies evolved with experience and changing economic circumstances. The role of the government was also different; less of a sponsor, and more of a counselor and process manager responding to specific problems and opportunities that emerged during the process of development. The development of wind power in Denmark was not staged or engineered by government authorities in an important way. Rather, the policies emerged out of a political process in which different parties had an input, and they reflected real-time experiences and changing economic and political circumstances in Denmark and the rest of the world.

As to the political actors, Jørgensen and Karnøe (1995) noted the importance of the antinuclear movement, organized in the Organization for information about Atomic Power (OOA). The OOA was important in positioning renewable energy sources as a necessary element in Denmark's future energy policy. It articulated the vision of local self-sufficiency and democracy, a vision in which renewables fitted but atomic power and large- scale fossil-fired power plants did not. Perhaps more important, the OAA aimed to present realistic alternatives to nuclear power. It joined forces with established physics researchers, which led to the "alternative energy plan" for Denmark, published in 1976. In this plan, wind power was selected as one of the future energy sources (among other alternative energy sources). Another important influence was the Organization for Renewable Energy (OVE) established in 1975. The OVE organized wind meetings, which brought together all kinds of people engaged in the construction of wind turbines. It also acted as an important forum for the diffusion of knowledge, experience and new ideas. Between 1974 and 1979, a great variety of wind turbine designs were constructed and tested by do-it-yourself builders and small companies.

Lauritsen et al. (1996) provided a somewhat different account than Jørgensen and Karnøe. They mark the establishment of the Wind Energy Committee by the Danish Academy of Technical Science council in October, 1974, as an important event in the development of wind power in Denmark. The task of the Wind Energy Committee was to investigate the possibility of wind energy in Denmark and a Danish wind turbine industry. The committee produced two reports in which it concluded that wind power was potentially economically viable and advocated a wind power development and demonstration program of 56,000,000 Danish kronen. Most of the money was for large unit wind power plants, as in the United States and the Netherlands.

The attention for wind power from both a political, social, and engineering point of view, stimulated the development of wind power. In the 1970s, all kind of experiments were done with different designs and unit sizes. Small size turbines were pioneered by grass-roots entrepreneurs and do-it-yourself builders in the 1974–1978 period. The first turbines were purchased by idealistic buyers whose early purchases help to build expertise. After 1977, a wind turbine industry emerged, consisting of companies that used industrial manufacturing techniques and standardized industry produced parts, using a design pioneered by Johannes Juul, a Danish engineer, in the 1950s. In 1978, a test station for small unit turbines was established in Risø, as part of the government's wind energy research program. According to Jørgensen and Karnøe (1995), the test station was very important in establishing a new industry. It worked as the common R&D department and played an important role in fostering cooperation among competing, antagonistic companies. Beyond this, the tests and accreditation scheme helped to establish a positive image for Danish wind turbines.

The development of wind power was supported by the decision of the Danish parliament to provide a 30% investment subsidy for wind power plants that were approved by the test station. The investment subsidy was lowered to 20% in 1981, increased to 30% in 1982, and reduced to 25% in 1983, and gradually reduced to 10% in 1988, the last year of investment support. In 1984, a 10-year agreement was signed between the utilities and the wind turbine owners regarding buy-back prices and grid connection costs. The utilities agreed to pay a price of 85% of the electricity sales prices for electricity delivered to the grid, on the condition that all electricity was delivered to the grid, if only a part of the production was delivered, the payment was 70% of the electricity price. The utility furthermore agreed to pay 35% of the grid connection costs. For private owners of wind turbines, there was a refund of the energy taxes and CO_2 taxes on top of the 70% to 85% buy-back rate. In 1985, an agreement was signed between the utilities and the Department of Energy to set up 100 MW wind power plants before 1990, followed by a second 100 MW agreement in 1990 to install an extra 100 MW wind power plants before 1994. The increasing institutionalization of Danish (wind) energy policy was underpinned by the Energy 2000 plan that came out in 1990, stating that by the year 2005,

10% of Danish electricity consumption should be generated by wind power. The government intervened in 1991 when the renegotiation of the 1984 agreement between the wind turbine organizations and utilities failed. Some limitations with respect to the size and ownership of private turbines were repealed in 1992.

The policies slowly emerged with the development of wind power technology; first they concentrated on research, development, and demonstration, the setting of quality and reliability standards; and later, when reliable technologies were available, a market was created through investment subsidy policies that were gradually phased out. The policies were more or less supportive; the government did not act as a sponsor or regulator but assumed the role of a catalyst and mediator by building on initiatives and exerting pressures in the strategic games between different groups, notably the utilities, private wind turbine owners, turbine manufacturers, and land use planning authorities. It was an example of a successful energy technology policy that helped to build a new industry (with exports worth $1,400,000,000 in 1990) and at the same time achieved environmental benefits in a cost-effective manner. Not all of its policies turned out to be a success; in retrospect, the decision to fund the development of large wind turbines proved to be wrong.[8] The gradual upscaling of small-scale wind turbines, from less than 55 kW in 1980–1983 to 600 kW in 1995, proved to be more economical.

From the perspective of regime management, the Danish policy is very interesting. It confirms our model of technological transitions about the importance of the coincidence of successful niche policies against the backdrop of changing regimes. It also shows the importance of learning, the creation of new actor networks, and changes in the institutional framework. More importantly, it demonstrates some of the advantages of a flexible, sequential policy aimed at modulating the dynamics of sociotechnical change into socially beneficial directions and using windows of opportunity within the evolving dominant regime. Unlike the California policy that merely created a market for wind power through tax credits and buy-back regulation, the Danish policy toward wind power was much more oriented to learning, fostering collaboration and institutional adaptation, and correcting undesirable outcomes. Before a decision to support wind power was made, the feasibility of wind power was investigated, the outcomes of which were used for a research and development program. A wind turbine test station was set up to weed out dismal products, the choice of turbine design was left to the market (and not perverted by adverse government incentives). Furthermore, the government did not underwrite a large part of business risks, and there were no radical policy changes that created false expectations and

---

[8]It was not necessarily a waste of money. Some of the knowledge that was generated could be used in building small turbines and could be usefully applied in developing today's big turbines.

business booms and busts as had happened in California. Protection of wind power was never too strong so as to dull incentives for technical improvement and cost reduction, and market stimulation incentives were gradually phased out when the technology became more mature.

The policies contributed to the unforeseen success of the Danish wind turbine industry, although the success owed a great deal to other, arguably more important, factors; in particular, the accumulated expertise in wind turbine technology, the work of grass-root entrepreneurs, the powerful antinuclear movement, public support for wind power, and the California market for wind power. What made the Danish policy so successful however (far more successful than the Californian policy or the Dutch policy that was quite similar to the Danish policy) is that it built on the ongoing dynamics of sociotechnical change and modulated these changes. It did not act as a sponsor nor as a regulator but more like a process manager, using both support and control measures.

Strategic Niche Management

From our discussion of continuity and change in technological regimes, strategic niche management (SNM) emerged as a possible strategy to manage the transition process into a different regime. The relative success of the Danish wind power policy showed the importance of process management, in general and as a de facto implementation of SNM. We now approach the issue from the other side, and set out an approach for SNM as we distilled it out of our studies of societal experiments with alternative motor cars and transport systems.

Strategic niche management is thus the planned development of protected spaces for certain applications of a new technology.[9] The SNM approach differs from the "technology push" approach by bringing in the knowledge and expertise of users and other actors into the technology development process. This generates interactive learning processes and institutional adaptation. The SNM approach differs from technology control policies by being aimed at the development of new technologies; the development aspect is an important aspect of strategic niche management.

The idea behind creating a protected space for a promising technology is that the new technology gets a chance to develop from an idea or prototype, or (in the case of motorcars) a showpiece on exhibition into a technology that is actually used. Such use is important for articulation processes to take place, to learn about the viability of the new technology and to build a supporting network around the product. SNM is more than just experimenting

---

[9]An example of the creation of a protected space at the company level is the formation of a team of designers by a car company intended to develop a new, environmentally friendly car concept. The design team is given a laboratory of its own and not hindered by a number of design restrictions that apply to the regular development of prototypes (such as cost restrictions and choice of materials).

with a new technology, however; it is aimed at making institutional connections and adaptations, at stimulating learning processes, both of which are necessary for the further development and use of the new technology.

More specifically, the aims of SNM are:

- To articulate the necessary changes in technology and in the institutional framework that are necessary for the economic success (diffusion) of the new technology;
- To learn more about the technical and economical feasibility and environmental gains of different technology options—that is, to learn more about the social desirability of the options;
- To stimulate the further development of these technologies, achieve cost efficiencies in mass production, promote the development of complementary technologies and skills, and stimulate changes in social organization that are important to the wider diffusion of the new technology;
- To build a constituency behind a product—of firms, researchers, and public authorities—whose semicoordinated actions are necessary to bring about a substantial shift in interconnected technologies and practices.

The management of niches can be done by firms, governments, and other social actors, but need not necessarily occur in a systematic and coordinated way. Actors have different interests, technological capabilities, powers, belief systems, and expectations. Moreover, there are usually several technological options that compete with each other.

How are technological niches created and managed? From a managerial rather than a process perspective, SNM consists of five interrelated steps: the choice of the technology, the selection of the experiment, the set-up of the experiment, the upscaling of the experiment by means of policy, and the breakdown of protection.[10] These steps should not be perceived as consecutive steps, but rather as overlapping and interrelated activities. The time dimension is important, however, insofar as the activities to be undertaken at a later stage build on earlier, often irreversible decisions in previous stages. This implies the need to anticipate the requirements of later stages in the initial design phase.

The Choice of Technology

There are usually different types of solutions for a problem, with different costs and benefits. A choice must be made as to which technology should be supported. Technologies particularly eligible for support through SNM are

---

[10]We have been developing the SNM perspective for cases in which technologies are promoted by governments for their relevance to solving environmental problems. This influences our framing of SNM. A workbook for the implementation of SNM is under development (Weber, Hoogma, Lane, & Schot, 1998).

technologies that are outside the existing regime or paradigm but that may greatly alleviate a social problem (like environmental degradation or road congestion) at a cost that is not prohibitively high. For contributing to a regime change, the technology has to meet four additional criteria, apart from the social precondition. It must:

- Have major technological opportunities embedded in it, have sufficient scope for branching and extension and for overcoming initial limitations (*Technological—scientific precondition*).
- Exhibit increasing returns or learning economies (*Economic precondition*).
- Be consistent with actual or feasible forms of organization and control and be compatible with important user needs and values of early adopters (*Managerial or institutional precondition*).
- Be attractive to use for certain applications in which the disadvantages of the new technology count less.

The social, technological–scientific, economic, and managerial (or institutional) preconditions are preconditions for regime shifts, already identified in the project "Technological Paradigms and Transition Paths" (Kemp et al., 1994). The fourth precondition is specific to the management of regime shifts through the creation and development of niches. Choice of technology implies a dilemma for SNM. To explore options for coevolution of technologies and their contexts, focusing on a specific technology is necessary but may lead to a bias in emerging patterns. Thus, SNM as transition tool rather than a market introduction strategy will have to allow for a variety of technological options and exploration of the options, while simultaneously working toward the embedding of at least one of them.

The Selection of the Experiment

In choosing an experiment, it is important to select a protected space in which the new technology may be used at relatively low cost and cause as little disruption and discomfort as possible. The space may be a certain application (for example, the use of solar cells for pleasure boats), a geographical area (a region or a city), or a jurisdictional unit. The heterogeneity of the selected environment means that there always areas and types of application for which the new technology is attractive, in which the disadvantages count less and the advantages are valued higher. For example, electric vehicles (EVs) that do not emit pollutants at the point of use are attractive for the use in cities with high levels of pollution. The disadvantages of EVs, such as the low range and the recharging of the batteries, are less of a problem for fleet owners (taxi companies, utilities, public transport companies) than they are for consumers. Adding these two considerations suggests the use of EVs by fleet owners in cities as a societal experiment.

There is a second important element with respect to the choice of experiment: its conduciveness to (appropriate) learning. The users involved in the experiment should not have too idiosyncratic requirements. If there is too great a gap between the requirements of early users and the larger population of potential users, the technology may never reach the larger group. One of the reason that electric vehicles never became a passenger car despite the wide use in milk cart fleets and golf carts is that the requirements of milk cart and golf cart users were completely different from the requirements of automobile drivers: range and speed and acceleration were not important, and technical change subsequently moved in a different direction,[11] using the internal combustion engine as the design for improvement. A related point about users is that users should also be critical, and able to communicate their requirements, so as to push suppliers to constantly improve the product.

The Set-Up of the Experiment

This is perhaps the most difficult step, as a balance must be struck between protection against and exposure to selection pressure. The choice of policy instruments can be based on the barriers that exist to the use and diffusion of the new technology. These barriers may be economic, as in when the new technology is unable to compete with conventional technologies, given the prevailing cost structure. They may be technical, in the form of a lack of complementary technologies, new infrastructure that is needed, or appropriate skills. And they may be social and institutional, having to do with existing laws, practices, perceptions, habits, and differences of interest. To deal successfully with these barriers, an integrated and coordinated policy is required. Possible elements of such a policy are: the formulation of long-term goals, the creation of an actor network, and the use of taxes, subsidies, public procurement, and standards. The policies should not attempt, however, to overcome all the barriers, making life too easy for the new technology. It should also be noted that the success of the experiment in terms of the original project goals is less important than setting interactive learning processes in motion and building new relationships between interdependent actors.

Scaling Up the Experiment

The next step concerns the scaling up of the experiment by means of policy. Even a highly successful experiment may require some kind of support from

---

[11]We owe this point to Robin Cowan (personal communication, 1998).

public policymakers, especially in the case of environmentally benign technologies (whose benefits are undervalued in the marketplace). Again this raises the question of how far governments should go in the support of a particular technology, and whether they should bear part of the costs or let others carry the costs. It also raises the issue of complementary policies that would underpin and reinforce SNM; for an exploration, see Schot, Elzen, and Hoogma (1994) and Schot and Rip (1997).

The Breakdown of Protection

The final step is the phased breakdown of protection. Support for the new technology may no longer be necessary or desirable when the results are disappointing and the prospects dim. Breakdown of protection is also needed to introduce full selection pressure and to reduce windfall profits. The breakdown is best done in a controlled, phased way in order not to cause heavy problems for the companies involved.

Experiences with wind power technologies and policies can be interpreted in terms of our discussion of SNM. We will examine two questions. First, to what extent can the policies be seen to conforming to the idea of SNM? And second, what is the usefulness of SNM in the light of the aforementioned policy experiences? For example, could some of the problems and setbacks encountered in the process have been prevented through SNM? Regarding the first question, the policies indeed can be seen as attempts to stimulate the development of new technology by protecting it temporarily from the myopia of the market. Both in California and Denmark, a technological niche was created by government policies, although the way in which it was different in each place. In California, protection was much stronger, with investment subsidies of 50% in the 1980–1985 period (in addition to guaranteed electricity rates), compared to 20% to 30% in Denmark. In Denmark, wind turbines were controlled through an accreditation scheme, there were also siting, capacity and ownership regulations. Buy-back rates were introduced only in 1984 at levels above those in California. The rates were not set by the government but the outcome of negotiations between utilities and private wind turbine owners.

Looking back, one might say that in California there was too much protection and too little control. The protection was also of an adversarial kind, being based on capital cost instead of performance. The government had little influence on both the microeffects of noise and visual intrusion and macrooutcomes of total capacity that was installed. According to Cox et al. (1991), the high number of wind turbines that resulted from the wind program was well above the number needed to develop the technology. A policy along the lines of SNM could have prevented this. Common sense might have achieved the same thing, but this would miss the important point that the Californian wind power policy was not specifically designed for wind power. The tax credits and PURPA requirements applied for all kinds of alternative energy sources—wind power just happened to obtain the greatest advantage from them. The Danish policies followed the scheme of SNM much better, by gradually phasing out protection and by focusing on learning and adaptation. This led to much better results, although, as already noted, this owed a great deal to other factors. We also should be careful in portraying policymaking as an exercise in logic and intelligence; policies are the outcome of bargaining and compromise between societal actors and coalitions in a context of multiple goals, conflicting interests, and uncertainty. The policies toward windpower in California and Denmark were highly politicized; they reflected different political ideologies and concerns. In the United States, the encouragement of renewables and cogeneration coincided with the political drive to break the monopoly powers of electric utilities. In Denmark, wind power was promoted on the basis of environmental benefits and the benefits of local self-sufficiency. In the United States, financial involvement from private investors was welcomed, whereas in Denmark such involvement was rejected by the public and by political parties. In Denmark, policymakers opted for a strategy of "controlled development," for political reasons rather than economic energy policy reasons. The success of the Danish policy thus owed a great deal to political expedience. This does not make the approach of SNM less useful. In a political world, there is still a need for policy models to inform political choices.

In what ways could policymakers have benefited from the SNM approach? What lessons can we draw from the aforementioned policy experiences in the light of the SNM model, which emphasises the need of flexible decision making and experiments? Do they attest to the need of using SNM (or elements of SNM) for promising technologies? The experiences with wind power show the importance of finding a balance between protection and selection pressures. This is a continuing task for policymakers on which much of the success will hinge. Policies should be flexible and adaptive, aimed at correcting undesirable outcomes and amplifying desirable ones (where needed). To be able to do so, monitoring, practice analysis, and policy evaluation should be an integral part of a government support program. As to the policies, SNM emphasizes the importance of learning. The principal aim of technology introduction policies should be learning—learning about problems and how they may be overcome, how the technology may be integrated into the existing system, for what users the technology is attractive, for what purposes the technology may be used, how it may be sold, and so on. Learning is far more important than achieving high sales. There are even some advantages in small-scale projects; they are less costly and more conducive to modification and change. Of course, small-scale projects do not help to achieve economy in production. But there is often more to be learned from a program lasting 10 years than a short program of 5 years that costs twice as much. Long collaborative projects also help to secure endurable collaboration to sustain a new type of development.

SNM also emphasizes that once the technology has been proved viable in the original domain of application, the scaling up of the user experiment or domain should be done in an ordered way. Jump starts will lead to inefficiencies, antagonism, and possibly obstruction by those who are harmed. Well-intended but ill-designed government support policies may very well harm the development of a new technology by giving the technology a bad reputation. In California, the sight of motionless wind turbines certainly damaged the reputation of wind power.

The setting of the experiment is also critical. Both in California and Denmark, locations were carefully selected on the basis of wind resource studies, but there were also some complaints about noise and visual intrusion that could have been prevented through the choice of different sites and the use of more silent turbines.

Finally, the breaking down of protection so as to subject the technology to normal market selection pressures; this may prove to be a difficult thing to do because of the investments that are made and the interests that have developed, and now press for continuation of the support program. The California program was expanded in 1984 for 2 more years despite the many problems and the downward revision of oil price predictions, which made wind power a less economical source of power.[12] The breakdown of support is best done in an ordered way so that firms get time to get used to the new situation by adjusting their business; again, the Californian policy stands out as a negative example.

REFLECTIONS ON SNM AS A STRATEGY TO CONSTRUCT TRANSITION PATHS

How far have we come in our attempt to develop a strategy to achieve regime shifts? One general point that structured our discussion deserves to be highlighted here: the strength of SNM lies in how it sets out to explore and exploit coevolution processes and patterns by reflectively intervening in them. Acting to create opportunities for learning, aligning goals and interests, shaping interactions—these are key features of SNMfe, atures that will guide its further development.

---

[12]According to Cox et al. (1991), the wind power program could not be justified on economic grounds after 1983. Continuing the program to the end of 1985 resulted in an estimated cost of $560,000,000 (in 1980 dollars) (Cox et al, 1991).

Thus, SNM is not just a useful addition to a spectrum of policy instruments. It is a necessary component of intentional transformation of regimes, multiactor processes in which government policy is only one component (although sometimes a decisive one). As a related point, SNM makes coevolution reflexive, perhaps creating new patterns for man to make his own (technological) history.

When SNM is specified in terms of "do" and "don't" guidelines, its basic strength is converted—and to some extent betrayed or degraded—into a method that can be followed. Thus, it is important to always locate the method in relation to coevolution processes and patterns. One way to do this is to point out problems involved in following SNM as a management tool.

First, one must find a balance between protection and selection pressure. Too much protection may lead to expensive failures but too little protection may inhibit an interesting alternative path of development. This calls for ongoing monitoring and evaluation of coevolution processes and of the support policies themselves.

Second, there is no guarantee for success; changing circumstances may render the technology less attractive and technological promises may not materialize. Hence, it is important to promote technologies with ample opportunities for improvement, and with a large cost-reduction potential that can be applied in a wide range of applications. A technology that does not yield immediate benefits may still turn out to be a useful technology in the long term. This means that it is important to take a long-term perspective. For example, government support of electric vehicles has been criticized (e.g., Wallace, 1995) on the grounds that the environmental gains are limited and that their performance is poor compared to internal combustion vehicles. But this need not be true in a long-term vision, where electricity is generated by solar energy and advanced batteries become available. Improved batteries may also pave the way for hydrogen fuel-cell powered automobiles and the wider use of solar energy.

Third, it may be difficult for government to end the support for a technology because of the investments that have been made and resistance from those who have benefited from such programs—the "angry technological orphans" (as Paul David has called them) whose expectations have been falsely nourished.

And fourth, it is important undertaken as part of a long term transition strategy and program, to create sufficient momentum or critical mass. To date, most experiments with alternative transport technologies have been rather small and covered a short period of time. Experiments should be of sufficient size and time span to allow for learning economies and to bring about institutional change. There is also a danger that the knowledge that is accumulated in the experiment will be lost once the experiment is over.

By pointing out problems if the method of SNM were followed too mechanically, we developed, at the same time, further guidelines at a higher level, guidelines that require some assessment of the overall dynamics of the path dependency that is being created. What remains is the question of whether a regime shift will actually be achieved, and if the new regime will reflect the hopes of the strategic niche "managers." This question cannot be answered in a definitive way; the results of coevolutionary processes are not determinate, and there is no linear causality between specific actions and outcomes at the collective level, which would allow one to reason back from desired outcomes to purposeful action. On the other hand, some guidance is necessary in constructing transition paths in order to increase the probability that there will indeed be a transition, and that it will be in the right direction.

We cannot address these complex questions here, but it is clear that they have to be confronted in any strategy to achieve regime shifts. SNM is important in constructing new paths and path dependencies, but a further step is necessary. Our general analysis on technical regimes and transition paths provides an entrance point. The stabilization of technological regimes as rule sets implies a reversal from a situation in which rules are tentative and not constraining, to one in which the rules guide action and interaction, having an existence of their own, independent of the actors (and in a sense they have, because they are located at the collective level of a regime, and cannot easily be changed at the microlevel of individual action). Thus, a strategy to achieve a regime shift must also attempt to bring about such a reversal. Such strategies are well known (although not well analyzed) for the creation of industry standards. An example of unintentional reversal is visible in Arthur's (1988) analysis of competing technologies, where minor differences and contingencies may tip the balance and create irreversible advantages. Technological regimes encompass more than just industry standards and differential learning effects, so the situation is more complex. The basic phenomenon, of stabilization through reversal, is still the same, however. Approaches to effect such a transformation will have an opportunistic element of triggering new linkages and alignments (see Molina, 1993), of seeing and using opportunities to shift the balance. Understanding the dynamics and patterns of coevolution processes will inform the opportunistic actions and allow the introduction of some "method in the madness."

ACKNOWLEDGMENTS

The authors wish to thank Rinie van Est, Bernhard Truffer, Christian Rakos, Robin Cowan, and Ron White for their comments on an earlier version of the chapter.

REFERENCES

Arthur, W. B. (1988). Competing technologies: An overview. In G. Dosi, C. Freeman, R. Nelson, G. Silverberg, & L. Soete (Eds.), *Technical Change and Economic Theory* (pp. 590–607). London: Pinter Publishers.

Beniger, J. R. (1986). *The control revolution.* Cambridge, MA: Harvard University Press.

Ceruzzi, P. (1996). From scientific instrument to everyday appliance: The emergence of personal computers, 1970–77. *History and Technology, 13*, 1–31.

Clark, K. B. (1985). The interaction of design hierarchies and market concepts in technological evolution. *Research Policy, 14*, 235–251.

Cox, A. J., Blumstein, C. J., & Gilbert, R. J. (1991). Wind power in California. A case study of targeted tax subsidies. In R. J. Gilbert (Ed.), *Regulatory Choices* (pp. 347–369).

Dosi, G. (1982). Technological paradigms and technological trajectories: A suggested interpretation of the determinants and directions of technical change. *Research Policy, 6*, 147–162.

Dosi, G. (1988). The nature of the innovation process. In G. Dosi, C. Freeman, R. Nelson, G. Silverberg, & L. Soete (Eds.), *Technical Change and Economic Theory.* London: Pinter Publishers.

Dosi, G., Malerba, F., & Orsenigo, L. (1993). Evolutionary regimes and industrial dynamics. In L. Magnusson (Ed.), *Evolutionary and neo-Schumpeterian approaches to economics* . Boston: Kluwer.

Edwards, P. N. (1995). From impact to social process: Computers in society and culture. In S. Jasanoff, G. E. Mrkle , J. C. Petersen, & T. Pinch (Eds.), *Handbook of science and technology studies* (pp. 257–285). Thousand Oaks, CA: Sage.

Elzen, B., Hoogma, R., & Schot, J. (1996). *Mobiliteit met toekomst. Naar een vraaggericht technologiebeleid* [Mobility with a future. Toward a demand-oriented technology policy]. Report to the Ministry of Traffic and Transport (in Dutch).

Freeman, C., & Perez, C. (1988). Structural crises of adjustment, business cycles and investment behaviour. In G. Dosi, C. Freeman, R. Nelson, G. Silverberg, & L. Soete (Eds.), *Technical change and economic theory* (pp. 38–66). London: Pinter Publishers.

Glasbergen, P. (Ed.). (1994). *Managing environmental disputes: Network management as an alternative.* Dordrecht, the Netherlands: Kluwer.

Grubb, M. (1992). Wind energy. In M, Grubb (Ed.), *Emerging energy technologies. Impacts and policy implications* (pp. 163–191). Royal Institute of International Affairs.

Hahn, F. (1989). Information dynamics and equilibrium. In F. Hahn (Ed.), *The economics of missing markets, information and games* (pp. 106–128). Oxford: Clarendon Press.

Hughes, T. P. (1979). The electrification of America: The role of system builders. *Technology and Culture, 20*, 124–161.

Hughes, T. P. (1983). *Networks of power: Electrification in western society, 1880–1930.* Baltimore, MD: The John Hopkins University Press.

Hughes, T. P. (1989). The evolution of large technological systems. In W. E. Bijker, T. P. Hughes, & T. J. Pinch (Eds.), *The social construction of technological systems: New directions in the sociology and history of technology* (pp. 51–82). Cambridge, MA: MIT Press.

Islas, J. (1997). Getting round the lock-in in electricity generating systems: The example of the gas turbine. *Research Policy, 26*(1), 49–66.

Jørgensen, U., & Karnøe, P. (1995). The Danish wind-turbine story: Technical solutions to political visions?. In A. Rip, T. Misa, & J. Schot (Eds.), *Managing technology in society. New forms for the control of technology* (pp. 57–82). London: Pinter Publishers.

Kemp, R. (1994). Technology and the transition to environmental sustainability: The problem of technological regime shifts. *Futures, 26*(10), 1023–1046.

Kemp, R. (1996). The transition from hydrocarbons: The issues for policy. In S. Faucheux, D. Pearce, & J. L. R. Proops (Eds.), *Models of sustainable development* (pp. 151–175). Cheltenham, England: Edward Elgar.

Kemp, R. (1997). *Environmental policy and technical change: A comparison of the technological impact of policy instruments.* Cheltenham, England: Edward Elgar.

Kemp, R., Miles, I., Smith, K., et al. (1994). Technology and the transition to environmental stability: Continuity and change in complex technology systems. For SEER research program of the Commission of the European Communities (DG XII).

Kemp, R., Schot, J., & Hoogma, R. (1998). Regime shifts to sustainability through processes of niche formation: The approach of strategic niche management. *Technology Analysis & Strategic Management*, 10(2), 175–195.

Kemp, R., & Soete, L. (1992). The greening of technological progress: An evolutionary perspective. *Futures*, 24(5), 437–457.

Kooiman, J. (Ed.). (1993). *Modern governance: New government–society interactions*. London: Sage.

Landau, R., Taylor, T., & Wrigh, G. (1996). *Mosaic of economic growth*. Stanford, CA: Stanford University Press.

Lauritsen, A. S., Svendsen, T. C., & Sørensen, B. (1996). *Wind power in Denmark*. Final Report of APAS/RENA project for EU, Denmark.

MacKenzie, D. (1992). Economic and sociological explanations of technical change. In R. Coombs, P. Saviotti, & V. Walsh (Eds), *Technological change and company strategies* (pp. 25–48). London: Academic Press.

Malerba, F., & Orsenigo, L. (1993). Technological regimes and firm behaviour. *Industrial and Corporate Change*, 2, 45–71.

Molina, A. H. (1993). In search of insights into the generation of techno-economic trends: Micro- and macro-constituencies in the microprocessor industry. *Research Policy*, 22, 473–506.

Nelson, R. R., & Winter, S. G. (1977). In search of useful theory of innovation. *Research Policy*, 6, 36–76.

Nelson, R. R., & Winter, S. G. (1982). *An evolutionary theory of economic change*. Cambridge, MA: Bellknap Press.

North, D. C. (1990). *Institutions, institutional change and economic performance*. Cambridge, England: Cambridge University Press.

OECD. (1992). *Technology and the economy: The key relationships*. Paris: Author.

Righter, R. W. (1994). Wind energy in California. A new bonanza. *California History*, pp. 142–155.

Rip, A. (1992). A quasi-evolutionary model of technological development and a cognitive approach to technology policy. *Rivista di Studi Epistemologici e Sociali Sulla Scienza e la Tecnologia*, 2, 69–103.

Rip, A., & Kemp, R. (1998). Technological change. In S. Rayner & E. L. Malone (Eds.), *Human choice and climate change: An international assessment, Vol 3* (pp. 329–399). Columbus, OH: Batelle Press.

Rip, A., Misa, T., & Schot, J. (Eds.). (1995). *Managing technology in society: New forms for the control of technology*. London: Pinter Publishers.

Rosenberg, N. (1976). The direction of technological change: Inducement mechanisms and focussing devices. In *Perspectives on technology* (pp. 108–125). Cambridge, England: Cambridge University Press.

Rosenberg, N. (1994). Path-dependent aspects of technological change. In N. Rosenberg (Ed.), *Exploring the black box: Technology, economics, and history* (pp. 9–23). Cambridge, England: Cambridge University Press.

Rosenberg, N. (1995). Uncertainty and technological change. In R. Landau et al. (Eds.), *Mosaic of Economic Growth* (pp. 334–353). Stanford, CA: Stanford University Press.

Schot, J. W. (1992). Constructive technology assessment and technology dynamics: The case of clean technologies. *Science, Technology and Human Values*, 17(1), 36–57.

Schot, J. W. (1998). The usefulness of evolutionary models for explaining innovation: The case of the Netherlands in the nineteenth century. *History of Technology*, 14, 173–200.

Schot, J. W., Elzen, B., & Hoogma, R. (1994). Strategies for shifting technological systems: The case of the automobile system. *Futures*, 26(10), 1060–76.

Schot, J. W., & Rip, A. (1997). The past and future of constructive technology assessment. *Technological Forecasting and Social Change*, 54, 251–268.

Schot, J. W., Slob, A., & Hoogma, R. (1996). *The implementation of sustainable technology as a strategic niche management problem*. Report for the Dutch National Program on Sustainable Technology (in Dutch).

Spier, F. (1995). Norbert Elias' theorie van civilisatieprocessen opnieuw ter discussie. Een verkenning van de opkomende sociologie van regimes. *Amsterdams Sociologisch Tijdschrift, 22*(2), 297–323.

Star, J. A., & MacMillan, I. C. (1991). Entrepreneurship, resource cooptation, and social contracting. In A. Etzioni & P. R. Lawrence (Eds.), *Socio-economics: Toward a new synthesis* (pp. 167–184). NY: Armonk.

Truffer, B. P., Dürrenberger, C. D., Jaeger, C., Rudel, R., & Rothen, S. (1999). Innovative social responses in the face of global climate change. In P. Cebon, U. Dahinden, H. Davies, D. Imboden, & C. Jaeger (Eds.), *A view from the Alps: Regional perspectives on climate change* (pp. 351–434). Cambridge, MA: MIT Press.

van de Ende, J. (1994). *The turn of the tide: Computerization in Dutch society, 1900–1965*. Unpublished PhD dissertation, Delft University.

van Est, R. (1999). *Winds of Change. A Comparative Study of the Politics of Wind Energy Innovation in California and Denmark*. Utrecht: International Books.

Wallace, D. (1995). *Environmental policy and industrial innovation: Strategies in Europe, the US and Japan*. RIIA, London: Earthscan.

Weber, M., Hoogma, R., Lane, B., & Schot, J. (1998). *Expanding technological niches: How to manage experiments with sustainable transport technologies*. Institute for Prospective Technology Studies, Sevilla/University of Twente, Enschede.

IV

Path Creation
As Mobilization

11

Show-and-Tell:
Product Demonstrations
and Path Creation
of Technological Change

Joseph Lampel
University of Nottingham

> *If a man makes a better mouse-trap, the world will beat a path to his door.*
> —Ralph Waldo Emerson (cited in Evans, 1968, p. 468)

Emerson's observation has often been used to reassure innovators that good technologies are destined to overcome the barrier of public indifference. Reality, as the research on innovation clearly shows, belies the optimism of this adage (Georghiou, Metcalfe, Gibbons, Ray, & Evans, 1986; Jelinek & Schoonhoven, 1990). The success of new technologies is often dependent on persuading nonexperts, such as investors and potential customers, of their merits. This persuasion process requires champions of new technologies to present their innovation to the relevant public in a manner likely to elicit their interest, approval, and ultimately, their support. History shows that simply communicating the characteristics of new technologies is usually insufficient to overcome the skepticism that confronts major innovations. This has frequently led innovators to stage the introduction of new technologies into the public arena in such a way as to dramatize their performance and application. Common venues for these staged introductions are trade fairs, international exhibitions, press conferences, or even stockholder meetings (Barhydt, 1987; Grove & Fisk, 1989; Kotler, 1973). These occasions are scripted as dramas in which the new technology is the central protagonist. Their intent is not so much to convey basic information as to highlight the potential of the new technology, while at the same time diverting attention away from limitations and potential difficulties.

What role do such spectacles play in technological innovation? In this chapter I argue that such spectacles are designed to defeat a key axiom of the rational paradigm of technological change. Namely, that the rate at which new innovations become established is contingent on reducing the initial uncertainty that potential backers and users experience when first confronted with the new technology (Rothwell & Robertson, 1973; Souder & Moenaert, 1992).

Uncertainty on the part of investors and consumers poses a threat to innovators; it delays the flow of resources needed to improve and promote the new technology. Innovators usually try to deal with this problem by providing investors and consumers with as much factual information as is necessary to reduce uncertainty. The major drawback of this approach is that new technologies are often crude and imperfect. More factual information, given the critical scrutiny that new technologies receive from investors and consumers, may be counterproductive—it raises more questions and only serves to create more uncertainty. Technological dramas can be seen as attempts by innovators to break out of this vicious circle by influencing the psychology and atmosphere surrounding the emergence of new technologies. The attempt is to manage the context in which new innovations are evaluated, moving it away from detached analysis, where evidence is weighed against claims, to one where uncritical enthusiasm prevails.

I argue that changing the routines that investors and consumers normally use to evaluate innovations is the key to this task. Garud and Rappa (1994) argued that researchers in technological communities employ evaluation routines that assign different feasibility estimates to nascent technologies. Investors and consumers also employ evaluation routines, but in their case, they tend to oscillate between two types of evaluation routines: critical routines and commitment routines. *Critical evaluation routines* focus attention on problems and limitations. They tend to foster a wait-and-see attitude on the part of investors and consumers. *Commitment evaluation routines*, by contrast, focus on the achievements and future potential of the new technology. They tend to produce a more positive interpretation of the new technology, and by extension, a greater willingness to back or adopt it.

The problem facing innovators is that investors and consumers normally rely on critical evaluation routines. This is because these groups only have a cursory understanding of the new technology, because information circulating about the new technology is ambiguous and contradictory, and because they are aware that claims made by innovators are self-serving. To get investors and consumers to see the new technology in a positive light, innovators must disable critical evaluation routines and substitute commitment routines in their place.

How is this task accomplished? I use a number of well-known historical vignettes to examine and illustrate the process. These include Eli Whitney's demonstration of interchangeable parts for rifles, Edison's unveiling of his illumination system, the maiden voyage of the first diesel locomotive, Steven Jobs' presentation of the NeXT computer to the public, and others. Analysis of these vignettes shows that technological dramas usually begin with a buildup process during which innovators structure public awareness in anticipation of the upcoming performance. The performances that follow employ different scripts, depending on the technology and the needs of the innovators. An examination of these scripts shows that they attempt to convey basic themes such as heroic conquests of time and space, revolutionary breakthroughs, magic acts, or the triumphal celebration of a new technology.

The basic premise of this chapter is that technological dramas are not only designed to produce lasting impression, but also intended to influence constituencies such as sponsors, investors, buyers, and competitors, whose opinions and actions are critical to the success of new technologies (Morus, 1988; Pfaffenberger, 1992). The staging of technological performances can therefore be seen as an element in a larger repertoire of persuasion tactics that innovators employ to mobilize and hold together sponsoring coalitions. This gives rise to two questions to which we turn in the latter part of this chapter. First, why is this type of persuasion tactics effective in recruiting the support of investors and customers? And second, do technological dramas fit into our current theories of technological change, and if so. how?

I argue that the answer to the first question is provided by recent bandwagon theories of collective action (Abrahamson & Rosenkopf, 1993; Granovetter, 1978). These theories suggest that large-scale change such as adoption of new technologies often begins with changes in the perception and behaviour of a small number of key actors. These changes impact the perceptions and then subsequently the behavior of other actors; creating a process that Oliver (1993) described as "threshold contagion": The actions of one group reduce the action threshold of other groups. In the case of technological dramas, this means reducing the action threshold that prevents investors and consumers from changing evaluation from critical to commitment routines. Threshold contagion must therefore take place if technological dramas are to succeed. But this will not occur unless a bandwagon process can be initiated; something that technological dramas do not always succeed in doing.

The second question turns our attention to current theories of technological change. Recent theories of technological change argue that technologies develop along trajectories, and that they compete for resources as they develop (Dosi, 1982). Thus, a differential advantage in resource recruitment during one stage of this development may subsequently translate into a decisive advantage. So-called path dependent theories of technological change push this point even further (Arthur, 1989); they argue that this differential advantage is often accidental, and in the long run may result in products that are inferior to their rivals. On the face of it, *technological dramas* appear to be events whose impact on technological trajectories belongs more to the realm of accident than design. Closer inspection, however, suggests that technological dramas cannot be seen as attempts by innovators to "tweak" an objective technological trajectory, but are in fact part of an enactment process in which the behavior of collectivities is changed through an envisioning process. This envisioning process does not conform to the calculative rationality that dominates path dependent theories of technological change. It does, however, accord well with more recent path creation theories (Karnøe & Garud, 1996). And going beyond this, the envisioning process of technological dramas also suggests that future theories of technological change should broaden their scope to include a richer and more nuanced view of human cognition and collective action.

COMMUNICATION AS COLLUSIVE PERSUASION

In an economy where new innovations are announced daily, acceptance and support for new technologies is often difficult to obtain (Bernheim & Shoven, 1992; Schwartz & Kamien, 1978). In an effort to mobilize resources for their ideas, innovators seek to communicate the viability and economic potential of their technologies to others in their own community, and beyond, to important financial and market constituencies.

Garud and Rappa (1994) argued that within technological communities, competition for resources revolves around scrutiny of claims made by innovators, and that the scrutiny depends on "evaluation routines" that technological communities employ to judge the merits of technologies. As Garud and Rappa (1994) put it:

> Researchers externalize their technological beliefs by creating routines that are then employed to evaluate the technology. The evaluation routines, in turn, filter data in a way that influences whether or not researchers perceive information as useful. Researchers with different beliefs attempt to sway each other with respect to the routines utilized to judge the technology. It is in this sense that technological systems are negotiated. (p. 347)

In line with the work of Kelly (1963) and Bateson (1972), Garud and Rappa see evaluation routines as filters screening out information not consistent with beliefs about the technology. The screening process is defensive and self-reinforcing, producing a "sealing" of belief systems against contrary evidence from technological paradigms. As Garud and Rappa (1994) put it:

> Data inconsistent with an individual's evaluation routines are either ignored or appear as noise. Data consistent with evaluation routines are perceived as information and cognitively rearranged in a manner that reinforces an individual's beliefs. Given bounds to rationality, this bracketing of perception occurs because individuals may be more interested in confirming their beliefs than in actively trying to disprove them. (p. 347)

This view of evaluation routines may be correct as far as actors in the technological community are concerned because members of technological communities are "initiates" in the broad sense of the term; they usually undergo a long process of learning and socialization that produces deeply held assumptions about technology. Evaluation routines create a defensive ring around these assumptions. What is more, the strength of this defensive ring depends on the cost of changing these assumptions. The more that members of a technological community invest in a technological paradigm—cognitively, professionally, or even economically—the stronger their evaluation routines will act to prevent inconsistent information from being admitted. Although this may have unintended dysfunctional consequences by delaying recognition of potentially important developments, it is important to bear in mind that strong screening action by evaluation routines also has important beneficial effects. They act to stabilize expectations about the future, thereby giving researchers and innovators the security they need to continue working on a given technological trajectory when evidence to the contrary would suggest that further effort is futile.

The same dynamics do not hold for potential users or investors who are outside the technological community. Their interaction with the new technology is not based on intimate understanding of the basic technological issues. Nor do they have deeply held assumptions about the merits of competing technological paradigms that would constrain their cognitive and resource investments. Their evaluation routines focus on utility, on potential profits, and possible risks. The information they use to make this evaluation does not come from their own acquaintance with the technology, but from other sources, including the innovators themselves. However, aware that innovators have an incentive to provide incomplete, or even distorted, information in order to solicit support, they are likely to be cautious when evaluating claims (Merrifield, 1981; Roussel, 1983; White & Graham, 1978).

Thus, the evaluation routines that dominate technological communities are likely to be different than those on which investors and users rely to make their judgment. The evaluation routines of members of technological communities can be characterized as "commitment routines," whereas those of external actors who are not intimately involved in transforming technological ideas into reality can be characterized as "critical routines."

If the focus of commitment routines is consistency and certainty, the focus of critical routines is inconsistencies and ambiguities—precisely because inconsistencies and ambiguities can signal problems and risks that make the technology a poor bet for use or investment (Boyadjian & Warren, 1987; Hawkins, 1986). Critical evaluation routines pose a major danger to innovators inasmuch as they breed a wait-and-see attitude among potential backers and users. Faced with uncertainty, investors may postpone support until they have more "hard" data about such issues as the potential size of the market and the ability of innovators to protect the proprietary advantage they currently enjoy (Dasgupta, 1988; Kamien and Schwartz, 1982). Users, for their part, may postpone purchases because they are aware of the dangers of being early adopters (Katz & Shapiro, 1985). Faced with this problem, innovators may decide to increase the intensity of factual communication, targeting key constituencies with more detail about their technology. Here, however, they face a paradox.

In literal communications such as press releases or technical reports, presenters of information indicate to their audiences that the information they are receiving must be taken at face value. The prediction of technical performance, projected costs, and sales contained in literal communication are intended to reduce uncertainty about the scope of the market, investment requirements, and financial returns. In other words, these predictions are intended to satisfy the screening criteria on which critical evaluation routines are based. Recipients of literal communications, however, are aware that an attempt is being made to influence them. They are likely to be vigilant, paying special attention to the weakest part of the case that innovators are attempting to make for their technology. If innovators can marshal an overwhelming array of objective data, they may overcome this problem using literal communication alone. However, if—as is often the case with new technology—many uncertainties persist about the technology, exclusive reliance on literal communication may have the opposite effect to the one desired. Instead of reducing uncertainty, it may actually focus attention on unresolved problems with the new technology. Instead of mobilizing support, it may make potential allies even more reluctant to provide investment and support for the technology.

Dramatic communications offer innovators a way out of this impasse. Whereas *literal communications* assume that the dominant mode of gaining acceptance for innovations is uncertainty reduction, *dramatic communications* look to certainty creation as a way of gaining acceptance. Certainty creation, however, is regulated by commitment routines. In technological communities commitment routines are the product of scientific and technological paradigms, and are internalized through a powerful process of training and socialization. Potential consumers and investors, however, have little more than a passing familiarity with these paradigms. Their commitment routines originate from deeply held social beliefs about the power and impact of technology—beliefs that are an amalgam of the factual and the mythical. The interaction of factual and mythical beliefs shapes the way that commitment routines are used. Whereas an emphasis on the factual discounts the credibility of information that is not directly verifiable, an emphasis on the mythical lends credibility to attribution about the capability of the technology. Thus, attention is not confined to technology as an artifact designed to perform a specific task; it extends to technology as a vehicle of social transformation capable of satisfying unmet desires and generating untold wealth.

Outside technological communities, the key to the operation of commitment routines is the premise that the specific technology one is observing will deliver on the promises made by technology in general. There is an obvious tension here between the reality and the promise; a tension that dramatic communications seeks to defuse. It does this by disavowing realism; by informing audiences that what they are about to see is a representation of reality, rather than reality itself. By making the admission that what is being shown is partial and selective (Speier, 1980), dramatic communications invalidate the most crucial claim of literal communications—the claim that what is presented is accurate and comprehensive. This may appear paradoxical if one subscribes to persuasion based on rational argument and systematic analysis of evidence. It becomes more logical if what is being attempted is a transition from uncertainty reduction to certainty creation; from persuasion based on literal communication to one based on dramatic communications.

Dramatic communications disable critical routines by enticing individuals into what Goffman (1974) called the "theatrical frame." Having done this, viewers abandon their stance of detached observers, and instead become participants who collude in the performance. Such collusion is at the heart of technological performances as "certainty creating events." As Weick (1995) pointed out, it is possible to persuade people by argument—compelling them by force of logic to see things one way as opposed to another. However, a far more effective way of persuasion is one in which both sides willingly collude in creating truth. The problem, of course, is that voluntary collusion should in principle be very difficult in a world of economic self-interest. It is difficult to understand at first sight why hardheaded investors or self-interested consumers will collude with their own persuasion. Difficult, that is, until we remember that as a society we are steeped in the conventions of artistic and cultural performances. Triggering these conventions increases the likelihood that an audience will abandon critical routines in favor of commitment routines. Triggering these conventions allows innovators to manipulate the vocabulary that their audience uses to interpret the structure and plot of the unfolding drama (Burns, 1972).

From a dramaturgical point of view, therefore, the innovators' task is as follows: They must lure viewers away from their stance of detached observation and make them into active participants. They must insinuate dramaturgical conventions into the staging so as to encourage viewers/participants into a collusion in which commitment routines predominate over critical routines. And finally, they must script the technological performance in such a way as to make the central message they seek to convey as powerful and persuasive as possible. From a practical point of view, this divides technological performances into three parts: setting the stage, building up anticipation, and finally, orchestrating the show.

SETTING THE STAGE

Getting an audience to abandon critical in favor of commitment routines requires a transition from a world where literal communications predominate to one where dramatic conventions hold sway. Doing this is largely a matter of redefining the space in which the performance takes place. In principle, innovators have two choices: If sufficient resources exist, innovators usually construct a setting dedicated to their own performances, creating what may be called a "one-machine show." However, if the scope of the technological drama exceeds available resources, innovators may opt for a "show within a show," enacting the technological drama in settings where other performances are taking place.

Attracting attention and creating awareness is the first step in constructing the public space necessary for an independent performance, such as a one-machine show (Arnot, Chariau, Huesmann, Lawrensonk, & Theobold, 1977). In the absence of areas or structures specifically designated for performance, the crowd itself can provide the means necessary to create a temporary performance space. The gathering of a crowd demarcates a performance space distinct from the rest of the environment. By focusing their attention on this space, the crowd produces a collective atmosphere that is qualitatively different from the mundane atmosphere that normally characterises public spaces. For example, after Peter Jensen and Edwin Pridham demonstrated their loudspeaker for an invited group in the stadium in Golden Gate Park in San Francisco, California, on December 10, 1915, they were asked by the *San Francisco Bulletin* to deliver a concert of recorded music on Christmas Eve in a plaza in front of San Francisco's new city hall. A crowd estimated at 100,000 showed up to listen to the recorded music that sounded as though it had been "uttered by a giant" (Lewis, 1990, p. 190).

Masters of the one-machine show invest considerable amount of resources in creating the unique settings needed to exert maximum impact. Thomas Edison periodically transformed Menlo Park into a stage on which technological dramas were enacted for the benefit of crowds lured by advance publicity. Steven Jobs rented the San Francisco symphony hall for the unveiling of the NeXT computer, and then used the press to hype the upcoming performance. In both instances, innovators expended a great deal of effort to create settings in which they could command the attention of their audiences.

The time and resources that Edison and Jobs invested in creating a public space are beyond the means of most innovators. In many instances, structuring public space for a dramaturgical event is much easier when the performance is part of, or ancillary to, a larger event. Traveling actors and acrobats often took advantage of country fairs to mount their performances. In modern times, we have become accustomed to film festivals, which are part trade fairs attended by industry insiders, and part spectacle, in which the larger public participates. Of course, trade fairs have been commonly used for decades in most industries to introduce new products. The problem of introducing new technologies within a trade fair context is that they often fail to stand out from other products that are being presented. Many innovators have therefore attempted to increase attention to their product by maneuvering them to the foreground of these forums. For example, the Pioneer Zephyr, the first diesel locomotive, was unveiled to the public in 1934 as part of a nonstop run from Denver to Chicago. Leaving Denver at dawn, the Zephyr arrived in Chicago on May 26, the opening day of the World's Fair. The train ran onto a great open-air stage as a grand climax to the transportation pageant known as the "Wings of a Century." Thirty years later, AT&T made a splash in 1964 at the New York World's fair with its introduction of Picturephone, the first video telephone system (Bode, 1971). The system was inaugurated by Lady Bird Johnson, who used the video connection to place a call to the White House in front of the assembled press.

BUILDING UP ANTICIPATION

To get an audience to suspend their habitual critical evaluation routines it is not enough to draw their attention to a forthcoming performance. It is also necessary to build anticipation about what that performance will deliver. Producers of technological performances have learned from their brethren in show business that building up such anticipation relies more on hints and allusions than on forthright reporting of factual information (Laemmle, 1976). The trick is to arouse curiosity first, increase this to suspense, and then transform the suspense into readiness to believe. The challenge of an effective buildup is to reveal a sufficient amount of information to build suspense, but not so much as to undermine the drama.

The balance between total withholding and full disclosure is often a function of the intrinsic shock value of the technology itself. Thomas Edison, for example, who was a master at staging technological dramas, often relied on the press to build up public anticipation and curiosity (Conot, 1980; Nye, 1983). In early December 1879, as the work on the lighting system was nearing completion, Edison gave reporter Marshall Fox of the *New York Herald* free access to Menlo Park (Conot, 1980; Nye, 1983). Fox spent 2 weeks gathering material and pictures for a story that appeared on December 21, 1879. The article described the lighting system in considerable detail, and was also sprinkled with glowing vignettes about the tortuous course that took Edison and his team to the threshold of discovery.

In the following days, the *Herald*'s article stimulated extraordinary public interest. Delegations of manufacturers and technical experts visited Menlo Park to see the lighting system. Edison arranged private demonstrations for their benefit but announced his intention not to mount a public demonstration until New Year's Eve. In the meantime, the press maintained its intensive coverage of comings and goings at Menlo Park. A constant stream of curious arrivals prompted the following observation from the *Herald*

> All came with one passion—the electric light and its maker. They are of all classes, these visitors, of different degrees of wealth and importance in the community and varying degrees of scientific ignorance. Few, indeed, are they who can approximately measure what has been done in this matter, and still fewer those who knowing its worth, admit it. But the homage of the mass to genius which they cannot comprehend makes up in quantity at any rate for any shortcomings in quality. (Jehl, 1936, p. 411)

Another example of how the selective release of information can generate enthusiasm for new technologies occurred almost 50 years later when the Pioneer Zephyr train made its "epoch making" run from Denver to Chicago. Unlike Edison, the managers of the Electro-Motive corporation provided glimpses of their new technology before it was even operational. Reporters who were shown scale models of the Pioneer Zephyr disseminated somewhat fanciful projections of the new train. As Reck (1948) described it:

> ... folks encountered such strange words as "Streamlined ... one hundred miles an hour ... articulated ... shining aluminium ... light, swift, steady ..." They saw photo-

graphs of a sleek tube of metal, so radically different from the conventional black train that it seemed like a Buck Rogers dream. (p. 73)

The build-up process that began with titillating news reports continued when the first prototypes were ready. Everywhere the Zephyr went it attracted large crowds. Reporters who rode the train compared the experience to that of "flight through the stratosphere" (Reck, 1948; p. 74). By the time the Pioneer Zephyr arrived at Denver, public mood had shifted from curiosity to eager anticipation. A crowd that included celebrities, writers, radio people, and top railroad officials swarmed to greet the locomotive. Crowds lined the tracks between Denver and Chicago and cheered the Zephyr as it passed. Meanwhile, loudspeakers at the Chicago fair recorded the progress of the train for the benefit of the awaiting thousands. When the Zephyr finally rolled onto the stage of the Wings of a Century Pageant, the "tense crowd, unable to restrain itself any longer, poured down from the amphitheater onto the stage, paying hoarse tribute to a train that had just made the longest and swiftest non stop run in railroad history" (Reck, 1948, p. 86).

A reliance on previews and selective release of information is effective when innovators are reasonably confident of the revolutionary character of their technology. However, when the innovation is an improvement rather than a breakthrough, glimpses of the new technology may dissipate the anticipation that is often a key feature of technological dramas, in part because they allow for advanced discussions and criticism by knowledgeable observers. A time-honored way of arousing curiosity is to deliberately cloak a performance in secrecy. Steven Jobs, co-founder of Apple Computer, resorted to this device prior to the introduction of the NeXT computer in 1988. Jobs, as the *New York Times* noted, was regarded in the computer industry as the "Andrew Lloyd Webber of product introduction, a master of stage flair and special effects" (Pollack, 1988). Shortly after Steven Jobs left Apple in 1985, he announced his intention to start a company that would equal, if not surpass, what his first venture had achieved. NeXT, the company Jobs formed, was slow to bring a product to market, but when the unveiling of its first computer was announced in October 1988, excitement ran high. What increased anticipation to a fever pitch was the blanket of secrecy that surrounded the new computer. In the past, Jobs had adroitly used the press, selectively leaking information to journalists (Lazzareschi, 1988). On the NeXT venture, however, Jobs went to extremes to avoid information leakage. Prospective recruits were not allowed to see the machine, but were told that they must join NeXT "as a leap of faith" (Newsweek, 1988). No outsiders were given an advance peek at the machine, and access was restricted to inside employees also.

These precautions created a feeding frenzy for any tidbits of information (Lazzareschi, 1988). By the time the 4,500 invitations to the Davies Symphony Hall in San Francisco were mailed, tickets were in such a premium that scalpers were rumored to be doing a brisk business. The audience lined up at 7:30 a.m., 2 hours before the doors opened, hoping for the best seat in the hall (Cringely, 1988). Media interest was so high that special pools of newspaper photographers and television cameramen were organized due to the limited availability of space. There were 260 reporters seated in the orchestra pit when the hall's lights dimmed at 9:39 a.m. The black curtain rolled back, and Steven Jobs emerged to enthusiastic applause.

THE MAIN ACT

Technological performances attempt to link two realities: the machine as it is, and the machine as it will be. The machine that the audience actually sees is an imperfect system, its purpose is often ill defined, and its markets are generally uncertain. By contrast, the machine that the innovators desire the audience to see is already fully developed; its use is clearly articulated, and its economic promise is about to be realized.

By exploiting the utopian impulses that technology evokes in our society, technological performances direct the audience's attention away from the imperfections of the specific technology. Although the latest innovation may fail, technology in general is triumphant. Therefore, the effectiveness of technological performances depends in part on the extent to which they can tap transformative and revolutionary archetypes (e.g., the transformation of darkness into light by electricity, and of ignorance into knowledge by the information revolution). In the cold light of inquiry, important constituencies often express doubts about the technical and economic viability of new technologies. These doubts activate critical routines, and once activated, these routines generate more questions that innovators must address. Technological performances seek to quell these doubts by exploiting the permissible ambiguity that exists between the performance text and its internal meaning, between the script that the spectators see and the message that the innovators are attempting to convey (Cole, 1975; Styan, 1975).

An analysis of technological dramas shows that, like their counterparts in the arts, they rely on conventional narratives and cultural archetypes. The narratives may consist of pioneering sagas that emphasize the trials and tribulations of new technologies (e.g., Edison's journey from humble origins to greatness, or the lonely band of General Motors (GM) pioneers developing the diesel locomotive in the face of corporate skepticism). Audiences who are versed in these sagas therefore arrive at the dramaturgical event as participants rather than as critical observers. The ensuing performance takes the audience from an initial state of suspense to triumphant finale that affirms the promising future of the new technology.

Innovators trying to launch radical technologies must often contend with highly visible efforts to discredit their claims. For example, a British parliamentary commission—seeking to calm panic in gas stocks resulting from news of Edison's experiments—consulted leading scientists of the day and then concluded that practical electric lighting was next to impossible (Jehl, 1936). A dramatic display of electric lighting, especially one that surpassed expectations, was clearly a very effective retort to this opposition. More generally, performances tailored to answer critics and transform skeptics into believers are often structured as a confrontation between the technology and a heroic task. The technology sets out to tackle formidable obstacles and go beyond known limits. The first diesel electric locomotive came under sustained attack from steam locomotive manufacturers, who claimed that it lacked the necessary power and reliability for hauling freight and passengers. To prove these critics wrong, GM's Electro-Motive corporation, in cooperation with Burlington Railways, announced that they would run their Zephyr from Denver to Chicago in 1 day, a distance of 1,015 miles. The fastest scheduled steam locomotive time on this run was around 26 hours; the Electro-Motive corporation proposed to make the same trip in 14 hours. The train arrived in Chicago 13 hours and 5 minutes after leaving Denver, making its travel the longest and fastest nonstop run in railroad history. However, it is worth noting that the Zephyr's feat was made possible by a special team of engineers that went along on the trip, fixing technical problems as they emerged. The diesel locomotive was clearly not fully prepared for normal railroad operation. Its dramatic performance, however, persuasively argued that it could overcome whatever problems still remained.

Dramatic performances of innovation are sometimes designed to demonstrate the feasibility of an idea, rather than the feasibility of an existing system. In such instances, innovators script the drama as an iconic presentation, stressing that performance is an illustration and not the "real thing." The power of the performance comes from the suggestion that although the new technology is not yet a practical reality, it heralds the dawn of a new age. Two demonstrations, more than 100 years apart, illustrate this utopian theme.

In early 1801, in front of an audience composed of members of Congress, President John Adams, and President-elect Thomas Jefferson, Eli Whitney used a simple screwdriver to fit ten different lock mechanisms to the same musket (Green, 1956; Smith, 1990). The event created a great deal of excitement, and was widely regarded as a revolutionary breakthrough that later came to be known as the "American system of manufacturing." Another demonstration that was widely perceived as the "dawn of a new age" took place on April 7, 1927. In front of an audience of top businessmen, government officials, and other luminaries, Secretary of Commerce Herbert Hoover used a rudimentary television system to communicate from Washington to New York. "Human genius", declared Hoover, "has now destroyed the impediment of distance in a new respect, and in a manner hitherto unknown. What its use may finally be no one can tell, any more than man could foresee in past years the modern development of the telegraphy or the telephone" (*New York Times*, April 8, 1927, p. 1). As if to emphasize his point, the demonstration suggested two potential uses for the new technology. In the first part of the demonstration, individuals in Washington and New York conversed using the system as a visual extension of the telephone. In the second part, a comedian stepped in front of the camera and performed a vaudeville act. The *New York Times* led with a report that reflected the enthusiasm generated by the event. Not surprisingly, the piece said little about the many obstacles that were yet to be surmounted before television could became a practical reality:

> The demonstration of combined telephone and television...is one that outruns the imagination of all the wizards of prophecy. It is one of the few things that Leonardo da Vinci, Roger Bacon, Jules Verne and other masters of forecasting failed utterly to anticipate. Even interpreters of the Bible are having trouble in finding a passage which forecasts television. (April 8, 1927, p. 5)

Although defusing doubts about feasibility is usually an important goal of the debut of a new technology, performances are often used to address other issues as well. In many instances, innovators are concerned about priority disputes that may hamper commercial exploitation of their innovations. A public demonstration of new technology does not preempt rival claims but it can shape public perceptions of priority. These public perceptions may deter rivals from taking legal actions or play a role in legal proceedings if and when they take place. For example, Edison was concerned about competitors who were working on the problem of electric lighting. The demonstration at Menlo Park was partly intended to stake his claim to being the first to have perfected a practical and safe system of electric illumination. And although Peter Jensen and Edwin Pridham sought to establish priority for their invention by mounting a demonstration in front of the San Francisco City Hall, they also attempted to keep the nature of their new system secret by operating the controls from a balcony and by concealing the loudspeaker behind an American flag (Lewis, 1990).

Finally, there are technological dramas that are unabashed celebrations of achievement. The attempt here is to dazzle, not simply to persuade. To convince an audience that the innovation operates in a realm where mundane technical concerns should not apply. The innovator's persona and achievements often becomes part of this celebration. The spotlight is not just on the technology, but on the innovator's creative genius as well. An example of a technological performance where the charisma of the innovator is just as important as the machine is Steve Jobs' unveiling of the NeXT computer. Jobs hired George Coates, a theater director with extensive experience in multimedia presentations, to stage the performance. Coates' stage design consisted of a vase of flowers and an unknown object hidden by a black shroud, against an austere black backdrop. When Jobs stepped on the stage, he did not unveil the computer immediately. Instead, he began with a graphic presentation of the history of the computer industry, and then recounted the saga of the NeXT venture since its inception. At last it was the machine's turn to perform. To increase the tension, Jobs followed the tradition of magicians and daredevils, by warning the audience that the performance they were about to witness could malfunction. The warning conveyed a sense that NeXT had truly moved into uncharted territory, but it also served to immunize the performance from the consequences of failure:

> Now what I'd like to do now is do some live demos for you.... I'd like to remind you of the first two laws of demoing. [Some laughter]. First law of demos is that demos will always crash. And the second law of demos is that their probability of crashing goes up with the number of people watching. [More laughter and applause]. So if something goes wrong today, have some compassion for the demo-er. (Stross, 1993, p. 178)

The performance began with a graphic display of quantum wave functions. A microphone was added, transforming the computer into an oscilloscope that displayed sound as visual data. From the computer's stored memory, Jobs called forward selections from Martin Luther King's "I Have a Dream" speech, John F. Kennedy's "Ask Not What Your Country Can Do for You," and a NASA voice transcript, "The Eagle Has Landed." After demonstrating the machine's digital library, Jobs turned to its musical capabilities. A digital signal processor conjured musical selections for harpsichord, strings, and Indonesian gamelan. The closing climax of the performance came when a violinist from the San Francisco Symphony Orchestra stepped forward to play a Bach violin duet with the NeXT Computer.

Objectively speaking, many of the capabilities that the Next computer performed during this demonstration were peripheral to the main uses of such computers. However, the purpose of the display was not to render a realistic equivalence, but to create a utopian image, stressing possibility over actuality. Response to the demonstration may have been enthusiastic but this alone begs the question that is surely of crucial importance: Do such performances influence the evolution of the technology, and if so, how?

COLLECTIVE DYANMICS AND PATH CREATION

Technological dramas persuade individuals to envision the future, and then bring the consequences of the future into the present. The process harnesses our ability as individuals to introspectively experience such vision, and it also is driven by collective social processes. In technological communities, this means discussion and debate in which innovators attempt to persuade others to use their preferred evaluation routines when judging claims in what Garud and Rappa (1994) described as "a highly negotiated political process" (p. 348). The interaction between innovators and key constituencies such as investors or customers is fundamentally different. Members of a technological community subscribe to a set of evaluation routines that are reasonably consistent, whereas external constituencies dispose of more eclectic set of routines. Furthermore, they are likely to use these routines differently, depending on their experience, interests, and personality.

What innovators would like to do is align the evaluation routines of key constituencies in their favor. The fundamental challenge facing innovators, however, is the heterogeneity of their key constituencies. Mobilizing support from one person or one institution at a time is a costly and time-consuming process. Technological dramas have two advantages in this regard. First, they reduce the time needed to communicate with the relevant constituencies. Second, when effective, they transform a loose aggregation of potential supporters into a cohesive group of backers. Edison's demonstration of his electric illumination system is a case in point.

The first system of illumination was put on the steamer, the Columbia. Anchored at the foot of Wall Street, the Columbia's full splendor was exhibited each night for all to see. Next, Edison electrified Fifth Avenue, making this prestigious address a showcase for the new technology. Soon requests for the system were pouring in, ensuring Edison's victory over competing systems. To quote Conot (1980):

> During the summer and fall [of 1881] a steadily increasing number of places were wired: hotels, machine shops, cotton, wool, and silk mills, railroad and locomotive works, flour mills, piano and organ factories, meat packers, printers and engravers, sugar refineries, and steamships. James Gordon Bennett had his newspaper plant and his yacht outfitted—and of course, if Bennett's yacht had electric lights, Gould's had to have them, too. (p. 223)

The effectiveness of dramaturgical events such as the demonstration mounted by Edison depended on the actions of a few rich tycoons such as Jay Gould, who made an open and highly visible commitment to his system. These actions influenced other customers who were hanging back; willing to believe, but unwilling to commit. When these customers became openly committed to the new technology, others who were until now sceptical shifted their evaluation from critical to evaluation routines. They were now believers, but not yet committed to action. And when they committed to action, the evaluation routines of other were affected, and they, in turn, shifted from a critical to a commitment stance.

Work by Granovetter (1978), Macy (1991), Coleman (1990), and Marwell and Oliver (1993) threw light on this process. They demonstrated that discontinuous collective change, such as adoptions of innovation by industries, may be the result of shifts in perception by a small number of actors, which then triggers a bandwagon process that takes a collectivity past a "critical mass." There is no need for the actors to negotiate this change; instead, the change comes about because actors behave in a serial rather than in a coordinated fashion. The calculations of each actor are influenced by the behavior of their peers. Collective dynamics that seem, in retrospect, to have been coordinated, in reality represent a convergence of distinct perceptions.

Technological performances attempt to trigger and then influence this convergence process. First by inducing pivotal actors to shift from critical to commitment routines, and then by using these actions to influence the evaluation routines of other actors. To successfully do this, technological performances must overcome resistance from two sources. To begin with, there is a perceptual threshold that prevents the transition from critical to commitment routines from taking place. And next, there is an action threshold that prevents cognitive commitment from being translated into action. For the process to attain collective critical mass, it must lower these thresholds sequentially, but not for all observers simultaneously. What has to take place is in effect a bandwagon process (Abrahamson & Rosenkopf, 1993). Technological performances seek to lower their action threshold of the collectivity by targeting the perceptual threshold of pivotal actors. If this attempt succeeds, it becomes easier to lower the action threshold as well. Open commitment on the part of pivotal actors, in turn, reduces the perceptual threshold of other actors, increasing the likelihood that they will switch from critical to commitment routines, which, in turn, impacts their action threshold.

The whole process was described by Oliver (1993) as a "threshold contagion": A shift in perceptions, attitudes, and actions which feeds on itself. Once this contagion process reaches critical mass, it is not difficult as a rule for innovators to keep it moving. The real difficulty is clearly at the crucial stage where it is necessary to get individuals and organizations that are in the best position to shift their evaluation routines, and then translate this increasing commitment into a willingness to back the new technology. These actors, however, are often likely to be the most resistance to influence. Their resistance may be due to the fact that they have more to lose if they make the wrong decision, or because their experience and knowledge gives them a better appreciation of the new technology. Regardless of which of these reasons is decisive, it may be strong enough to stop the process from moving toward its desired conclusion.

This failure is especially striking in technological performances that are very successful in purely dramatic terms. Steven Jobs' demonstration of the NeXT computer is a case in point The press gave the demonstration an enthusiastic review. "I arrived a nonbeliever, and I came away a convert" (Elmer-Dewitt, 1988), said Richard Shaffer, editor of the Technological Computer newsletter. In a *Los Angeles Times* (1988) article, of the eight industry experts, investors, competitors, and potential customers quoted in length, seven believed that the machine would be a success. Interestingly, their technical analysis of the computer was in general accurate; it was the conclusions that they drew from the data that, in retrospect, proved to be wildly optimistic. The sales that they, and NeXT, forecast for the computer did not materialize. In the 7 years of its existence, NeXT sold only 50,000 machines. Poor financial performance finally forced Jobs to discontinue manufacturing and concentrate on its software line. In fact, Jobs was the main backer and financier of the venture. It was his resources and reputation that have kept the company afloat. In 1996, Apple computers acquired NeXT computers for $400,000,000. Their interest, however, was not in NeXT as envisaged by Jobs, but in its potential application on the rapidly developing internet (Burrows, 1997). Thus, in spite of the tremendous success of the NeXT demonstration, the product itself as a stand-alone computer system did not create the revolution that Steven Jobs envisaged.

FROM PATH DEPENDENCE TO ENVISIONING TECHNOLOGICAL PATHS

Following Kuhn's (1970) theory of scientific revolutions, a number of researchers have argued that technological change is governed by competition among alternative technological paradigms (Dosi, 1982). Graud and Rappa (1994) suggested that technological paradigms that succeed in recruiting supporters are more likely to prevail over their competitors. Their theory, however, focuses on the recruiting process within technological communities. The same argument can be applied to external constituencies; technologies compete for the resources provided by investors and consumers. Those that succeed in getting this support early in the development of their trajectory are more likely to become dominant, whereas technologies that fail to get this support will lack the resources necessary to solve their key problems, and hence their trajectory will fail commercially.

The progressive amplification of an early differential advantage in resource recruitment is also an important feature in recent path dependent theories of technological change (David, 1985). Path dependent theories, however, argue that this differential advantage is usually a by-product of other activities. It is rarely, if ever, the result of a consciously thought out strategy on the part of innovators. However, by postulating that small events during the emergent phase of technological change can have a disproportionate impact on their subsequent development, this view suggests a role for technological dramas.

Technologies, argue path dependent theorists, move along a path that at selected points is vulnerable to intervention by social and ideological processes. The success or failure of a technological drama is therefore likely to have a marginal impact unless it comes at a time when technology is at a crossroad, or to use the language of catastrophe theory, if it emerges at a bifurcation point. When these points occur, small events can produce large effects. Such interventions are particularly important if they lock-in technological choices, making further technological change less malleable to social or political action (Arthur, 1989; Granovetter, 1992). Prior to lock-in, events may be influential because they trigger or accelerate the accumulation of choices that push technologies past a certain threshold that represent a "point of no return" for technological evolution (Arthur, Ermoliev, & Kaniovski, 1987).

Do technological performances act in this fashion? Do they constitute events that push technologies past a critical threshold? A path dependent explanation of technological performances would focus primarily on the event as a trigger in a chain reaction. What would be missing is the dynamics of collective sensemaking without which innovators cannot mobilize resources. *Technological performances* represent an attempt by innovators to change how innovations are evaluated. The intent is to replace critical evaluation routines that are focused on uncertainty reduction, with commitment routines that are oriented toward certainty creation. Claims become credible because they are evaluated from the perspective of future promise, rather than present reality. This is done on two levels; at one level, there is a sustained effort to disable critical evaluation routines by surrounding the new technology with an aura of a "sure bet"—the belief that a technology is bound to win in the forthcoming competitive struggle. At a deeper level, however, the aura of a sure bet is reinforced by a more powerful evocation of a future in which the technology in question plays a transformative role (Garud, Jain, & Phelps, 1997). Thus, the electric light, the diesel locomotive, the loudspeaker, the NeXt computer, are all presented as shining visions of future reality; Not quite here, but well within our reach!

Of course, there is usually considerable distance to be covered and numerous gaps to be bridged before the technology becomes an everyday reality. The innovators do not deny this. What they do instead is rely on the intrinsic property of commitment routines. Critical routines search for gaps, whereas commitment routines seek to "connect the dots," to imaginatively conceive the path that links the present and the future. The propensity is rooted in the cognitive response to ambiguity. In this case, the ambiguity of how the technology will get from "here" to "there.". It is well established that the mind responds to ambiguity by spontaneously seeking a deeper order even when there is none there to be had (March, 1994; Ross & Nisbett, 1991; Watzlawick, 1976). One way in which this is done is by building expectations of how the ambiguity will be resolved; this tends to restrict a person's attention to information that confirms what is expected. Certainty creation plays on this. As Weick (1995) puts it, "When perceivers act on their expectations, they may enact what they predict will be there. And when they see what they have enacted, using their predictions as a lens, they often confirm their prediction" (p. 153).

It is this aspect of human cognition that path dependent theories of technological dramas ignore. Although path dependent theories of technology make the important point that small events can disproportionately shape trajectories, the cognitive models that individuals use to interpret these events is objective and calculative. Events change the costs/benefits tradeoffs. Actors examine these tradeoffs and act according to their objective needs and local conditions. There may be uncertainty about these tradeoffs but no ambiguity; no room for interpretive flexibility. In other words, given the same information, all rational actors with the same needs and in the same local conditions should arrive at the same decision.

What is probably the best known example of path dependent theories, the QUERTY keyboard, illustrates this model well (David, 1985). The keyboard is the result of a historical accident but this accident gave rise to the current keyboard design. The keyboard is by no means the best design possible, and switching to a better design would clearly benefit everybody. However, doing this would impose costs on individuals that they are willing to pay. An external actor, for example, a government, could step in and remedy this situation by offering to pay individuals a fee. What this actor would be doing is shifting the costs/benefits structure, and hence changing the calculus of keyboard users. What it is not doing is creating a qualitatively different interpretation of the situation; there is no ambiguity here, and, in fact, if the measure is to have its intended effect, it is imperative that it should be as free of ambiguity as possible.

The QUERTY example has been used to illustrate the gap that exists between individual and collective rationality. What is rational for individuals may not be rational for the collectivity, and vice versa. The problem of bridging the gap between individual and collective rationality is also approached rationally; for example, by changing the situation's payoff structure. It would make no sense from a path dependent point of view to bridge this gap by appealing to the noncalculative part of the human mind—to affect, imagination, or fantasy. Yet, this is precisely what technological dramas attempt to do. And they try do this precisely because innovators realized that technological change is not a purely an objective and calculative process; that it depends on a commitment to the future, as much as a proper evaluation of the present.

What technological dramas are in effect trying to do is to envision a path between what the machine can do at present and what it can do in the future, and then trigger a process that can help bring about this future. The path envisioned is created in the province of the imagination, but its consequences are expected to be real. If the drama fails to stimulate a shift from critical to evaluation routines, actors are likely to adopt a wait-and-see attitude of "Let others take the lead, I will make my commitment once the promise becomes reality." Of course, if enough actors adopt this attitude, the technological trajectory will most likely fail, exactly as predicted by path dependent theories. On the other hand, if the path is evaluated using commitment routines, there is greater probability that key constituencies will commit to the new technology and that their behavior will initiate a bandwagon process that will ensure the successful creation of the technological trajectory.

The difference between path dependence and path creation therefore comes down to a different views on how the mind works. In path dependence theories, the human mind is essentially a very powerful computer, whereas in path creation the human mind is conceived more broadly as having the capacity to feel and imagine. The difference is crucial when it comes to how actors see and evaluate the future. Actors in path dependence theories see the future as a set of scenarios that they evaluate, using the tools and vocabulary of accounting and probability. Actors in path creation see the future as a complex whole that they evaluate using a vocabulary drawn from the social and cultural milieus in which they operate (Simons, Mechling, & Schreier, 1984). When it comes to explaining technological change, both these views can be useful. However, when it comes to explaining why technological dramas are used in the first place, only the latter provides a satisfactory framework.

CONCLUSION

To catch a mouse—make a noise like a cheese.
                                                    —Lewis Kornfeld (1992)

Throughout most of the 20th century, scholars and commentators argued for a sequential relationship between science, technology, and the commercialization of new products and processes (Freeman, 1974). Science comes first, technology builds on science, and new products are derived from generic technologies once they are ripe. Technological dramas are clearly peripheral to such a linear model because there is little in such dramaturgical events to explain how they can substantively influence the development of a technology.

The last 20 years has seen the sequential model fall from grace, to be replaced by a complex multilevel picture of innovation in which science, technology, and commercialization interact in a nonlinear and nonsequential way. The image of innovation that we have today is nondeterministic. It is more subject to the influence of contingencies and more likely to be swayed by historical accidents than the one that prevailed earlier. If technological dramas were a curiosity in the old model, the question arises as to whether they have a more important place in the new one.

The answer seems to be *yes*, but it is a qualified yes. At times, technological dramas do play an important role in the development of new technologies but at other times, they are clearly peripheral events. Given the uncertain influence of technological dramas on outcomes, the question naturally arises as to what practical lessons they communicate? What advice can innovators take from the experience of their predecessors?

The answer to this question must contend with the uncertain nature of technological change. There are few facts that are not open to conflicting interpretations when questions are raised about a new technology: Does the technology contain the promise of a large market or is it destined to marginality? Can the new technology be brought to market relatively quickly, or must it go through a lengthy and expensive process of improvement? Economic theory has traditionally argued that markets resolve these questions in a rational manner, using the objective information provided by innovators. In short, if you make a better mousetrap and announce it to the world, the news will eventually reach those who stand to profit from your invention, and they, once they have calculated the potential profits, will beat a path to your door.

Of course the reality of the market is quite different than the one portrayed in economic theory. At no time do decisionmakers have a clear and stable image of the terrain in which they are operating; the information they receive is usually fragmentary and frequently contradictory. Potential investors and customers may be rational, but when rationality drives their analysis they are likely to subject new technology to highly critical scrutiny. Thus, attempting to persuade these actors by providing more detailed information will usually fail precisely because critical evaluation routines focus attention on the disadvantages and problems.

Making believers out of disbelievers by appealing to objective evidence is sometimes blocked by the very rationality that drives the process. The more one pushes, the more the resistance grows (David, 1991). Breaking this rational impasse may call for a strategy that radically reframes the situation. This reframing may appear counterintuitive at first sight. Imitating the "noise" made by cheese is a prescription that appears paradoxical, even absurd, until we realize that switching from a chase to a lure strategy often makes sense. Technological dramas arguably do precisely this: They work by luring audiences away from their habitual critical stance into an evaluation routine that predisposes them to commitment. A luring strategy, however, depends on attracting and molding attention, and this, in turn, points to artful use of symbol and spectacle. So in the final analysis, if we insist on finding a place for technological dramas in the process of technological change, we must look to the area where technology as artefact interacts with technology as narrative and representation. In this area, human imagination and wish fulfillment may be just as important to shaping technology as economic calculations and technical constraints, and at critical juncture in the evolution of technology, even more important.

REFERENCES

Abrahamson, E., & Rosenkopf, L. (1993). Institutional and competitive bandwagons: Using mathematical modeling as a tool to explore innovation diffusion. *Academy of Management Review, 18,* 487–517.

Arnott, J. F. Chariau, J., Huesmann, H., Lawrensonk, T., & Theobald, R. (Eds.). (1977). Theater space: An examination of the interaction between space, technology, performance, and society. Munich, Germany: Prestel-Verlag.

Arthur, W. B. (1989). Competing technologies, increasing returns, and lock-in by historical events. *Economic Journal, 99,* 116–31.

Arthur, W. B., Ermoliev, Y. M., & Kaniovski, Y. M. (1987). Path-dependent processes and the emergence of macro-structure. *European Journal of Operational Research, 30,* 294–303.

Barhydt, J. D. (1987). *The complete book of product publicity.* New York: American Management Association.

Bateson, G. (1972). *Steps to an ecology of mind.* San Francisco, CA: Chandler Publishing.

Bernheim, B. D., & Shoven, J. B. (1992). Comparing the cost of capital in the United States and Japan. In N. Rosenberg, R. Landau, & D. C. Mowery (Eds.), Technology and the wealth of nations (pp. 151–174). Stanford, CA: Stanford University Press.

Bode, H. W. (1971). *Synergy: Technical integration and technological innovation in the Bell System.* Murray Hill, NJ: Bell Laboratories.

Boyadjian, H. J., & Warren, J. F. (1987). *Risks: Reading corporate signals.* New York: John Wiley.

Burns, E. (1972). *Theatricality: A study of convention in the theater and social life.* London: Longman.

Burrows, P. (1997, January 13). How Apple took its NeXT Step. *Business Week,* 36–37.

Cole, D. (1975). *The theatrical event.* Middletown, CT: Wesleyan University Press.

Coleman, J. S. (1990). *Foundations of social theory.* Cambridge, MA: The Belknap Press.

Conot, R. (1980). *Thomas A. Edison: A streak of luck.* New York, NY: De Capo Press.

Cringely, R. X. (1988, October 17). Next question: Are those Gates quotes erasable? *Infoworld,* 94.

Dasgupta, P. (1988). Patents, priority, and imitation or, the economics of races and waiting games. *Economic Journal,* 98, 66–80.

David, P. A. (1985, May). Clio and the economics of QWERTY. *American Economic Review,* 75, 332–337.

David, P. A. (1991). *The hero and the herd in technological history: Reflections on Thomas Edison and the battle of the systems.* In P. Higonnet, D. S. Landes, & H. Rosovsky (Eds.), Favorites of fortune: Technology, growth, and economic development since the Industrial Revolution (pp. 72–119). Cambridge, MA: Harvard University Press.

Dosi, G. (1982). Technological paradigms and technological trajectories. *Research Policy,* 11, 147–162.

Elmer-Dewitt, P. (1988, October 24). Soul of the next machine. *Time,* 80–82.

Evans, B. (1968). *Dictionary of quotation.* New York: Delaorte Press.

Freeman, C. (1974). *The economics of industrial innovation.* Harmondsworth, England: Penguin.

Garud, R., Jain, S., & Phelps, C. (1997, August 19–22). Transmutional change: Path dependence and creation in the web browser market. Paper presented at the Path Creation and Path Dependence Conference, Copenhagen, Denmark.

Garud, R., & Rappa, M. A. (1994). A Socio-cognitive model of technology evolution: The case of cochlear implants. *Organization Science.* 5, 344–362.

Georghiou, L., Metcalfe, J. S., Gibbons, M., Ray, T., & Evans, J. (1986). *Post-innovation performance: Technological development and competition.* London: MacMillan.

Goffman, E. (1974). *Frame analysis: An essay on the organization of experience.* New York: Harper & Row.

Granovetter, M. (1978). Threshold models of collective behavior. *American Journal of Sociology.* 83, 1420–1443.

Granovetter, M. (1992a). Economic institutions as social constructions: A framework for analysis. *Acta Sociological,* 35(1), 3–11.

Green, C. McLaughlin. (1956). *Eli Whitney and the birth of American technology.* Boston: Little, Brown, & Co.

Grove, S. J., & Fisk, R. P. (1989). Impression management in services marketing: A dramaturgical perspective. In R. A. Giacalone & P. Ronsenfeld (Eds.), *Impression management in the organization.* Hillsdale, NJ: Lawrence Erlbaum Associates.

Hawkins, D. F. (1986). *The effectiveness of the annual report as a communication vehicle: A digest of the relevant literature.* Morristown, NJ: Financial Executives Research Foundation.

Jehl, F. (1936). *Menlo Park reminiscences.* Dearborn, MI: Edison Institute.

Jelinek, M., & Schoonhoven, C. B. (1990). *The innovation marathon: Lessons from high technology firms.* San Francisco, CA: Jossey-Bass.

Kamien, M., & Schwartz, N. (1982). *Market structure and innovation.* Cambridge, England: Cambridge University Press.

Karnøe, P., & Garud, R. (1996). Path creation and dependence in the Danish wind turbine field. In J. Porac & M. Ventresca (Eds.), *The social construction of industries and markets.* New York: Pergamon.

Katz, M., & Shapiro, C. (1985). Network externalities, competition and compatibility. *American Economic Review,* 75, 424–440.

Kelly, G. A. (1963). *A theory of personality.* New York: Norton.

Kornfeld, L. (1992). *To catch a mouse—make a noise like a cheese* (3rd ed.). Fort Worth, TX: Summit.

Kotler, P. (1973). Atmospherics as a marketing tool. *Journal of Retailing,* 49, 48–64.

Kuhn, T. S. (1970). *The structure of scientific revolutions.* Chicago, IL: University of Chicago Press.

Laemmle, C. (1976). The business of motion pictures. In T. Balio (Ed.), *The American film industry.* Madison: The University of Wisconsin Press.

Lazzareschi, C. (1988, October 12). Secrecy increases allure of Jobs' next computer venture. *Los Angeles Times*, 1.

Lewis, W. D. (1990). Peter L. Jensen and the amplification of sound. In C. W. Pursell, Jr. (Ed.), *Technology in America: A history of individuals and ideas* (2nd ed., pp. 190–210).

Los Angeles Times. (1988, October 23). Will Steven Jobs' computer sell? (p. 4).

Macy, M. (1991). Chains of cooperation: Threshold effects in collective action. *American Sociological Review, 55*, 730–747.

March, J. G. (1994). *A primer on decision making*. New York: Free Press.

Marwell, G., & Oliver, P. E. (1993). *The critical mass in collective action: A micro–social theory*. New York: Cambridge University Press.

Merrifield, D. B. (1981). Selecting projects for commercial success. *Research Management, 24*, 13–18.

Morus, I. R. (1988). The sociology of sparks: An episode in the history and meaning of electricity. *Social Studies of Science, 18*, 387–417.

Newsweek. (1988, October 24). *Steve Jobs comes back, Vol. 112*, 7, pp. 46–51.

Nye, D. E. (1983). *The invented self: An anti-biography, from documents of Thomas A. Edison by D. E. Nye*. Odense, Denmark: Odense University Press.

Oliver, P. E. (1993). Formal models of collective action. *Annual Review of Sociology, 19*, 271–300.

Pfaffenberger, B. (1992). Technological dramas. *Science, Technology, and Human Values, 17*, 282–312.

Pollack, A. (1988, October 10). NeXT Inc. Produces a gala (and also a new computer). *New York Times*, D1.

Reck, F. M. (1948). *On time: The history of General Motors locomotives*. Detroit, MI: General Motors Corporation.

Ross, L., & Nisbett, R. E. (1991). *The persona and the situation*. New York: McGraw-Hill.

Rothwell, R., & Robertson, A. B. (1973). The role of communication in technological innovation. *Research Policy, 2*, 204–225.

Roussel, P. A. (1983, September/October). Cutting down on the guesswork in R&D. *Harvard Business Review, 154–160*.

Schwartz, N. L., & Kamien, M. I. (1978). Self-financing of an R&D project. *American Economic Review, 68*, 252–261.

Simons, H. W., Mechling, E. W., & Schreier, H. N. (1984). The functions of human communication in mobilizing for action from the bottom up: The rhetoric of social movements. In C. C. Arnold & J. W. Bowers (Eds.), *Handbook of rhetorical and communication theory* (pp. 792–867). Boston: Allyn & Bacon.

Smith, M. R. (1990). Eli Whitney and the American system of manufacturing. In C. W. Pursell, Jr. (Ed.), *Technology in America: A history of individuals and ideas* (2nd ed., pp. 45–61). Boston: MIT Press.

Souder, W. E., & Moenaert, R. K. (1992). An information uncertainty model for integrating marketing R&D personnel in new product development projects. *Journal of Management Studies, 29*(4), 485–512.

Speier, H. (1980). The communication of hidden meaning. In H. D. Lasswell, D. Lerner, & H. Speier (Eds.), *Propaganda and communication in world history. Volume II: Emergence of public opinion* (pp. 261–300). Honolulu: The University of Hawaii Press.

Stross, R. E. (1993). *Steve Jobs and the NeXT big thing*. New York: Atheneum.

Styan, J. L. (1975). *Drama, stage and audience*. London: Cambridge University Press.

Watzlawick, P. (1976). *How real is real?* New York: Vintage Books.

Weick, K.E. (1995). *Sensemaking in organizations*. Beverly Hills, CA: Sage.

White, G. R., & Graham, M. B. (1978, March/April). How to spot a technological winner. *Harvard Business Review, 99–109*.

12

Innovation as a Community-Spanning Process: Looking for InterAction Strategies to Handle Path Dependency

Bart Van Looy
Koenraad Debackere
Rene Bouwen
University of Leuven, Belgium

INNOVATION AS A PROCESS SPANNING COMMUNITY BOUNDARIES: SETTING THE STAGE

Research into the management of technology and innovation has high-lighted the many pitfalls and problems that are usually encountered during the innovation journey. At different levels of attention and analysis, early work in the field (e.g., Allen, 1966, 1977; Allen & Frischmuth, 1969; Allen & Marquis, 1963; Myers & Marquis, 1965; Pelz & Andrews, 1967) to name just a few, has pointed to the importance of problem-framing and prob-lem-solving activities to accomplish the innovation task at hand. In es-sence, this rich and diverse stream of research has pointed to the central role of handling, and from a performance point of view, reducing uncer-tainty during the various phases of the innovation process. Information and information exchange were considered and shown to be critical elements in this endeavor.[1]

A closer look at this research program reveals that at various stages the innovation process benefits enormously from boundary-spanning information exchanges and insights. In this way, information asymmetries are reduced. Some scholars, including Thomas Allen (1977) and Michael Tushman (1977), have pointed to the important roles played by gatekeepers or boundary spanners during innovation processes. Not only does this boundary-spanning activity play an important role during the implementation and problem-solving phases of the innovation process, but, as these and other authors suggest and demonstrate, the problem-framing or gestation phases of the innovation process may benefit from these boundary-spanning interactions as well. It is not our aim to review here all of the findings and evidence at hand to support this boundary-spanning notion. However, we would assert that—following the French saying that *"du choc des idées jaillit l'esprit"*—the concept of *boundary-spanning* has received widespread attention and support as one of the key phenomena that occur or should occur during any innovation effort.

Using the concept of boundary-spanning as a starting point for the research reported in this chapter, though, immediately reveals the problematic nature of the concept as well. Throughout the innovation literature, boundary-spanning activities have been most often investigated at the level of the innovation project. At that specific level of analysis, boundary spanning is important and problematic at the same time because it points to the necessity of confronting and integrating different functions (e.g., marketing, research and development (R&D), engineering) or disciplines (e.g., mechanical engineering, electrical engineering, chemical engineering) or grammars (e.g., algorithmic or symbolic reasoning in the area of artificial intelligence vs. adaptive learning approaches in the area of neural networks) within and across organizations during the development of new technological knowledge and/or artifacts. As described and argued by such scholars as Constant (1980), Nelson and Winter (1982), Garud and Rappa (1994), each of those distinct disciplines, functions, or grammars is represented by communities of individuals who create and find their rallying point around a complex interaction of beliefs, evaluation routines, and artifacts.

This complex interaction is at the origin of the genesis of nontrivial path dependencies that, in turn, inhibit or impede this boundary-spanning activity In summary, innovation requires the spanning of boundaries across communities, though at the same time, communities create important impedance ef-

---

[1]An excellent (summary) reflection on the various aspects and focal points of this research program can be found in an article written by Brown and Eisenhardt (1995).

fects that prevent and imperil the boundary-spanning activity. This duality has been reported at different levels of analysis and in different study contexts. To name just a few:

- In their important studies of productivity of scientists in industrial R&D (situated at the microlevel of the individual scientist), Pelz and Andrews (1967) pointed to the existence of different creative tensions, some of which can be reduced to the basic dilemma or paradox just described: How does a scientist handle the balance between striving to become still better embedded in the expertise area (community) in which he or she already excels versus breaking away from these established routines, beliefs, and artifacts, to embrace new research trails? Similar phenomena have been reported by Allen and Marquis (1963) and Allen and Frischmuth (1969) when they investigated the "biasing" effects of prior engineering experience on the problem-solving strategies pursued by engineers. This phenomenon is succinctly described in Garud and Rappa's study (1994) on the evolution of cochlear implant technology:

  "Researchers develop specific competencies over time. These competencies accumulate in a path-dependent manner as earlier technological choices direct future options and solutions (Cohen & Levinthal, 1990; Arthur, 1988; David, 1985). As competencies become more specialized, researchers find it increasingly difficult to redirect themselves to pursue other paths. This is similar to the notion of the accumulation of "sticky" resources (Ghemawat, 1991). As a consequence, there are powerful incentives for researchers to persist with a path. In this way, beliefs are externalized as artifacts, which in turn shape the beliefs of the researchers associated with the development of these artifacts."

Werth's 1994 description of the development of rational drug design technology raises similar issues and dilemmas. In other words, the confrontation between path dependent activities versus boundary-spanning activities has been present at least in an implicit (but sometimes also in a more articulated) manner in much of the writing on the problem-solving activities of innovative professionals.

- At the project level, research on the "not-invented-here" syndrome (Katz & Allen, 1982) pointed to the same tension between the path dependency of the project team versus the openness of the project team toward new signals from other stakeholders both inside and outside their respective organizations. These boundary-spanning dilemmas have been amply documented ever since (Wheelwright & Clark, 1992). Katz's (1993) case study on the development of the alpha-chip at Digital provided yet another illustration of the paradoxical nature of the boundary-spanning activity between different communities within one and the same organization;

- At higher levels of analysis, scrutinizing the development of new technologies, the analyses by Constant (1980) on the development of the turbojet, by Thomson (1998) on the development of mechanized shoe production, by Garud (1994) on cochlear implant technology or by Burgelman (1994) on Intel's exit from the DRAM-industry all point to technological paths and directions taken or to be taken as the result of interactions between different communities. In line, Bijker's (1994) social-constructivist approach to decision making in technology development and technological evolution highlights this as a process where the confrontation of the beliefs, routines, and artifacts characteristic of and held by various communities is reevaluated, renegotiated, and fused into a synthesis around which a new community coalesces; such a synthesis may partially or completely include the members of the communities previously present in the complex negotiation and fusion. It is quite comforting to find that studies on the development of new scientific disciplines have long been confronted with these dynamics and can therefore serve as useful signposts to any student trying to approach the development of new technology and its subsequent embedding in an innovation endeavor from this same perspective. To name just a few: Ben-David's (1960) and Ben-David and Collins' (1966) work on the development of innovations in medicine and the origins of psychology as a "new" science, Edge and Mulkay's (1976) study of the development of radio astronomy, or Lemaine, MacLeod, Mulkay, and Weingart's (1976) still interesting work on the origins of new scientific disciplines.

CROSSING COMMUNITY BOUNDARIES
AS A WAY OF TRANSCENDING PATH DEPENDENCY

Hence, the dilemma raised in this introduction is not new and can be recognized in different settings at different levels of analysis. Innovation, whether studied as the development of new technological or scientific knowledge or the creation of new artifacts, is a process that requires boundaries between communities to be spanned or crossed. But, communities have their own fixed set of beliefs, evaluation routines, and artifacts that create powerful path dependencies inhibiting the very boundary-spanning activity that is at the heart of many successful innovations as they have been described and documented in the extensive amount of research that has just been briefly touched on. Hence the question turns into the direction of finding relevant action strategies that allow one to transcend these path dependencies.

The notion of path dependency reminds us of the historical antecedents of current situations: "A path-dependent sequence of economic change is one of which important influences upon the eventual outcome can be exerted by temporally remote events, including happenings dominated by chance element rather than systematic forces." (David, 1985, p. 332). Although the notion of path dependency has only recently gained interest in literature oriented toward economics, within other disciplines the idea that "history matters" is more accepted (Hirsch & Gillespie, 1997). Although the main message of David was one of creating historical awareness when studying economic processes, the idea that historical events might result in less than optimal technical trajectories has created a lot of animosity among economic scholars, because this questions the effectiveness of market mechanisms (see for instance Liebowitz & Margolis, 1990, 1995). Distinguishing between different kinds of path dependency and acknowledging arrangements that allow for path creation might settle this dispute (Liebowitz & Margolis, 1995). That path dependence cannot exist separated from its counterpart, path creation, becomes conceptually clear when turning to the notion of structuration, conceived by Giddens to arrive at a more inclusive understanding of social phenomena (Giddens, 1986; for an application in the domain of technology, see Orlikowski, 1992). Within the notion of structuration, agents and structure are not seen as independent notions but represent a duality. The structural properties of social systems can be seen as both medium and outcome of the practices of interacting actors these structural properties recursively organize. As such, actors reproduce and partially transform social systems. This (partial) reproduction is historically situated and achieved by knowledgeable actors. So when introducing the notion of path dependency, its counterpart, path creation, is implied.

This interdependence is further exemplified in Morison's (1966) study. The innovation described by Morison, brought about by Lieutenant Sims at the same time extrapolated and rejuvenated existing military practice. In this way, the case is an excellent illustration of Karnøe and Garud's (1997) statement that:

> An understanding exclusively based on path dependence could, both in theory and in practice, trap us, and thereby limit our abilities to maneuver into the future by reproducing the past. The evolution of field must be understood as emerging structures and processes that are shaped by initial conditions, random historical events and a transformative process that entails partial reproduction and partial creation.[2]

One of the other main characteristics and strengths of the path dependence model, besides historicity, lies in the insistence of its practitioners on the specific importance of microlevel events. The issues raised within this chapter are situated exactly at this level. Although the co-presence of path

---

[2]Although we endorse the main message of both authors, namely that processes imply past, future, as well as transformative processes, the notion of "random historical events" might be replaced by "historically situated events." The debates on Qwerty v. Dvorak's keyboard both illustrate this point, as well as the work done within the array of studies related to the social construction process of technology (for an overview see Bijker, Hughes, & Pinch, 1987; Bijker & Law, 1992; and further, Rip, Misa, & Schot, 1995).

dependence and creation becomes intuitively apparent on a more conceptual level, nonetheless, a lot of issues remain vague when one turns to the daily practice of the innovation process. Hence, our contribution can be seen as an attempt to create a more profound insight into the complexity of path dependency and creation at this level, by acknowledging the historical and situated nature of the actors involved in the development process. As such, the intention in this chapter is to add to our understanding of how to deal with the dilemma described by examining how path dependencies present themselves within the development process and how relevant action strategies for overcoming these path dependencies might look.

A starting point in this endeavor is the fact that our understanding of the innovation process can be further enhanced by framing the innovation process as a community-spanning process. As a consequence, the first task of this chapter is to show and describe how we can study, identify and clarify the innovation process from a community-spanning perspective. Portraying the innovation process as resulting from the involvement of different communities also means that ambiguity or asymmetries of interpretation enter the stage. As already mentioned, communities are characterized by a complex configuration of opinions, beliefs, routines, and artifacts. This causes *interpretation asymmetries* to come into being. Interpretation symmetries arise because different communities use different frames of reference to interpret and to evaluate technological phenomena. Hence, as communities are characterized by different beliefs, routines, and artifacts once they become involved in an innovation endeavor, community-spanning interactions will necessarily be marked by ambiguity. Ambiguity or equivocality implies much more than uncertainty. *Ambiguity* means the innovation effort is not clearly determined both in terms of its relevant parameter space and in terms of the relationships among the parameters. *Uncertainty*, on the other hand, implies that the parameter space and its interrelationships are determined, but their values or specifications are lacking and need further problem-solving activity. Uncertainty, then, is seen as a measure of the ignorance of a value for a variable, whereas *equivocality* is a measure of the organization's ignorance of whether a variable exists. Uncertainty relates to finding answers to well defined questions, equivocality or ambiguity implies that one is searching for the adequate questions (Daft & Lengel, 1986). This distinction underlies the difference made by Allen (1977) between problem definition and problem solving.

As both uncertainty and ambiguity are present in the innovation process, actors belonging to different communities are confronted with finding ways to handle this variety of beliefs, evaluation routines, and enabling artifacts. Stated otherwise, they have to deal with a situation of interpretation asymmetries. So far, we have only reached a limited understanding of how we might handle this ambiguity during technological innovations (whereas the relationship between information, information exchange, and uncertainty has received much more attention and has been much better articulated). As a consequence, any study of community-spanning strategies will have to focus on the way in which particular strategies for coping with ambiguity can be developed. This, though, immediately raises a second, more important issue. In order to successfully accomplish an innovation effort, community-spanning activities and processes cannot go on indefinitely, although they need to converge into a particular outcome by means of closure, fusion, or synthesis. Hence, any study of the innovation process needs to address the critical issue of community-spanning processes and strategies as they will have an immediate impact on the level of path creation versus the almost "linear" continuation of the path dependence.

These are the broad issues we want to address in this chapter. Our investigation begins at one particular level of analysis that presented itself as a very convenient starting point for our current research interest: the development of a new chemical process for metallurgical refining. We therefore first need to tell the story of the development of this process. Our ideas and assumptions on community-spanning processes and the underlying complexities as well as characteristics of relevant interaction strategies will be developed along this narrative.

NARRATING AND RECONSTRUCTING THE "PURIFICATION"[3] PROCESS: A TYPICAL STORY[4] OF ENGINEERS AND SCIENTISTS AND THEIR BELIEFS WHEN CONSTRUCTING AN INNOVATION

In this section, we present an account of a typical development process. Given the character of the issues we mean to address, the case-study approach has been selected as the most appropriate (Yin, 1984). The innovation process presented here is documented by reconstructing the sequence of events taking place from the start of the process to the final technical completion, spanning a 2-year period. The following constructs and their evolution over time, were documented by means of interviews and document analysis (Abbott, 1990; Pettigrew, 1990; Van de Ven & Scott Poole, 1989):

---

[3]*Purification* refers to a stage in the production process of Metalloy, which in its turn is a fictive name for a well known metal derivative. Fictive names are used to protect the names of the company, people, and processes involved.

[4]This "story" is a first reporting on an ongoing research project focusing on interaction strategies during the development process. The quotes interwoven in the description of the different events all stem from the R&D project manager. The process described here is situated within a European-based multinational group active in the process industry. The group has production facilities in Europe and the United States.

- The technical options taken into consideration: these reflect the uncertainty and ambiguity confronted;
- The actors involved and the interaction characteristics during each phase of the process: these are related to the interaction strategies in use.

Defining the Project

On October 19, 1994, an R&D-project concerning the purification phase of the production of Metalloy was presented and approved by a cross-functional committee. Both the R&D department and the business unit involved agreed on the objectives and the approach to scout, examine, and evaluate opportunities to improve the purification step in the existing process. This approval signaled the coalescence and translation of the various ideas on purification-process improvement that had been circulating for awhile among the different stakeholders into a dedicated budget, an action plan, and the assignment of experts to work on the chosen approach.

This new project definition was not unheard of, however. Already in the period 1992–1993, various groups (communities) within the company had started to look at the purification process. The main motivation for these efforts (at that time) resided in the evolution of world market prices for Metalloy. At the business unit, though, these efforts awakened mixed feelings. The hope for a truly pathbreaking result were almost nonexistent. After all, 85% of the world's Metalloy is produced under license, using the very process they themselves had developed 20 to 30 years previously. Obviously, the refining technology and process they had developed had paid off nicely for the company.

The payoff had been handsome, not only for the company as a whole; personal careers had benefited as well. The process champions of 30 years ago had all risen to senior management positions in the business unit. They were running the show; they knew what was important in further developing the refining process. Still more important, they had realized a process that over 20 years had earned their corporate parent a payback ratio of about 400% on the initial R&D investment required to develop the refining process. Would a path-breaking view of the refining process ever be possible? Surely not! At best, the improvements would come from improved process control and instrumentation. This was a logical cognitive step to this pioneering development community. Indeed, when they first pioneered the process in the 1950s and 1960s, electronics and instrumentation did not exist. At best, they were theoretical leaps at the time, with no engineering implications at all. So the pioneering community felt that the younger generation of their profession, who had been raised with microprocessors and computers, could best look at this type of process improvement. In this way, the technological trajectory envisioned and created several decades before would be extended, and the success story would further be reinforced.

As a consequence, during this first period, the efforts of the refining project were devoted to the first stages of the process; no clear prospects for improvements or breakthroughs were defined or anticipated. The only certainty existed in that the business unit advocated that a rigorous approach be followed. In 1994, the need to do something more specific, leading to more tangible short-term results, was becoming acute to the business unit (BU). Fierce competition threatened the company's position in the product market. In addition, major competitors spread rumors about realizing productivity gains in the range of 20% to 50% by applying more efficient processes. This was done, so they claimed, - to a large extent by better process control. The improvement leap was doubted by the BU (weren't *they* at the origins of the world's leading process?); however, the instrumentation avenue, as already described, was taken seriously enough to consider the possibility. So the "5-year approach" was redefined and resulted in the project as approved on October 19, 1994.

Designing a Problem-Solution Approach

As a first step, the project manager (PM) assigned to the project started a large information campaign during the first months of 1995. The project manager was an experienced R&D collaborator. He visited everybody within the company assumed to possess knowledge relevant to purification. This information round consisted of bilateral and informal talks.

> Bringing everybody together would be inefficient and time consuming. Moreover, as I was trying to establish for myself and my team a more profound insight in the process under study, I didn't want to become involved in a "power game" between different experts, all with their own agendas and preferences. You must not forget that this organization has grown out of several independent companies that used to be competitors; old rivalries are still present from time to time, especially on occasions where the question "who is the real expert here?" is posed. Bringing them all together opens the risk of an escalation. (excerpt from interview)

At this stage, the main objective for the project manager was to get a clear insight into the underlying chemical processes, the methodology to be followed, and the possible options for improvement. In the project definition, an exhaustive list of possibilities was listed, totaling in the neighborhood of 20 different options or solution possibilities. The PM was looking for clues to organize and prioritize these different options. On the side of the BU, this broad exploratory stage was not needed to know where possible profits could be found; one specific option, recycling, was considered the most fruitful, besides the instrumentation strategy. This point of view was communicated to the PM.

Coordination between the BU community and the R&D community in the company at this stage was experienced by the PM as rather directive:

> They wanted me to just look into the recycling option; everything else was considered of minor importance or even a waste of time. So I had no choice but to spend the first 3 or 4 months of the project on doing experimentation related to recycling. Although I did not believe it was really going to lead to something worthwhile. (excerpt from interview)

So, the many alternative options described in the project plan stayed on ice for a while. In March 1995, a first intermediate report on the recycling option was presented to the BU. The results of the first months of experimentation made clear that working on recycling was not a viable option:

> This step won me time; I spent the first 3 months only on recycling, doing experiments, and so on. Now the results made clear that solutions would have to be found elsewhere. By writing this report I got them off my back. I agreed that R&D should be customer oriented; on the other hand, a researcher needs certain degrees of freedom to explore new possible options even if their outcome is unknown. The first months the BU was really directive toward our work; they did not hesitate to tell us what to do and how to do it. Now that we had proven that recycling would not work, they lost interest; they left us doing the project without dictating how to proceed. (excerpt from interview)

Exploring Different Options—Pursuing Viable Options

The project was not ended at this stage though; the project definition included a whole range of options and a project plan for 2 years. The findings on recycling did not lead to an abrupt ending of the project: "Commitments were made; the project definition at the start was broader than mere recycling. So now we could work systematically on the whole range of possibilities that was defined (excerpt from interview)". The manager of the R&D department fully understood this need and further shielded his R&D collaborators from undue BU meddling. In the months to come, a full experimental setup was constructed and adequate process control equipment was installed. During the summer, further reports were produced containing preliminary results on the different options. These reports certainly had an impressive management aspect to them; "we were communicating that we were busy with a whole range of viable options. So at the BU level they regularly noticed that we were busy; that the project was not in a dead-end street" (excerpt from interview).

During this period a "strange" series of events was registered when the experimental runs were being rigorously monitored.

> By coincidence, we noticed that minor shifts in the temperature created some unexpected side effects. Having a rigorous research method allowed us to detect this. You

are nowhere without a rigorous research approach. Examining this phenomenon more closely revealed that the presence of [chemical substance] had some influence. However the nature of the impact on the purification process was unclear. (excerpt from interview)

During the summer, a new intermediate report was made, resulting in a September meeting between R&D and the BU. An overview of different action strategies was presented. They were organized in terms of priorities as perceived by the BU.

If you look at the presentation of September 29th … we described and organized the five different options in order of their preference; and read this … "parallel with the five described options we will explore the possibilities of stabilizing the process by adding [chemical substance] to the process." We were not sure that it would lead to anything at that stage but the observations done so far made us believe this could become something "big." However we did not want to come out with this yet; the chances were good that the idea would be killed right away. By stating it in this way, we managed to continue our explorations in that direction as well. (excerpt from interview)

The next 6 months were spent working rigorously on the different options. The project manager devoted a lot of effort to the sixth (bootlegging) option as well. During this period, no official interaction with the BU took place.

The observations we made on these strange "side-effects" just kept me busy. It could be the case that the same mechanisms were applying here as were very well known in the Power[5] area. So I started to discuss this possibility with [name], a real expert in Power. He selected a list of about 25 powders; we started to test systematically whether they had any impact on the process. And two of them really did. So, after 2 months, we knew we had hit upon something really good. We had identified two elements; and the best effects were obtained when combining them. (excerpt from interview)

This did not mean the other options were neglected:

We worked on all options quite simultaneously. However—while progressing—it became clear that the major breakthrough would lie with the sixth option, implying savings of several millions of dollars a year. But we also worked out some serious improvements for the agitation part of the process; this resulted from the development of the second option. (excerpt from interview)

In March 1996, the first results related to this option were presented at an R&D meeting. It became clear that the findings were plausible and viable. However, there was still no one from the BU present at that meeting.

When we knew that the addition of [chemical substance] and [chemical substance] could lead to a serious improvement of the process, in terms of effectiveness as well as

---

[5]Fictive name; refers here to an expert area related to power processes and applications.

efficiency, we still needed to define the optimal doses. If this might sound like an easy question, finding the answer is definitely something else. We would not go to the BU before we had determined the optimal doses and had done a whole series of experiments so that the robustness of the phenomenon was indisputable. If we had not had this kind of data, they would have blown us away. (excerpt from interview)

Drawing Conclusions Toward Implementation

On June 19th, the findings were presented at the BU. What was planned to be a 1-hour meeting between 4:00 and 5:00 p.m. lasted until 9:00 p.m. "They could not believe that these results came out of it, but the evidence was there. We had to explain it over and over again" (excerpt from interview). A final report was written in November; the findings were presented at the corporate level and during the 1996–1997 budget meetings as well. The implementation phase is being worked out right now.

Of course I would like to have worked together with the BU in another way. But collaboration in such a project means other things to me. We needed feedback, not directives. If I had been working with other people, the interaction might have been completely different. Working with [other BU manager's name] for instance is completely different. I keep him informed on all findings in each stage. He takes time to listen and to brainstorm, to explore. (excerpt from interview)

Whether the outcome would be different if the interaction with the BU had been otherwise? "I guess so, but we can not know this for sure, can we?" (excerpt from interview).

Looking at the Purification Process
From a Community Perspective

This process description indicates yet again that interaction and cooperation during development processes entail particular requirements. Although on the one hand, cross-functional cooperation might be beneficial in terms of final outcomes, on the other hand, realizing this cooperation is a complex enterprise. The diversity needed to create new insights hampers at the same time an open and constructive exchange of ideas or, as already stated, hampers the crossing of boundaries between communities. This is a dilemma that can be identified as well within the field of organizational learning or knowledge creation. As "learning" and "exploring" relationships can be characterized by trust, openness, reciprocity, support, and recognition (Argyris, 1992; Argyris & Schön 1974), the antecedents of these interaction characteristics refer to homogeneity or a common identity (see e.g., Festinger, Schachter, & Back, 1960; Tajfl & Turner, 1986; and more recently McAllister, 1995). Hence the requirements for developing constructive co-operative relationships are in sharp contrast with the idea of diversity or requisite variety so prominently present within the innovation literature (see, e.g., Myers & Marquis, 1965; and more recently, Nonaka, 1990, 1992; Nonaka & Takeuchi, 1994).

The development in the designing of relevant interaction strategies to deal with these paradoxical requirements starts, however, with understanding their underlying dynamics. As described in the introduction, the notion of a community is highly relevant for looking at the evolution of scientific disciplines and technological developments. Communities and communal behavior can be relevantly defined at different levels of analysis. Communities are *collectivities*, sharing beliefs, hopes and search heuristics. As a consequence, communities as a concept are to be linked to the notion of "problem domains" as developed by Trist (1983): "Functional social systems which occupy a position in social space between the society as a whole and the single organization" (p. 57). *Technological communities*, for instance, can be seen as the group of scientists and engineers, who are working toward solving an interrelated set of technological problems and who may be organizationally and geographically dispersed but who nevertheless have a shared interest and hence communicate with each other (Debackere & Rappa, 1994). Underlying the relevance of the notion of community is the idea that scientific and technological developments are inherently social as well: "While it is true that that scientists grapple with nature, they also grapple with each other" (Rappa & Debackere, 1995, p. 324). Or, as pointed out by Medawar (1967), science is the art of the soluble. Good scientists work on the problems they believe they can solve and that are relevant by their peer community. This idea comes close to the cycle of credibility as it has been developed by Latour and Woolgar (1981).

However, the notion of communities can be extended to other, microlevels of analysis too; it then refers to collectivities that share an understanding of a problem domain. Whereas Lave and Wenger (1991) defined communities mainly in terms of stability,[6] a certain degree of homogeneity, in terms of problem approach and understanding seems to characterize them as well. One can speak of homogeneity not only at the level of cognitive frames (Miller, Burke, & Glick, 1996), but as well in terms of identity. It is in this latest sense that we use the notion of community. This collectivization of identity is closely related to Garud's shared beliefs, evaluation routines, and artifacts embraced by the community's actors. As a consequence, communities are powerful devices for path dependent thinking. Another consequence is that communities are powerful devices capable

---

[6]"A community of practice is a set of relations among persons, activity, and ,world over time and in relation with other tangential and overlapping communities of practice." (Lave & Wenger, 1991, p. 98)

more of exploitation than exploration (March, 1991). This is well illustrated by the attitudes and reactions of the BU pioneer community to the new purification development. Extending the dilemma of organizing versus learning, one might hypothesize that new communities are born as the result of a learning experience (Weick & Westley, 1996). However, once communities organize and institutionalize, emphasis shifts from creation to proliferation. This proliferation engenders routinization and conformity, thus generating powerful path dependencies or lock-in phenomena, making it increasingly difficult to break away from the path.

As mentioned, we argue that pathbreaking innovations necessarily imply an interaction spanning the boundaries of different communities within (and even external to) the organization. This community-spanning interaction figures as the pathway to understand the optimization of the purification process. We also suggest that this interaction has important implications as to the strategies that are to be deployed when attempting to span different communities. During the innovation process, boundaries between communities need to be crossed while at the same time, new frames of understanding are being built—generating eventually new communities. This implies interaction strategies of a more complex nature because building new frames of understanding touches on the identity of the actors involved. Given the attention paid by Brown and Duguid (1991) to the notion of community to illuminate the social dynamics at play in the processes of working, learning, and innovating, we shall take their work as a starting point to further explore the complexities underlying community spanning strategies.

Brown and Duguid (1991) built on the research of Orr (1996) to illuminate some central characteristics of work practice. Taking into the account the tension between canonical practice (the explicit or "official" knowledge and way of acting) and noncanonical practice (the actual way people perform their activities), Orr's (1990) ethnographic work allows one to derive three central features of work practice: narration, collaboration, and social construction. As for learning, the concept of legitimate peripheral participation is brought in (Brown & Duguid, 1991). This notion, developed by Lave and Wenger (1991) denotes the particular mode of engagement of a learner who participates in the actual practice of an expert, but only to a limited degree and with limited responsibility for the outcomes. Learning is seen as a process of participation in communities of practice; learning as legitimate peripheral participation involves becoming an insider. One becomes member of a community—be it a community of physicists, classmates, or scholars in philosophy or organizational behavior. So learning implies not only a relation to specific activities but also a relation to social communities; it implies becoming a full participant, a member, a certain kind of a person. (Lave & Wenger, 1991; see also Giddens, 1979, 1984). In both the process of working and learning, identity and identity formation play a central role. The same holds true when it comes to innovating.

When discussing innovation, Brown and Duguid (1991) introduce Daft and Weick's (1984) framework, which characterizes different kinds of organizations, each typified by its relationship to its environment.[7] Two of them can be seen as innovative organizations. The *discovering organization* is the archetype of the conventional innovative organization, responding to changes it detects in its environment. The *enacting organization*, on the other hand, is depicted as proactive. It does not only respond to its environment, but also creates the conditions to which it must respond. *Innovation*, then, is not simply a response to empirical observations of the environment, because the source of innovation lies in the interface between an organization and its environment. The process of innovating involves actively constructing a conceptual framework, imposing it on the environment and reflecting on their interaction. Brown and Duguid (1991) argued that the actual noncanonical practices of communities are continually leading to new interpretations of the world because they have a practical connection to the world. Closure is the likely result of the rigid adherence to formal practice. Rejecting a predetermined view and constructing an alternative view through narration are both seen as essential to innovation. By so doing, an enacting organization must be not only capable of reconceiving its environment, but also its own identity, as the two are in a significant sense mutually constitutive.

> This re-conceptualization is something that people who develop non-canonical practices are continuously doing, forging their own and their community's identity in their own terms so that they can break out of the restrictive hold of the formal descriptions of practice. Enacting organizations similarly regard both their environment and themselves as in some sense unanalyzed and therefore malleable. They do not assume that there is an ineluctable structure, a "right" answer, or a universal to be discovered; rather they continually look for innovative ways to impose new structure, ask new questions, develop a new view, become a new organization. (Brown & Duguid, 1991, p. 54)

Such a reconceptualization might occur as well when communities characterized by different opinions and views interact or try to cooperate or coalesce. However, such reconceptualization also introduces a difference between working and learning on the one hand, and innovating on the other. Learning and working can be depicted as evolving within a community, whereas innovation often implies fusing the views and opinions held between different communities.

In the purification case, the tension between the "prescribed" or canonical view of the purification process—as developed within the BU commu-

---

[7]These relationships are labeled: undirected viewing, conditioned viewing, discovering, and enacting.

nity over a 20-year time period—and what was actually observed during the first experimentation round triggered development efforts within the R&D team. The specific way of collaborating during this process, however, shows some remarkable characteristics. In this case, a variety of interactions between different communities was essential. The development efforts, especially with regard to the sixth option, resulted in a breakthrough because of the joint effort of two communities: the R&D team and the specialists within the application department. Note that people belonging to the BU community were left out during this particular phase and that this temporary withdrawal from interaction was experienced as crucial to advancing the project.

Thus, community-spanning interactions do not seem to present themselves as developing in a straightforward, linear sequence of steps and actions. Neither can they be portrayed as open and constructive during the whole process. Still, the R&D community perceived this particular way of collaborating as the most relevant way to advance during the innovation process. Why this is the case becomes understandable by connecting the dilemma sketched in the introduction with the conceptualization of the organizing processes as embedded within communities of practice and its implications in terms of identity. On the one hand, the innovation process can benefit from bringing in new perspectives by spanning the boundaries between communities. At the same time, introducing new perspectives challenges the existing order of beliefs, routines, and embraced artifacts. As identity is at stake, this could be seen as a profound and even painful process implying reactions of denial, rejection, and aggression (see Lewin, 1948). Paradoxically, withdrawal then sometimes becomes necessary to allow advancement. Balancing between both ends of the spectrum—openness and closure—seems to be an inherent characteristic of more complex strategies that allow for spanning of the boundaries between communities.

A NEW LOOK AT THE PURIFICATION PROCESS AS A SEQUENCE OF OPENNESS AND CLOSURE BETWEEN COMMUNITIES

In Table 12.1, the different stages, actors, options as well as interaction characteristics are summarized. Three communities can be distinguished along the purification process: (a) the engineering community responsible for running the operations within the business unit; (b) the research community, consisting of the R&D team responsible for working out the project; and finally, (c) the application community, playing a role in the development of one of the options. It becomes clear that the ways in which the communities interact and collaborate alters from phase to phase. Different assembly rules are used at different stages (Weick, 1979); not only the perceived equivocality related to the task influences the choice of these assembly rules, the experiences during the different interactions also lead to changing the applied rules. On experiencing the interaction between his team and the BU Metalloy as directive and one-sided—and hence destructive for his attempts to take a fresh look at the purification process—the project manager shifted toward an "impression" management type of communication with this community.

In so doing, he avoided in-depth discussions and a potentially damaging confrontation with the dominant logic of a powerful community. The early observations, when developing a full experimental process set-up, led to interactions with another community. This community was regarded as experienced with the phenomena under observation. Also, because of the experiments conducted in isolation by the R&D team, the level of ambiguity between both communities was tolerable so that cooperation became a viable community-spanning strategy. Eventually this community-spanning cooperation resulted in a serious breakthrough. Hence, the sequence of community-spanning interactions might be summarized as follows:

Stage 1: Open information exchange at the start; after awhile, BU starts to direct R&D project team. The community-spanning strategy almost has a confrontational character.

Stage 2: R&D temporarily withdraws from interaction with the BU, experiments "in isolation" help to reduce the uncertainty related to the "novel" observations without having to struggle with the ambiguity present in the broader interaction.

Stage 3: Application community is drawn into cooperation with R&D community at acceptable levels of ambiguity.

Stage 4: R&D community confronts the BU community with its results. Both communities arrive at reducing ambiguity.

Stage 5: Emergent cooperation between the R&D and the BU community.

It seems nonproductive to start guessing about what might have happened if the project manager and his team and the BU had interacted in a different way at the start. However, we do find it worthwhile to explore the dynamics underlying the interaction sequence just analyzed.

We already referred to Garud's work on technological evolution. We argued that communities are collectivities of individuals sharing beliefs, evaluation routines, and artifacts. In this way, communities are like Plato's prisoners: They create their own reality, their own identity, their own truths or paradigms. As a consequence, interactions between communities that do not share the "core set" become increasingly difficult the greater the distances between these core sets of beliefs, routines, and artifacts. These dis-

TABLE 12.1

Revisiting the "Purification" Story: Phases, Actors, Options and Interaction Characteristics

Period:	10/24/94	10/25/94 to 3/95	3/95 to 9/25/95	9/29/96 to 6/18/96	6/19/96 to 12/96
Phases:	Defining the project	Designing a problem solution approach—Exploring recycling option	Exploring different options	Pursuing viable options	Drawing conclusions towards implementation
Actors involved:	Mainly BU Metalloy & R&D competitors: Rumors of performance improvements as a trigger for focus/redefinition	R&D, PM and his team BU Metalloy, R&D, and "local" experts	R&D, PM and his team BU Metalloy BU Power	R&D, PM and his team	R&D BU Metalloy corporate levels
Options taken into consideration:	Broad range of options and ideas, related to different concerns and viewpoints	List of +/–twenty possible options	Grouping of actions/possibilities into six "broad" classes	Working out technical steps for different options Focusing on "sixth" option as well as on possibilities for agitation	Two options found worth implementing: adding [element] and [element] + agitation
Interaction characteristics:	Constructively combining efforts based on parallel interests	Directive: One-sided between R&D, BU Metalloy Informal, open between R&D and local experts	R&D providing BU Metalloy with feed-back as to give the impression that "everything is running well" Between R&D and BU Power open and constructive	R&D remaining "silent" to build "strength of argument" before entering the "presentation" stage	R&D convincing BU by means of an "expert" approach.

tances are both at the origins of the installment of path dependencies and the ambiguity that exists between different communities.

The notion of truth, in a pragmatic sense, makes it understandable why bringing in new perspectives is often a hazardous enterprise, as this notion of truth is inherently linked to the notion of identity or "absorptive capacity" at the microlevel. It became clear that there is more involved here than cognition (Cohen & Levinthal, 1990; Langlois, 1996), because persons as a whole are involved in these processes. In this respect, the work of James (1907/1995) turns out to be rather up-to-date. James (1907/1995), building on the insights of Dewey and Schuler, provides us with the following definition of truth: "'Truth' in our ideas and beliefs means the same thing that it means in science. It means that ideas (which themselves are but parts of our experiences) become true just in so far as they help us to get into satisfactory relation with other parts of our experience" (p. 24).

This view of truth is derived from examining situations whereby individuals settle into new opinions, as is the case in innovation efforts. Analyzing this phenomenon leads to the demarcation of the following dynamics (according to James, 1907/1955):

> The individual has a stock of old opinions already, but he meets a new experience that puts them to a strain. Somebody contradicts them; or in a reflective moment he discovers that they contradict each other; or he hears of facts with which they are incompatible; or desires arise in him which they cease to satisfy. The result is an inward trouble to which his mind till then had been a stranger, and from which he seeks to escape by modifying his previous mass of opinions. He saves as much of it as he can, for in this matter of belief we are all extreme conservatives. So he tries to change first this opinion, and then that (for they resist change very variously) until at last some new idea comes up which he can graft upon the ancient stock with a minimum of disturbance of the latter, some idea that mediates between the stock and the new experience and runs them into one another most felicitously and expediently. This new idea is then adopted as the true one. It preserves the older stocks of truth with a minimum of modification, stretching them just enough to make them admit the novelty, but conceiving that in ways as familiar as the case leaves possible. (p. 24)

The notion of inertia, or path dependency, figures prominently in this account of truth. James stresses the importance of older truths as they might hinder the adoption of new insights (cf., the first confrontation between the BU community and the R&D community). Bringing in perspectives too ambiguous and hence too distant from existing knowledge will result in denial or rejection (James, 1907/1955):

> An outré explanation, violating all our preconceptions, would never pass a true account of novelty. We should scratch round industriously till we found something less eccentric. The most violent revolutions in an individual's belief leave most of his old

order standing. Time and space, cause and effect, nature and history, one's own biography remain untouched. New truth is always a go-between, a smoother-over of transitions. It marries old opinion to new fact so as ever to show a minimum of jolt, a maximum of continuity. We hold a theory true just in proportion to its success in solving this problem of *maxima* and *minima*. But success in solving this problem is eminently a matter of approximation. We say this theory solves it on the whole more satisfactorily than that theory; but that means more satisfactorily to ourselves, and individuals will emphasize their points of satisfaction differently. To a certain degree, therefore, everything here is plastic. The point I now urge you to observe particularly is the part played by the older truths. Their influence is absolutely controlling. Loyalty to them is the first principle—in most cases it is the only principle; for by far the most usual way of handling phenomena so novel that they would make for a serious rearrangement of our preconceptions is to ignore them altogether, or to abuse those who bear witness for them. (p. 25)

Returning to the purification process, one observes that the R&D team and the project manager had their own opinion about a suitable approach and were looking for new ways to conceive of the process. Once this became apparent efforts were undertaken by the BU to impose their view. When these attempts to influence the action of the R&D team in the direction of their own opinions—that is, creating homogeneity in terms of the view of what the process was about and how to approach the project—failed, the interaction became minimal. The BU lost interest and started to ignore the development efforts. From the side of the R&D community, this loss of interest was not really seen as a problem. In the light of their previous experience related to the collaboration with the BU, looking for shelter was seen as the next thing to do. Within the R&D community, the risk of "being abused when bearing witness of new conceptions" was clearly acknowledged: "We would not go to the BU before we had determined the optimal doses and had done a whole series of experiments so that the robustness of the phenomenon was indisputable. If we had not had this kind of data, they would have blown us away" (excerpt from interview).

Obviously, changing paths or trajectories in technological development is a major challenge. In the context of organizational innovation, modification of opinions has been documented as a profound process (Steyaert, Bouwen, & Van Looy, 1996). This is what occurs along the technological development trajectory as well. Path-breaking innovations indeed require a rethinking of community members' perceived truths and hence, a reconfiguration of their own identity as persons who are totally involved in, and devoted to, their practice. It is not astonishing then that a powerful, almost cumulative inertia is at work, which probably increases exponentially the better and the longer people are embedded in their respective communities (Rappa, Debackere, & Garud, 1992). Action strategies are then required to cross boundaries between communities, notwithstanding these powerful inertial forces.

DISCUSSION: TOWARD MORE FINE-GRAINED INTERACTION STRATEGIES FOR INNOVATION PROCESSES

As the beginning of this chapter showed, the innovation process benefits from gatekeeping; community-spanning activities are beneficial in terms of problem framing as well as problem solving. This dual nature of boundary spanning brings us to the paradoxical requirements of the interaction strategies with which one is confronted. The innovation process can be seen as a process in which one needs to address and reduce uncertainty as well as to handle ambiguity. Working on uncertainty is achieved most efficiently by working within the boundaries of one community. Developing novel solutions, and hence addressing ambiguity, benefits from crossing boundaries. Each activity hampers the other; hence, the paradoxical nature of the innovation process as a community-spanning process and the need to address the issue of relevant action strategies.

When discussing the case, it became clear that communities do not only create important forces of inertia; their interactions and confrontations are at the origins of breaking away from well-known paths and trajectories as well. These confrontations are not random, though. Actors in a particular community cross the boundaries of their community by looking for partners in other communities who are believed to have an affinity with the ideas or insights to be developed. The different community-spanning strategies discussed and developed in this chapter demonstrate that actors look for the construction of temporal "zones of proximal development" (Vygotsky, 1986). If this common ground cannot be found (temporarily), then withdrawal is a strategy to be advised. The central focus of these community-spanning strategies is thus on the variety of potential modes for handling ambiguity. This ambiguity stems from the diversity of beliefs, evaluation routines, and enabling artifacts that create community-specific path dependencies. Finding ways to relate and fuse this diversity forms the core of these strategies. As such, this process implies working on the tension between continuity and novelty, and pacing the transition between both. (Bouwen & Fry, 1988; Bouwen, Van Looy, Debackere, & Fry, 1998).

As shown in the brief case development, these community-spanning strategies, like DNA, may consist of a limited set of building blocks: (a) temporary withdrawal from communities that become too belligerent, (b) confrontation between communities, and (c) co-operation and joint problem framing and problem solving. Just as the four basic DNA building blocks, by their spatial sequence and interaction, are at the basis of an almost unlimited variety of living species; so can the spatial and temporal sequence of the three community-spanning mechanisms create an almost unlimited range of options available to engage in community-spanning interactions. Time figures here as a crucial ingredient, allowing the reduction of the present ambiguity. In conjunction with time, third parties that play the role of go-between constitute another major ingredient for action strategies (Steyaert & Janssens, 1998; Van De Ven & Poole, 1988) that imply path creation. Given the profound character of certain episodes of the innovation process—in terms of rethinking one's own preconceptions of what holds true—these interaction strategies will imply "go between" episodes and/or persons." Moreover, technical artifacts can here play this go-between role as they provide a common ground around which to interact, as the role of the rigorously conducted experiments during the final stages of the process suggests.

To conclude, we have pointed to the relevance of applying the notions of path dependence and creation in understanding innovation processes. Fine-grained interaction strategies, described as microlevel social phenomena, have been shown as the primary vehicle to bridge the gap between path creation and path dependence. In doing so, we demonstrated the need to unravel the notion of path dependence and path creation at a microlevel of analysis by acknowledging the historically situated nature of any development process.

ACKNOWLEDGMENT

The authors would like to thank Koen Heyrman for his assistance with data collection.

REFERENCES

Abbott, A. (1990). A primer on sequence methods. *Organization Science, 1*(4), 375–392

Allen, T. J. (1966). Performance of information channels in the transfer of technology. *Industrial Management Review, 8* 87–98.

Allen, T. J. (1977). *Managing the flow of technology.* Cambridge, MA: The MIT Press.

Allen, T. J., & Frischmuth, D. (1969). A model for the description and evaluation of technical problem solving. *IEEE Transactions on Engineering Management, 16*, 58–64.

Allen, T. J., & Marquis, D. G. (1963). Positive and negative biasing sets: The effect of prior experience on research performance. *IEEE Transactions on Engineering Management, 11*, 158–162.

Argyris, C. (1992). *On organisational learning.* Oxford, England: Blackwell.

Argyris, C., & Schön D. (1974). *Organisational learning: A theory of action perspective.* Reading, MA: Addison Wesley.

Arthur, B. E. (1988). Self-reinforcing mechanisms in economics. In P. Anderson (Ed.), *The economy as an evolving complex system.* Reading, MA: Addison-Wesley.

Ben-David, J. (1960). Roles and innovations in medicine. *American Journal of Sociology, 65*, 557–569.

Ben-David, J., & Collins, R. (1966). Social factors in the origins of a new science: The case of psychology, *American Sociological Review, 31*, 451–465.

Bijker, W. E.(1994). *Of bicycles, bulbs and bakelite.* Cambridge, MA: MIT Press

Bijker, W., Hughes, T., & Pinch, T. (Eds.). (1987). *The social construction of technological systems.* Cambridge, MA: MIT Press.

Bijker, W., & Law, J. (Eds.). (1992). *Shaping technology/building society: Studies in socio-technical change.* Cambridge, MA: MIT Press.

Bouwen, R., & Fry, R. (1988). An agenda for managing organisational innovation and developement in the 1990's. In M. Lambrecht (Ed), *Corporate revival: Managing into the nineties* (pp. 153–172). Belgium, Leuven: Leuven University Press.

Bouwen, R., Van Looy, B., Debackere, K., & Fry, R. (1998, July). *Assembling trajectories: A process view on innovation and change efforts.* Paper presented at Egos Conference Maastricht: The Netherlands.

Brown, J. S., & Duguid, P. (1991). Organizational learning as communities-of-practice: Towards a unified view on working, learning and innovating. *Organization Science, 2,* 40–57.

Brown, S., & Eisenhardt, K. (1995). Product development: Past research, present findings, and future directions. *Academy of Management Review, 20*(2), 343–378.

Burgelman, R. (1994). Fading memories: A process theory of strategic business exit in dynamic environments. *Administrative Science Quarterly, 39,* 24–56.

Cohen, W. A., & Levinthal, D. A. (1990). Absorptive capacity: A new perspective on learning and innovation. *Administrative Science Quarterly, 35,* 128–152.

Constant, E. (1980). *The origins of the turbojet revolution.* Baltimore, MD: Johns Hopkins University Press.

Daft, R. L., & Lengel, R. H.(1986). Organizational requirements, media richness and structural design. *Management Science, 32,* 5.

Daft, R. L., & Weick, K. (1984). Toward a model of organizations as interpretation systems. *Academy of Management Review, 9*(2), 284–295.

David, P. (1985). Clio and the economics of QWERTY. *Economic History, 75,* 227–232.

Debackere, K., & Rappa, M. (1994). Technological communities and the diffusion of knowledge: A replication and validation. *R&D Management, 24,* 355–371

Edge, D. O., & Mulkay, M. J.(1976). *Astronomy transformed: The emergence of radio astronomy in Britain.* New York: Wiley.

Festinger, L., Schachter, S., & Back, K. (1960). *Social pressures in informal groups.* Stanford, CA: Stanford University Press.

Garud, R., & Rappa, M. A. (1994). A socio-cognitive model of technology evolution: The case of cochlear implants. *Organisation Science, 4,* 344–362.

Ghemawat, P. (1991). *Commitment: The dynamics of strategy.* New York: The Free Press.

Giddens, A. (1979). *Central problems in social theory: Action, structure and contradiction in social analysis.* New York: MacMillan.

Giddens, A. (1986). *The constitution of society.* Cornwall: Polity Press.

Hirsch P., & Gillespie, J. (1997, August). *Unpacking path dependence: Differential valuations accorded history across disciplines.* Paper presented at Path Creation and Path Dependence Workshop held at Copenhagen, Denmark.

James, W. (1995). *Pragmatism.* New York: Dover. (Original work published 1907)

Katz, R. (1993, Summer). How a band of technological renegades created the Alpha chip. *Research Technology Management.*

Katz, R., & Allen, T. J. (1982). Investigating the not-invented-here syndrome: A look at the performance, tenure and communication patterns of 50 R&D project groups. *R&D Management, 12,* 1, 7–19.

Kramer, R., & Tyler, T. (1996). *Trust in organizations: Frontiers of theory and research.* Thousand Oaks, CA: Sage.

Langlois, R. N. (1996). Cognition and capabilities: Opportunities seized and missed in the history of the computer industry. In R. Garud, P. Nayyar, & Z. Shapira (Eds.), *Technological innovation: Oversights and foresights.* New York: Cambridge University Press.

Latour, B., & Woolgar, S. (1986). *Laboratory life: The construction of scientific facts* (2nd ed.). Princeton, NJ: Princeton University Press.

Lave, J., & Wenger, E. (1991). *Situated learning: Legitimate peripheral participation.* Cambridge, England: Cambridge University Press.

Lemaine, G., MacLeod, R., Mulkay, M., & Weingart, P. (Eds.). (1976). *Perspectives on the emergence of scientific disciplines.* The Hague, The Netherlands: Mouton.

Lewin, K. (1948). *Resolving social conflicts.* New York: Harper and Row.

Liebowitz, S. J., & Margolis, S. E. (1990). The fable of the keys. *Journal of Law and Economics, 33,* 1–25.

Liebowitz, S. J., & Margolis, S. E. (1995). Path dependence, lock-in, and history. *The Journal of Law, Economics and Organization, 11*(1), 205–226

March, J. (1991). Exploration and exploitation in organizational learning. *Organization Science, 2,* 1.

McAllister, D. (1995). Affect-and cognition-based trust as foundations for interpersonal co-operation in organizations. *Academy of Management Journal, 38*(1), 24–59.

Medawar, P. B.(1967). *The art of the soluble.* London: Methuen.

Miller, C., Burke, L., & Glick, W. (1996, August). Cognitive diversity among upper-echelon executives: Implications for strategic decision processes. Paper presented at the Annual Academy of Management Conference, Cincinnati, OH.

Myers, S., & Marquis, D. F. (1965). *Successful industrial innovations.* Washington, DC: National Science Foundation.

Nelson, R. R., & Winter, S. G. (1982). *An evolutionary theory of economic change.* Cambridge, MA: Harvard University Press.

Nonaka, I. (1990, Spring). Redundant, overlapping information: A Japanese approach to managing the innovation process. *California Management Review,* 27–38.

Nonaka, I. (1994). A dynamic theory of organisational knowledge creation. *Organisation Science, 5*(1), 14–36.

Nonaka, I., & Takeuchi, H. (1995). *The knowledge creating company.* New York: Oxford University Press.

Orlikowski, W. (1992). The duality of technology: Rethinking the concept of technology in organizations. *Organization Science, 3*(3), 398–427.

Orr, J. (1996). *Talking about machines: An ethnography of a modern job.* Cornell University, Ithaca, NY: IRL Press.

Pelz, D. C., & Andrews, F. M. (1967). *Scientists in organisations.* New York: Wiley.

Pettigrew, A. M. (1990). Longitudinal field research on change: Theory and practice. *Organisation Science, 1*(3), 267–291.

Pinch, T., & Bijker, W. (1987). The social construction of facts and artefacts: Or how the sociology of science and the sociology of technology might benefit each other. In W. E. Bijker, T. Hughes, & T. Pinch (Eds.), *The social construction of technological systems.* Cambridge, MA: The MIT Press.

Rappa, M., & Debackere, K. (1995). An analysis of entry and persistence among scientists in an emerging field of science: The case of neural networks. *R&D Management, 25*(3), 323–341.

Rappa, M. A., Debackere, K., & Garud, R. (1992). Technological progress and the duration of contribution spans. *Technological Forecasting and Social Change, 42*(3), 133–145.

Rip, A., Misa, T. J., & Schot, J. (Eds.). (1995). *Managing technology in society: The approach of constructive technology assessment.* London, New York: Pinter Publishers.

Steyaert, C., Bouwen, R., & Van Looy, B. (1996). Conversational construction of new meaning configuration in organizational innovation; a generative approach. *European Journal of Work and Organizational Psychology, 5*(1), 67–89.

Steyaert, C., & Janssens, M. (1999). The world in two and a third way out? *Scandinavian Journal of Management, 15,* 121–139.

Tajfl, H., & Turner, J. C. (1986). The social identity theory of inter-group behavior. In S. Worchel & W. G. Austin (Eds.), *Psychology of inter-group relations,* pp. 33–47. Chicago: Nelson-Hall.

Thomson, R. (1988). *The path to mechanised shoe production in the U.S.* Chapel Hill, NC: University of North Carolina Press.

Trist, E. L. (1983). Referent organisations and the development of inter-organisational domains. *Human Relations, 36*(3), 247–268.

Turner, J. C. (1987). *Rediscovering the social group: A self-categorisation theory.* Oxford, England: Basil Blackwell.

Van de Ven, A., & Poole M. S. (1988). Paradoxical requirements for a theory of change. In R. E. Quinn & K. S. Cameron (Eds.), *Paradox and transformation: Towards a theory of change in organisation and management.* Cambridge, MA: Ballinger.

Vygotsky, L. (1986). *Thought and language.* Cambridge, MA: MIT Press.

Weick, K. (1979). *The social psychology of organising.* New York: Random House.

Weick, K., & Westley, F. (1996). Organizational learning: affirming an oxymoron. In S. Clegg, C. Hardy, & W. Nord (Eds.), *Handbook of organization studies* (pp. 440–453). Beverley Hills, CA: Sage.

Werth, B. (1994). *The billion dollar molecule.* New York: Touchstone Books.

Wheelwright, S. C., & Clark, K. B. (1992). *Revolutionising product development.* New York: The Free Press.

Yin, R. (1984). *Case study research: Design and methods* (Sage applied social research methods series, Vol. 5).

13

Technologies of Managing and the Mobilization of Paths

Jan Mouritsen
Niels Dechow
Copenhagen Business School, Copenhagen, Denmark

CRAFTING THE "DOMINANT DESIGN"

In their chapter (chap. 8, this volume) Porac, Rosa, Spanjol, Saxon, and Nielsen make the important point that for the U.S. minivan market a certain dominant design emerged at a certain point in time as buyers and sellers developed a mutually reinforcing understanding of the idea of a minivan. This created a new state of coherence, which reflected a gradual extinction of alternative designs of minivans that were not competitive.

Our chapter is also concerned with the development of dominant design. However, rather than focusing on the final effect of the process of development work, we study the process itself and arrive at conclusions, which in certain respects are dissimilar to those of Porac et al. Rather than suggesting that a dominant design is the optimal result of a process of distillation, we suggest that a dominant design reflects a juxtaposition and realignment of elements already in place.

This difference may be due to a difference in empirical domain; Porac et al. are concerned with technical solutions in the product area, whereas we are concerned with techniques of management. It is also due to a difference in perspectives on how technologies and competencies are identified and demonstrated. We study technologies while they are in the making and in the process of becoming something; here the focus is on the clashes between potential technological solutions as they emerge over time. Our study illustrates how a design can become dominant through an alignment of competing alternatives. Through juxtaposing alternative supplier–management mechanisms, managerial technologies are involved in mobilizing firm-specific competence as it points it out, stabilises it, and crafts its relationship to corporate performance. A dominant design may not be mere "competence uniformation" as seems to be Porac et al.'s view but also possibly an organization of various modes of technological competencies vis-à-vis each other. Coherence is made possible here, not only by the extinction of alternative possibilities, but through their alignment in a new system of fragile and possibly provisional relations of dependence and autonomy made possible and governed by managerial technologies. That is to say, we analyze the work to manufacture and identify competencies by studying strategy in the making rather than the after-the-fact strategy, and we discuss how managerial technologies are mobilized to identify, stabilize and govern competence.

Path Creation and Path Dependence

We propose to study competence and coherence as a relationship between path creation and path dependency, which are two interrelated dimensions of history. The former directs attention to the creative forces of agency, and the latter emphasises the continuity of collective arrangements toward a future. The tensions between creation and continuity, however, are not antithetical, because to craft new visions and versions of competence and cohesion, a historically forged repertoire of rules and resources have to be mobilized and applied anew. Therefore, creation is not independent of history, but neither is history a mechanistic determinant for the future. Thus, we should be concerned not to exaggerate path dependency, although, again, trajectories are drawn on, mobilized, and transformed to make the future amenable to control. The "routines" in place (Nelson & Winter, 1982) and the accompanying competencies and capabilities (Hamel & Prahalad, 1994; Wernerfelt, 1984) may enable firms to exploit certain established orders. The creative destruction of routines, itself requiring routines, may allow the formation of new constellations of capabilities in an explored world (March, 1991). There are multiple interpretations of history mobilized to identify a trajectory that is brought forth to condition the present and future.

Structure Agency Framework

A strong, paradigmatic opposition between path creation and path dependence may be unfortunate. Agency and structure may be not dualisms but dualities (Giddens, 1984; Latour, 1987). Following Giddens, structure—organised as stocks of rules and resources—is both medium and outcome of interaction. Rules and resources are means (knowledge, methodical procedures) that people mobilize as integral to interaction and at the same time, in being mobilized, they are also reproduced. Structure is thus historical, but also in a sense "outside time and space, save in its instantiations and co-ordination as memory traces, and is marked by an 'absence of the subject.' The social systems in which structure is recursively implicated ... comprise the situated activities of human agents, reproduced across time and space" (Giddens, 1984, p. 25). *Path dependence* here is historically forged patterns of knowledge and methodical procedures, which in being drawn on and reproduced form a pattern of continuity with the past. Structure is not only constraining; it is also enabling because any piece of interaction requires procedures for its mobilization. Likewise, *path creation* depends on agency, or the potentiality to act otherwise. *Agency* presupposes a creative and critical incorporation of the past in order reflexively to accommodate the potentiality of a new world. According to such a view path creation and path dependence are both located in any piece of interaction. Crafting new paths and reproducing existing paths are both part of a project whose ontology incorporates both the view that humans are both history conscious species and that continue to transform history to make and bring about the future.

For Latour (1991), the world is continually provisional and fragile. The courses of action that can be mobilized are numerous, all depending on complex networks of interacting elements. If history may in some situations seem to be continuous, this is an effect of technologies mobilized to make the world durable (Latour 1991). Technologies, however, are constructs where humans and nonhumans engage with each other in a mutually reinforcing way. There is, according to Latourian theory, no overall historical motor that requires the world to take particular paths, but on the other hand, certain paths may be reproduced and reinforced by the existence of certain technologies that bind, as long as they are not problematized, in continuous fashion by humans and nonhumans together in reproduced networks. Path creation is concerned with the provisional and fragile character of social life where, in principle, any form of continuity is dependent on technologies, which for a period have been "black boxed," but that are continually possible to reopen. Likewise, path dependence reflects the longevity of black boxed relations made durable by technology.

From the perspectives of both Giddens and Latour, path creation and path dependence are always related to each other. They presuppose each other and form two moments in social life that cannot be separated. Giddens' (1984) duality of structure captures this nicely, because:

> the duality of structure [implies that] the constitution of agents and structures are not
> two independently given sets of phenomena, a dualism, but represent a duality. Ac-
> cording to the notion of the duality of structure, the structural properties of social sys-
> tems are both the medium and outcome of the practices they recursively organize....
> Structure is not to be equated with constraint but is always constraining and enabling.
> (p. 25)

From this perspective, we are concerned with the dialectic between path
dependence and path creation. Drawing on parts of the methodology actor
networks (Latour, 1987; Law, 1994) we investigate how two firms produced
an account of "strategy-in-the-making" (see also Mintzberg, 1994). Specifically, we seek to study how two firms attempted to introduce "world class
supplier relations" (Ooster, 1990; Welch & Nayak, 1992), and we do this by
analyzing the way such an aspiration is performed. By that we mean to identify the heterogeneous media that bring world class forth and make it a stable representation of firms' core competence. We do not assume that a priori
world class has a certain immutable content. On the contrary, we assume
that world class has to be mobilized, and we study how it is made to perform,
how it is drawn on to produce effects, and thus how it is situated in social settings where it is both a medium to install certain practices and a result where
its "content" is determined. World class, to paraphrase Giddens (1984), is
both medium and outcome. It is through using world class as a resource that
firms enable inquiry as it presents itself as a mechanism to be explored. And
it is through exploring world class that it generates its particular content and
meaning. World class thus has to be performed; it is given voice, meaning
and effects in being drawn on. It does not exist as an immutable fact, but
gradually it gets defined and made to work in its application. Obviously, in
this sense, it is an open, fragile, and shaky resource.

STRATEGY

This view contrasts to other theories of strategy (e.g., Mahoney & Pandian,
1992; Porter, 1985; Prahalad & Hamel, 1990) that appear to suggest that
advantages and competencies exist as "facts." We suggest, in contrast, that
they may become facts, if their mobilization is successful. But before this has
been accomplished, they are fragile potentialities in search of historical
paths and trajectories to which they can cling to make their position stronger, and in search of allies that can protect them from becoming extinct. It is
this process of solidifying world class that we propose to study. In doing this,
we do not presume that advantages and competencies exist; we suggest that
they may be fabricated. Neither do we presume that there is only one trajectory in a firm; several heterogeneous candidates may exist in a kaleidoscopic
of organizational capabilities and competencies.

This may be counterintuitive to a manager who considers his or her business and whose view of its competencies may be clear and obvious. They are already solidified. However, the process of solidification is a complex one, drawing together from various organizational resources certain—and different—records of organizational trajectories. The trajectory, which for a moment has been elevated to representing the firm's core competence, however, has been through a process of distillation where the heterogeneous activities that make up organization are rendered simple and solid. The seeming strength of a certain core competence—with a durability of decades, to follow on from Prahalad and Hamel's argument—is there primarily for reasons of convention. It is a social product. Competencies and resources are composites. It has been suggested to map them in terms of a number of resources and capabilities. Resources can be identified as financial, physical, human, and structural and also in image and information resources (Amit & Shoemaker, 1993; Best, 1990; Itami & Roehl, 1987). On the basis of that resource pool, simply the capabilities of the individual firm is what it can do more effectively than its rivals (Grant, 1996). Although not denying this position we would, however, be concerned to underline that such a view is that of the victorious, and that there is little, if any, debate as to how the war was fought. We suggest studying the war and getting to the mechanisms that pointed out the trajectories that were given the ability to speak for the firm's competencies.

EUROOIL AND SUPERDESIGN

Through case studies of two firms—EuroOil and SuperDesign—we illustrate how path dependence and path creation are related. We show how world class was mobilized via debates on various trajectories of history, where managers at different locations in the firm mobilized their own local knowledge and attempted to make it universal to the firms' problems. We compare the two firms in order to show that the same basic idea of world class can often be *very different things* in different firms and *different things* within each firm. In this way, we show that world class is an "icon," opening the process of testing current routines and capabilities against general best practices, allowing a debate on the competence base of the firm.

This process of presenting the firms' competencies is one where internal organizational participants seek to stage themselves as spokespeople for world class and who seek through a set of devices to construct everybody else as "appendices" to their competence and capability. To do this, they have to demonstrate that their proposed translation of world class is workable. *Workability* refers to the probability of its practical application. It has a paradoxical position because, in a sense, it is impossible to demonstrate that a practice works until it has been put to the test. In paying attention to workability, organizational actors find historical episodes and moments where their preferred method has succeeded. In effect, what they do is show that a procedure already in place, albeit some times with marginal visibility in the firm, has been able to construct successful organizational action. In a sense, workability presumes a prior technology of managing, which has already proven its power in history. When managers attempt to mobilize world class, they do so by pointing out certain more or less well rehearsed procedures, media, organizing principles (in short, managerial technologies), that in certain historical episodes and glimpses have been called on to solve problems. Therefore, when world class is mobilized, it is often via a voice speaking from the point of a managerial technology already in place, and the process to arrive at competencies is not merely a dissemination of a grand vision of the possible functioning of the firm; it is also a matter of its determination and translation into possible practices as they may be envisaged by their presence in procedures, paper, charts, and techniques—often quite mundane and ordinary—which are used or have been used before. World class mobilizes managerial technologies that in turn, if successful, define what world class is all about.

This is obviously no innocent process; in translations between world class and managerial technology and back, actors attempt to re-form the firms. They attempt to re-group networks and create a new topology of centers and peripheries where the work to be done is to incorporate others' managerial technologies into their vocabulary (albeit as appendices, of one's own technologies). It is also, and perhaps more importantly so, a process of subsumption where one is subordinated by others even if these still do their original work. The translations between world class and managerial technologies and back is a process of destillation where no managerial technology is kicked out; most of them, however, are made parts that implement the work of networks where others are partners with a voice to influence, condition and transform networks.

To illustrate this thesis and develop the argument, we draw on a study of two firms, both of which embarked on a voyage toward world class supplier relations. We traced the translations between world class and managerial technologies but only while they were being shaped. That is, we followed strategy while it was invented, and thus we were able to see some of convolutions that strategy debates construct.

THE STUDY: CORE COMPETENCIES IN THE MAKING

We draw on Latour's (1987) sociology of translation and suggest that strategy making is best studied before a strategy is made. In other words, in this chapter, we study core competencies in the making. It is based on a study of two companies, SuperDesign and EuroOil, that attempted to achieve world class supplier relations. We studied the translations made between the overall strategic agenda based on general best-practices, and the work to make them organizational through a core-competence development process. The study shows how material and managerial knowledge was employed to find premises for supplier relations in both firms, and in both firms, the strategic agenda and the meaning of core competencies were on the move.

In both firms, multiple translations of the firm's history were introduced to justify what the firm's present and future strategic problems could be. These vignettes of history showed strategy as a process where organizational strengths and weaknesses, core competencies, and relations of power were pointed out, defined, defended, and indeed performed. We saw that fragments of existing organizational routines were called forth and reinterpreted as solutions to strategic problems still in the making in top management's offices.

We paid particular attention to the rhetoric used to present different possible managerial technologies as responses to the strategic agenda. In this debate, each actor engaged in the fabrication of the relevance of their particular managerial technology. To do this, each had to show how it could mobilize world class. This was done, if successful, first by, a conceptualisation of world class and a presentation of its effects in plain, ordinary, mundane organizational procedures. Second, the actor had to mobilize a set of steps by which to proceed and make it practical; and third, he or she had to show that this technology could subsume other managerial technologies. All three circumstances pointed to multiple organizational trajectories, each of which could be mobilized vis-à-vis world class. Multiple trajectories—or historical experiences—could be called forth and termed strategic core competencies.

We begin our discussion by presenting the two companies; we then present our analysis of the translations between world class and its signification in the firms. The last part of the chapter is oriented toward explicating the mechanisms that connect world class with managerial technologies ,and thus, with the relationship between path dependence captured in the mobilization of existing managerial technologies and path creation, as the new configurations of parts and partners that managerial technologies call forth in organizational and interorganizational networks.

TWO FIRMS—ONE STRATEGY

EuroOil and SuperDesign worked in different markets. Within their respective industries, both companies were highly respected. *EuroOil* was in the industrial market, more specifically in the oil industry, and recognized for its innovative production techniques. *SuperDesign* was in the consumer market, more specifically in the high-tech consumer industry, with electronic household products known for their superior design. Both companies had a strong reputation for their technical competence. In a broad sense, the image of both companies was that they both exhibited superior engineering skills.

Both companies had traditionally been using large pools of suppliers and had typically not considered any individual supplier of special strategic importance. Rather, on a simple make/buy basis, the market had been squeezed for the best price. The strategic imperative to attempt world class supplier relations in accordance with espoused best-practices was seen as a condition to future growth as a consequence of enormous costs incurred on technical tailor-made production equipment and products. Although the two companies were located in quite different industries, their top management teams each had identified similar strategic imperatives for supplier management: First, the total number of suppliers was to be reduced and organized in three layers, the first being the most important one. Second, supplier relations, primarily for the first tier, were to be managed on a long term basis. Third, all suppliers would participate in continuous improvement programs, and fourth, first tier suppliers would be engaged in development work.

Both EuroOil and SuperDesign were about to introduce what was termed a revolution in their respective organizations with the set of imperatives just stated. Neither of the firms had yet developed routines that reflected imperatives concerning widespread dialogue with suppliers. Although these four characteristics reflected "best practices," they did not indicate by what measures these practices should be realized. It was clear that best practices were conceptual and inspirational as well as aspirational. They were milestones of a standard of supplier management to be found in the future of both firms.

In SuperDesign, the new agenda implied a debate concerning superior design versus quality to cost. Historically the agenda had been to focus on the best design possible. When introducing world class, one top management representative suggested "We decided not to make an estimate of the costs. This is about creating a path into the future!" implying that this leap into the future had to be predicated upon trust and faith. It was a strategic venture!

In EuroOil, the outline of a new policy followed similar reasons. Although the companies work in different markets, and even though EuroOil had a strong tradition of using suppliers, both companies had managed their supplier relations at arms length. EuroOil had about 50 times more suppliers than SuperDesign. Both firms adopted the same formula for their strategy. When world class was introduced at EuroOil, a top manager made the observation: "It's interesting, when you write something [strategy] it seems that everybody agrees with you. However, many still don't understand how it is to be realised!" Also here, strategy was to be made. Consequently, strategy generated widespread debate in both organizations.

STRATEGIC IMPERATIVES

The strategic imperatives outlined in the two companies are represented in Table 13.1. It reflects a set of intentions with a certain visionary quality. The rhetorical qualities of the table's statements are clear, and it seems possible that they are perhaps more observable on paper than in relation to practice. The statements in the table are quotations from central managers.

This table shows two firms where the strategic imperatives are worded slightly differently but where scenarios appear to be consonant. It illustrates that the strategic imperatives based on world class identify two firms committed to interorganizational linkages designed for the long run; that substantial interaction with suppliers is made an important objective; that contracting is not the primary way to structure relationships to suppliers; that results are more important than specified inputs and behavior controls; and that the financial realm is to be controlled in the long run via frankness and openness in sharing internal information with others. This thesis is the semantic of world class. It is the vision that managers sent out in their firms and asked how to implement it.

REACTION POSTPONED

In both firms, reaction to these strategic imperatives was complex. The idea of world class is mainly an imperative for change but in itself it renders little insights about how the change process should be managed. World class was positioned as an icon yet to be pressed and employed. As the respondent just quoted above suggested: "nobody knew what to do."

Rather than implementing or diffusing world class into the firms, actors started to invent a series on contexts for understanding world class. They started to contextualise to relate it to organizational concerns as they were emanating from below. They started to develop frames of reference through problematization to craft the focus for organizational changes in the name of world class. This contextualization was accompanied by the mobilization of an actual or potential particular change program, which was staged in the organizational market place for solutions. We will look at contextualization and mobilization in turn.

TABLE 13.1

Strategic Imperatives at SuperDesign and EuroOil

SuperDesign	Strategic Imperative	EuroOil
We aim at building up a network of relatively few but stable key-suppliers with whom we can do direct long-term business. We avoid purchasing on the basis of spot transactions.	View of own company	We aim at a cross-organizational cooperation in view of the total life and use of products and services. We aim at an image of standardized practices.
Suppliers will be selected in advance of new product development and invited into integrated product development.	View of suppliers	Cross-functional-teams with EuroOil and suppliers will be defined prior to a new construction.
Contracts are to be kept as simple as possible.	Attitude to contracts	With suppliers, we want to focus on the opportunities for cost minimization rather than complex contract issues.
Quality cooperation shall be based on an agreed relationship, and goods are expected to be delivered at a zero-defect-level.	Product(ion) control	Functional specifications will abandon the prior regime of detailed specifications.
To be able to point out the elements in the pricing, which need optimizing, it is necessary to work with open calculations.	Financial controls	In relation to suppliers, EuroOil will, right from the start of a new project, introduce life cycle-costing

CONTEXTUALIZING THE STRATEGIC IMPERATIVES

World Class

In both companies, positions held by the purchasing and technical departments formed an important debate. What we saw was that the purchasing departments used world class to raise a set of conditions that would make purchasing more important because supplier relations were argued to be about more than merely ordering a set of specifications. This would pay but little attention, it was argued, to let suppliers engage in development work. The purchasing departments in SuperDesign and EuroOil both saw the strategic imperative as an opportunity to reorganize and abandon existing rules regarding the management of supplier relations. In both firms, the traditional situation had not left the purchasing departments with strategic influence over the suppliers as they had had merely to coordinate contracts and supervise supplier relations.

The technical departments saw the new agenda as a threat to their position because they typically had been called on to specify at great detail what the firms wanted suppliers to do. They were, however, not inclined to immediately see this as a problem, because they had suspected that their position vis-à-vis suppliers would be stable and immutable.

In both firms, the uncertainty about the implications of the strategic imperative led the purchasing and the technical departments to search for prior, historically informed solutions to the management of suppliers.

THE POSSIBLE NEW ROLE OF PURCHASING

The purpose of this search was to set up a context for the strategic imperative, which would later enable the technical departments and purchasing to influence the prospective position of suppliers. This led to the paradoxical situation that, in SuperDesign, the purchasing department presented the possibility to formulate a set of broad quality standards vis-à-vis suppliers, whereas in EuroOil, the purchasing department suggested that quality standards, which were used at that time, had to be abandoned. As a result of this, the purchasing departments attempted to draw in the specific translation of world class to suit a prospect of a more central position:

- At SuperDesign, one purchasing officer stated: "Traditionally, product development has been done without any involvement of suppliers—and with using the purchasing officers knowledge."
- At EuroOil, likewise another purchasing officers mentioned: "We in the purchasing department cannot accept to be involved only on the basis of already made decisions."

In contrast, the technical departments attempted to show that their conventional access reiterate the firms' positions vis-à-vis suppliers should be continued because they were the ones that possessed the insights needed to interact with suppliers on complex technical projects.

- As response to the purchasing department's move, an engineer at SuperDesign stated: "Although it was a nice proposal in theory, it wouldn't work in practice since the capability to develop superior product design was something inherent to the corporate culture—the skills of craftsmen, which could not be outsourced."

- At EuroOil, the comment by the technical department simply was "that although in theory it would be much cheaper not to specify subcontracted projecting in detail, in practice it was necessary. The idea of functional specifications would leave too much space for uncertainty, which could not be accepted as long as EuroOil would carry project liability."

In addition to the struggle between the purchasing and technical departments, in both firms, there was a set of company specific questions that differed between the firms. In SuperDesign, a third response to world class was a technology-mapping project oriented towards representing the skills located within the firm and at its suppliers. This part of SuperDesign wanted a visualization of the subcontractors' competencies that they could build on to make new products. This required a representation of the knowledge that could be drawn on, either inside or outside of the firm. This was an attempt to create an interorganizational business system that integrated suppliers and firm in a virtual mode where all competencies could be aligned at the drawing boards of SuperDesign. Here, the prospect of a pool of suppliers crafted to be of particular importance to the development work going on in the firm was mobilized.

At EuroOil, this approach was split between two different organizational utilities, A and B, that tried to engage world class in two opposing ways. Unit A was a relatively new plant in the beginning of its life cycle, whereas Unit B was an old one presumably toward the later stages of its life cycle.

- Unit A saw the strategic imperative as a response to the problem, that EuroOil did not trust its subcontractors and therefore incurred unnecessary costs on management.
- Unit B saw the strategic imperative as a opportunity to solve the problem that EuroOil did not have formal contracts that provided rules for when and how subcontractors could be punished or paid a bonus, depending on their ability to reduce costs.
- Unit A would try to develop the idea of workable specifications on the basis of conducting considerable dialogue with subcontractors.
- Unit B would like to refine the technical standards to minimize unnecessary and thereby costly discussions with suppliers.

A and B each used world class to address problems that were specific to their particular business. A wanted to start a dialogue, whereas B demanded the end of dialogue. A was about the start up a new production area employing new production technology; B would in a few years stop the production. Compared to the general debate between purchasing and technical departments mentioned, these two positions were not as visible, precisely because they only related to the specific business of each profit center. Nevertheless, in addition to the most visible debate between the purchasing and engineering departments, other attempts were made to relate the strategic rules to other site-specific problems.

Debating World Class

Summing up, in both companies, different departments reacted to the imperatives. The technical departments, currently at the core of supplier management, were challenged mainly by the purchasing departments. But also other contextualizations were attempted.

At SuperDesign, we saw that the purchasing department's proposal was first countered by the engineers, and secondly by a proposal to create an interorganizational information system that—as it described the suppliers' capabilities—was a medium to link internal and external competencies directly under supervision by top management. At EuroOil, we identified one managerial proposal by the purchasing department, countered by the engineering departments, and by two minor managerial responses located within two business units with quite different apprehensions of the problem. (See Fig. 13.1)

It appears that world class was staged in several places. They all had their focus on the question about what kind of supplier relations would be needed in the future. Through a debate on the boundaries of the firm, the world-class icon was given properties. The contextualization of world class

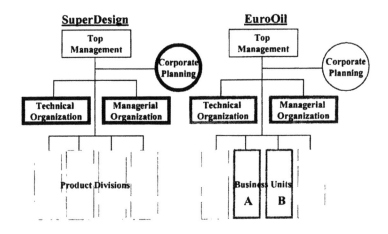

FIG. 13.1.   Places for debate in SuperDesign and EuroOil.

was a set of attempts to define the frame of reference by questioning the firms' boundaries and how they were to be managed.

Clearly, these contextualizations were forceful. In questioning interorganizational relations, even the very competencies of the firms were made debatable. By the contextualizations, at least two different lists of requirements defining core competence could be defined. One saw it as technical specifications implying a centrality of engineers. Another saw it as coordinating mechanisms handled by the purchasing departments. The core competence in these two situations was product and production technology on the one side and the ability to organize others on the other. At this point, there existed different sets of visions about the firms' competencies.

MOBILIZING DIFFERENT STRATEGIES

Purchasing Departments Mobilizing World Class

The struggle to define the practical principles of managing was not exclusion strategy, where a fundamental change in the operations of the firm was attempted; rather, it was an attempt at inclusion and realignment of certain relations of dependence and autonomy between organizational utilities within the firm.

In SuperDesign as well as in EuroOil the purchasing departments were eager to make proposals that could enable them to have supplier relations based on structure, where their problems could be solved by functional specification rather than by technical specification or by identifying and controlling for outputs rather than by detailed specification of inputs.

These propositions must be seen against the background of traditions in both companies. In SuperDesign, the purchasing department had never established and managed supplier relations before. As explained by a purchasing officer, the rule more than the exemption had been that "All the time, technicians in this company seemed to find most pleasure in travelling around making arrangements with new suppliers."

Likewise in EuroOil, due to the use of technical quality standards, the purchasing department had never even attempted the management of suppliers. They had always been managed by the engineering departments via the immense number of quality standards that had been imposed to control the work of suppliers. Therefore the game was about how to position internal and external actors in relation to each other. The purchasing departments' two possibilities concerned cross-functional teams (EuroOil) and integrated product development (SuperDesign).

The individual purchasing departments in SuperDesign and EuroOil individually had either presented the idea to introduce quality standards for supplier management or proposed to abandon historical standards of detail, specifying how subcontractors should perform their work. These proposals mobilized a strong role for purchasing departments.

- At SuperDesign, purchasing wanted to define a number of supplier teams. Then, before planning new products, the purchasing department would invite and head a team of suppliers, and encourage them to come up with their best ideas for cost-effective product development.
- At EuroOil, purchasing presented more modest ideas, explicitly stating that they would not be able to manage suppliers alone. Matters concerning choice of technology should be under supervision of technicians; however, purchasing departments would be managing the suppliers.

Technical Departments' Responses

At SuperDesign and EuroOil, the technical departments reacted similarly. They started to explain engineering routines and pointed out what criteria were employed, when and on what basis. At SuperDesign, technical departments erased a number of headlines that all specified different kinds of quality criteria. In fact, what had previously been characterized by the "loose" term culture, could be identified in a number of basic rules, ranging from technological criteria to quite subjective criteria regarding handling and products' required weight. As an example, a rationale behind these subjective criteria was that a customer buying expensive products from SuperDesign should literally feel the difference when handling the product. Products by SuperDesign were known for their superior quality products, which in fact can be regulated regarding mathematical ranges for noise, speed, and so on.

At EuroOil, the technical departments disclosed that the enormous number of technical quality standards provided logistical visibility. In this sense, engineers could realistically perform project management through applying standards because any change could be introduced via respecification of the standards; this would leave the purchasing departments marginal.

Contesting Mobilizations

Thus, we have seen the strategic imperative mobilized in a number of ways. Each of the purchasing departments argued that they wished to engage in supplier management—ranging from contract negotiations to integrated product development. In SuperDesign, one approach was based on introducing a new concept—the measurement of internal and external productivity of technology. At EuroOil the profit centers mobilized solutions reflecting among other things the life-cycle-stage of the plants.

The purchasing departments' attempt to establish themselves as centers for supplier management was contested by the technical departments in both firms. This contention did not question the attempts of the purchasing departments to mobilize expertise in managing suppliers as such. Rather the importance of the technical issues were represented as a competence essential to the management of suppliers. This representation was mobilized by unraveling certain elements of the technological complexity, thereby signaling the valuable potential of the technical departments. By increasing complexity, the technical departments had intervened against the purchasing departments' attempts to simplify the principles of supplier management based on arms-length principles at a distance where the functional specification was among our detailed technical specification. Thereby the technical departments had not only reduced the risk of becoming a part of the network proposed by the purchasing departments. Also the technical departments had contributed to new insights concerning complexity in defining the boundaries of the firm in relation to suppliers in a new and more aggressive move toward internalizing technical knowledge as strategic competence. Thereby the purchasing departments were challenged anew to proceed mobilizing solutions.

The contextualizations and mobilizations produced in this process were all challenged as to their workability. Could they deliver the goods in practice ? Would they work? They were rarely directly compared with the explicit strategy statement but obviously, world class was always an important metaphor. During this debate, potential future paths were identified and scanned. Concurrent reflections initiated by the game between departments, an ongoing process to incorporate the others' arguments, drove the creation of possible paths and attention to their realization, or workability.

Throughout this process of iterative contextualizations and mobilizations, the knowledge resources of the firm were refueled. EuroOil and SuperDesign were about to create a path of development. Although dependent on their history we see that the development process was not predetermined. In both companies, what the core competence had been and would be was only about to be constructed, as managerial and material knowledge was made explicit.

Workability is an important, if not the only, dependency factor of these firms' development paths, and this is why the process was more important than the resulting definition of the distinct core competence. In the following section, we discuss further the conditions of mobilizing workable solutions.

THE WORKABLE SOLUTION

A Process of Enlargement

The scenarios just presented above illustrate the difficulties of reaching solutions to strategic challenges. What we have illustrated is that different organizational entities compete to define how a certain strategic vision is to be presented, and thus to be constructed as relevant and meaningful to corporate concerns.

To prove workability, technologies of managing were employed. Most of the inscriptions provided by interested parties were technologies of managing in clear continuity with those they had been previously applying. Although molded to fit the new strategy, they were brushed up to support an argument about why the then present knowledge and procedures were, in fact, very appropriate if only given the chance to move into a freer space where their power could be released fully.

In other words, strategy resided at best partly in technologies of managing. The mobilizations only marginally touched on the imperative of world class in itself. They were not justified primarily by its own rhetorical force. Rather, in all circumstances, they were justified by arguments that touched on elements of a political economy of organizational relations. Good justifications made it possible for agents to somehow appropriate others' efforts; these included profits, knowledge, control, and investments. What this illustrates is that the fate of a strategy is not a mere process of implementation but also a process of enlargement. In the firms studied, this process of enlargement consisted of at least three elements. The first was concerned with who produced the mobilization of certain solutions. The second was concerned with the question how others' actions could be taken into account, and the third was concerned with the technologies of managing that connected a solution to the firms' problems in a workable manner.

First, each of the seven solutions mobilized in the firms were defined from different positions, each of which drew on the particular managerial and technical competencies located in these firms. This shows that supplier relations were not only down-stream activities; they were also involved in the configuration of the firms themselves as to how they should cope with, and indeed identify, core competencies.

Second, we illustrated that the way this occurred was to incorporate the network of people and procedures within and outside the firm in one mechanism. By this move, actors attempted to place themselves as centers in networks able to organize others' work. In this sense, the Latourian actor-network helps describe how the network was constituted. All actors were concerned with the same elements of the network, but disagreed about their centrality. In the different positions within EuroOil and SuperDesign, middle managers attempted to create others as supplements to their work. Others were made parts of their network without being partners. It was by negotiating world class best practices that each position attempted to use them in a way that made it difficult, if not impossible, for others to evade incorporation. Such ability to act on others without them being able to react on oneself is a Machiavellian form of strategizing, which has as prime purpose to incorporate others' knowledge. The network, as such, was not given up; it was the relative positions of power within it that were at stake.

Third, to place themselves as centers, actors "carve out" technologies of managing that enable action at a distance (Latour, 1987, Law, 1994) and serve as disembedding mechanisms (Giddens, 1990, 1991), which enable them to control organizational practices in time and space. Such technologies of managing were essential as they provided the justification that a particular inscription would be able to hold the network together, irrespective of its spread in time and space. In EuroOil and SuperDesign, these technologies of managing came in several forms. Open books which allowed access to suppliers' information systems, and life cycle costing which created insights are important here because they translated between corporate profitability and the nodes within the network. But other managerial technologies also acted as immutable movables: standards, functional specification, outsourcing, etcetera, which tended to make stable relations within the network. These were disembedding mechanisms, which stood out as interventions in a flow of activities, highlighting certain aspects and subjecting others to silence. Thereby, the technologies of managing enabled centers to bring back knowledge and effort by means that only marginally recognized these nodes as the producers of that effort.

Partners and Parts

In EuroOil and SuperDesign, the promotion of the best practices was a piece of translation where situated actors attempted to align their competence with the ethos of world class supplier relations. The work of translating world class best practices into workable solutions involved creating a network of people and procedures internal and external to the firms. Each of the translations provided its particular network, that was connected, not by excluding others, but by including others albeit in new capacities. An important distinction is between inclusion as network-partners and as network-parts.

- A *partner* is a node, which accords actorhood and presents as one who speaks and communicate in free language, in multidimensional modes about the grand design involving internal functions and external suppliers

- A *part*, in contrast, is a node within the interorganizational network, which is allowed to respond to the partner only in ways that support the overall framework suggested by the partner.

For example, the purchasing departments sought to introduce technologies of managing by which suppliers could be governed by outputs, thus understanding and evaluating them by the presence or absence of deviations from expected results. Here suppliers were give a voice and made partner of an interorganizational network, whereas the technical departments were rendered to do a set job. The technical departments, in contrast, attempted to make purchasing departments perform a marginal role. The distinction between the solutions of the two departments impacted not only on the direct work of the individual entity; it also re-created the organizational significance of other departments' work.

A *part* is a node in a network, which is rendered one-dimensional and made into a cybernetic element in the functioning of the firm. This is a situation not of exclusion but of designing organization such that certain parts are made peripheral. As partner, in contrast, the *entity* is constructed as multidimensional and with a voice to be heard. *Partners* are central, and they have to be minded in a general dialectic of control where their possible dissatisfaction is constructed as serious and has to be addressed. Parts' rebellion, even if potentially important, is not contemplated.

The construction of the network is not a matter of excluding nodes; it is more concerned with crafting the status of a certain node, such as suppliers. Are they to be as partners with a voice to be used to co-develop the business, or are they to be regarded as parts that will perform a task? The *network* is constituted by the interrelationships between actors within the firms and outside the firms and a set of devices that make their interrelationships durable. These devices, or *managerial technologies*, co-produce the network, establishing the rules by which nodes are made parts or partners. For example, the discussion in EuroOil about specification of supplies (functional or technical) was not merely a matter of a straightforward transition vis-à-vis suppliers from a rules to a results-orientated form of management. It was also an internal discussion in EuroOil about how suppliers were to be incorporated in the firm's management activities, and in turn it was an internal matter of what type of managerial knowledge should count the most. Purchasing would prefer functional specification and control by output- or results-oriented measures, whereas technical departments would prefer technical specification and control by inputs.

Managerial Technology

Strategy works if it is packaged in managerial technology; if not, strategy would be a free floating exercise, not only because it would not be organized, but more importantly, because it would not be serious. Managerial technology points out how strategy should be procedurized. Procedure may be thought to be trivial to strategy, but as EuroOil and SuperDesign illustrate, procedure is a precondition for the existence of a strategy. World class did not have any organizing effect until procedures were specified that pointed out what is inside and what is outside strategy. Strategy thus is nothing before it is procedurized. And the procedures come in large measure from internal historical trajectories, where actors attempt to mobilize their specific small bit of history to be the ones that can counted on as the firm's core competence. To do this, they have to have procedure; otherwise they cannot point out what they mean.

If, for example, in EuroOil functional specification was chosen, a managerial knowledge centering around results and outputs could point out what is strategic and what is not. Here, managerial technologies that emphasized a distanced view toward suppliers' activities would be put in place, management competence that relied on high level inscriptions, such as effect measures, would be put in place, and an emphasis on effects for the firm's costs and revenues would be in place. In contrast, if technical specification was adopted, a managerial knowledge concerning the engineering aspects of EuroOil's production would be emphasized, and then management competence would center on detailed insight into the technical transformations mobilized in the particular projects. In this situation, managerial technologies would center on the production of a design blueprint. Its form would be the forecast of technological effects for the project itself. Its effects would be engineering effects, whereas relations to the business side of the firm would be more marginal, if not effaced.

This example illustrates that the production of the network between firm and supplier depends on managerial technologies that make it durable. The managerial technologies to a certain extent may preconstitute the network because it is through the form of these technologies that engineering projects are made visible. But this technology is less an effect of a detached analytical activity of resources and capabilities than a matter of promoting certain forms of managerial technology and knowledge.

In SuperDesign, there were similar examples. The question whether to adopt a cost approach or a technological competence approach to suppliers was predicated on where in the firm the management of suppliers was to be located. Cost was important to purchasing that already had in place a system that ranked the suppliers by their efficiency. Here, the relevant managerial knowledge was the ability to compare standard cost, and the managerial technology preferred was the cost–management systems that could provide detailed, albeit one-dimensional, analyses of individual suppliers and compare these with others. This was management at a distance, where the inscription of the supplier in a standard cost would make any concrete understanding of its production process superfluous. To get to world class the language here would be productivity and efficiency.

In contrast, the technology competence approach would highlight the particular capabilities of the suppliers and their ability to cooperate on issues with a bearing on SuperDesign's product development problems. Here, managerial knowledge was much more concerned with the process of technology development itself and it required co-operation about technical details just as it required the firm to specify its plans to the supplier. Here managerial technologies were ones that inscribed suppliers in a more complex set of dimensions, of which cost was only one. This complexity made interaction more important than in the cost approach and comparisons between suppliers was made multidimensional. Here, the partners would be technologies and the language preferred for developing world class relations would be technical, and, to a certain degree, local.

Workability

Both firms illustrate that there is a distinct relationship between managerial knowledge, managerial technology, network effects in the form of a distinction between parts and partners, and mobilizations of world class. Indeed, world class was in search of meaning, a search that was constituted not merely as an analytical determination of optimal content. It is more likely constituted by the interests that make people market their competence and suggest managerial technologies that can support it. In mobilizing managerial technologies, actors also mobilize issues, problems, and concerns that define and redefine what supply chain management is all about. It did not, namely, have content; in contrast, it was given meaning through struggles for significance, and it had to be reestablished constantly by managerial technologies that defined the procedures by which problems were to be solved, and thus defined how organizational decisionmaking was to be executed. It was the managerial technologies—that produced durability, and thus produced strategy.

This is where workability comes in. For a certain idea of world class to be credible, it had to be able to withstand challenges on its probability to yield results. Such demonstration was predicated on a certain history of "tests," which could be pointed out or at least demonstrated to have been put in motion. *Workability* is premised on the possibility to demonstrate that the particular back and forth translations between world class and managerial technologies works for top management. It is not necessarily top management who uses the technologies directly, but they have to be confident that the organizational principle adopted has power; therefore, the demonstration of workability is important. This demonstration can only be achieved if it is mobilized against certain examples, however, rudimentary, of how managerial technologies would work to support corporate interests.

The Juxtaposition of Path Dependence and Path Creation

This discussion illustrates how history is important in developing organizational paths. Drawing on Giddens' (1984) duality of structure, the case studies illustrate how stocks of knowledge located in organizational and social rules and resources are constantly being mobilized and put to work. In a sense, historical experience only exists in being drawn on and inserted into a reproduced set of processes. It is the reproduction that mimics path dependence, but such reproduction is always fragile and provisional, because any agent could choose to act otherwise. This is path creation, where history is mobilized and used as a resource to condition interrelations between humans and managerial technologies in extended networks.

Path Creation Identified by Managerial Technology

Managerial technologies override the metaphorical utterance of strategy. By the case study, we show that strategy was only given content by the employment of specific managerial technologies fitted to alternative contextualizations. The managerial technologies were employed to explore world class imperatives. And this exploration was contextualized by questioning the firm's structure and relations to their suppliers. As a consequence, it was the managerial technologies that seriously pointed out what direction strategy should take; indeed, strategy could go in directly opposing directions, depending on what managerial technologies were mobilized to inscribe it.

   Enacted strategy is different from planned strategy (Mintzberg, 1994) simply because planned strategy is an imperative with little pragmatic content. However, neither is enacted strategy merely standard operating procedure. Enacted strategy is a result of how different organizational actors have been able to "fit" routines of the firm into the strategic imperative and been able to prove the fit of those routines and present these as workable solutions; workability here being a composite of workable change and workable practice. In principle, this has to do with path dependency.

Strategy—Mandatory Semantic Space

Managerial technologies reach out toward certain procedurized practices; the case studies illustrate that managerial technologies' work were reshuffling how the logic of historical experience should be made real in a firm. In this way, a small managerial technology is part of and, indeed, critically contributes to, a much larger project of strategy formulation.

The irony seems to be that strategy may be uttered and thus carve out for itself a semantic space. It is only, however, when attached to a managerial technology that it gets pragmatic content. Another irony is, that these managerial technologies could possibly not be employed to the same effects without a strategy carving out this very semantic space; this is not a trivial exploitation of past experience. Rather, the realization by certain procedurised practices was a complex process of exploration, and it is within the span of ongoing contextualizations and mobilizations that the strategic imperative was explored.

History was drawn in to produce strategy, but in the strategy-making process, the administrative routines were not employed simply within the confines of past experience. The routines were employed against the context of the strategic imperatives. As a semantic space the strategic imperatives allow the managerial technologies to be mobilized in ways that are different to the conditions of present administrative routines. Prior to their employment, imperatives symbolize only an empty space; but maybe it is exactly this emptiness that facilitates new path creation, where history clearly is drawn in—as the constraining as well as an enabling factor.

CONCLUSION

Drawing on and contrasting two firms' work with implementing strategy on the basis of a vision of world class supplier relations, this chapter discusses how firms mobilize their histories to generate knowledge about their core competencies. In this chapter, this theme is explored against explicit attempts to change the firm according to a new strategic agenda based on general industry knowledge of world class supplier relations.

The chapter discusses the translations mobilized within two separate business organizations—SuperDesign and EuroOil. In these two firms, the translation of world class was mobilized against possible multiple future paths and past histories in particular ways. We have attempted to show that strategy has to be translated into manageable organizational procedures; otherwise it is not realistic. Managerial technologies are the means of translation that make the claim that a certain vision of world class is probable.

These technologies, however, depend on capabilities developed through history to create history. As there are multiple forms of managerial technologies present in any organisational history, the struggle to translate from strategic vision to strategy implementation is conditioned on the way certain managerial technologies (to a certain extent already in place), are staged and marketed. The strength of the translation depends on how it is able to make itself center (and others periphery) in organizational and interorganizational networks. But indeed, the power of a (potential) translation depends not only on its ability to support top managers' interests but also on its ability to prove the workability of the specific combination of change and practice.

In translations, the position of other parties in the firm are rendered parts or partners. It is through translation of the strategic agenda, that position, actors, and function attempt to create others as appendices to their centrality. They do this by giving others a role as part of a whole network of organizational activities, or attempting to enroll them as vocal partners in the network.

The processes of translation involve giving space to others in such a way that their activities can be seen to be a supplement to one's own position. Therefore, the translations that make strategy real are based on procedures, controls, and technologies of managing already in place to an extent.

The strategy is nothing before it has latched onto a certain set of managerial technologies; it is only there that it generates content in new configurations of path creation predicated on the assemblage of old, historical trajectories.

REFERENCES

Amit, R., & Shoemaker, P. J. H. (1993). Strategic assets and organisational rent. In *Strategic Management Journal, 14*, (1) pp. 33-47.
Best, M. (1990). *The new competition.* Cambridge, England: Polity Press.
Giddens, A. (1984). *Constitution of society. Outline on the theory of structuration.* Cambridge, England: Polity Press.
Giddens, A. (1990). *The consequences of modernity.* Oxford, UK: Polity Press.
Giddens, A. (1991). *Modernity and self-identity—Self and society n the late modern age.* Oxford, UK: Polity Press.
Grant, R. (1996, Spring). The ressource-based theory of competitive advantage. *California Management Review, 33*, pp. 114–136.
Hamel, G., & Prahalad, C. K. (1994). *Competing for the future.* Boston: Harvard Business School Press.
Itami, H., & Roehl, T. (1987). *Mobilizing invisible assets.* Cambridge, MA: (Harvard University Press.
Latour, B. (1987). *Science in action: How to follow scientists and engineers through society.* Milton Keynes UK.
Latour, B. (1991). Technology is society made durable. In J. Law (Ed.), A *sociology of monsters, essays on power, technology and domination.* London: Sage Publications.
Law, J. (1994). *Organizing modernity.* Oxford, UK: Blackwell.
Mahoney, J., & Pandian, J. R. (1992). The resource-based view within the conversation of strategic management. *Strategic Management Journal, 13*, 363–380.
March, J. G. (1991). Exploration and exploitation in organizational learning. *Organizational Science, 2*(1), pp. 71-87.
Mintzberg, H. (1994). *The rise and fall of strategic planning.* New York: Prentice-Hall.
Nelson, R. R., & Winther, S. G. (1982). *An evolutionary theory of economic change.* Cambridge, MA: The Belknap Press of Harvard University Press.

Ooster, S. (1990). *Modern competitive analysis*. New York: Oxford University Press.

Penrose, E. T. (1972). *The theory of the growth of the firm*. Oxford, England: Basil Blackwell. (Original work published 1959)

Peteraf, M. A. (1993). The cornerstones of competitive advantage: A resource-based view. *Strategic Management Journal, 14*, 179–191.

Porter, M. E. (1985). *Competitive advantage*. New York: The Free Press.

Prahalad, C. K., & Hamel, G. (1990, May/June). The core competence of the corporation. *Harvard Business Review*, (pp.79–91).

Robson, K. (1992). Accounting numbers as "inscription": Action at a distance and the development of accounting. *Accounting, Organizations and Society, 17*(7), pp. 685–708.

Welch, J., & Nayak, P. R. (1992). Strategic sourcing: A progressive approach to the make or buy. Academy of Management Executive, 6 (1) pp. 23-31.

Wernerfelt, B. (1984). A resource-based view of the firm. *Strategic Management Journal, 5*, 171–180.

14

Why You Go to a Music Store to Buy a Synthesizer: Path Dependence and the Social Construction of Technology

Trevor Pinch
Cornell University

If you ask the man on the street "What's a synthesizer?" He will reply "A synthesizer is a keyboard instrument".... If you go into a retail store and say I want to see some electronic instruments, they'll send you to the keyboard department, because a synthesizer is a keyboard instrument by default.

—Don Buchla, synthesizer pioneer (interview)

I don't think I, or anybody else in the company, went into a music store before ... March of 1971.

—Bob Moog, synthesizer pioneer, talking about how the synthesizer he invented in 1964 was marketed (interview)

As Don Buchla, one of the inventors of the commercial electronic music synthesizer, stated, electronic music synthesizers today are essentially electronic keyboards with a range of special effects. By pressing the correct button, it is possible to emulate virtually any other instrument or reproduce a range of pre-stored specially selected electronic sounds for live performance. Often referred to simply as keyboards, such instruments are commonplace. They are to be found in children's bedrooms, on stage as part of a band, or in recording studios. To buy one, you simply visit the piano section of your local music store.

It has not always been like this, as the other synthesizer pioneer quoted, Bob Moog, indicates. In the spring of 1964, when Moog and Buchla independently developed the first voltage-controlled modular synthesizers, no one knew what the synthesizer would be used for or foresaw a mass market. The notion that a synthesizer might be like other portable keyboard devices to be sold in music stores was something that only slowly developed. Indeed, before 1971, Moog had conceived of his synthesizer as something only of interest to avant garde composers, academic musicians, and electronic music studios. He did not envisage a popular market for his instrument at all.

In this chapter, I use the early history of the synthesizer (1964–1971) as a way of exploring the notion of path dependence or how past actions affect specific technological trajectories (Hirsch & Gillespie, chap. 3, this volume). Economists Paul David (David, 1985) and Brian Arthur (Arthur, 1988) argued that the QWERTY keyboard is a good example of path dependence. Typewriter and computer manufacturers became locked in to a particular (possibly suboptimal) keyboard design and the choice of QWERTY continued to dominate keyboard design long after the initial circumstances leading to its adoption had vanished. In the synthesizer story, keyboards also play an important part; in this case, it is the multioctave keyboard common to the piano and the electric organ. Why has this keyboard, with individual black and white keys and its division of octaves into the notes of the 12-tone scale become the standard way of manipulating a new source of sound—electronic sound? Has synthesizer development, like that of the typewriter and computer keyboard, resulted from what the economists call lock-in?

The goal of this chapter is to try and use ideas in the social construction of technology to enrich our notion of path dependence. The social construction of technology (SCOT) is now well established (e.g., Bijker, 1995b; Bijker, Hughes, & Pinch, 1987) and is starting to influence work in management studies and organizational theory (e.g., Garud & Ahlstrom, 1997; Garud & Rappa, 1994; Porac, Rosa, Spanjol, & Saxon, chap. 8, this volume). From the outset, SCOT has opposed both technological and economic determinism. Stress has been placed on radical contingency and the different meanings to be found in technology by a diversity of heterogeneous actors or social groups. Such actors often contest the meaning of technologies. *Interpretative flexibility* is the term used to describe how a technology can be given a different meaning by different groups. For instance, in the case of the bicycle, Pinch and Bijker (1987) showed how different social groups could contest the predominant meaning of the penny farthing or ordinary bicycle. For one group, young men of means and verve, this bicycle was the macho bike to be used primarily for sport and recreation. For another group of elderly men and women, who mainly wanted to use the bike as a form of transport, this same bicycle was the unsafe bike.

Interpretative flexibility occurs, not only at the initial design stage of a technology's life, but also later on at the use stage. For instance, in a study of the early motor car in the US, we have shown how different meanings were given to the car when it was taken up in rural life (Kline & Pinch, 1996). Some farmers used the car for transport, but others found new meanings in the car in its use as a source of stationary power from which to run farm appliances.

Social constructivists recognize that despite "interpretative flexibility" technologies also exhibit long periods of stability. "Closure" or "stabilization" is also described as a social process. In the early history of the bicycle (Pinch & Bijker, 1987) we looked at specific closure mechanisms that started to limit the interpretative flexibility of the bicycle and which led to the predominant path of development—the safety bicycle (at least until the mountain bike came along (see Rosen, 1993)). Two closure mechanisms were outlined: redefinition of the problem and rhetorical closure. *Redefinition of the problem* was used to describe how closure was achieved around a component of the bicycle, the air tire. The air tire was introduced by Dunlop to solve vibrational problems, but was eventually adopted because it was seen to solve another problem, that of speed. Bicycles using air tires started to win races and this led to the widespread adoption of the air tire. *Rhetorical closure* was used to describe the power of advertisements in persuading people to purchase new bicycles.

In early SCOT, the emphasis was placed on how specific artefacts could be understood to be "socially constructed." In later work, it is not only technologies that are treated as being socially constructed but also user groups. In the case of the rural car, we argued that not only was the car a social construct but also this construct in turn helped reconstruct users' identities and practices (Kline & Pinch, 1996). It is this "mutual shaping," "coconstruction," "coproduction," or "coevolution" of technology and society that is at the core of today's social constructivist work on technology and that informs this chapter (Pinch, 1996). This opens the door to a broader framework; for instance, some authors have found it useful to talk about "sociotechnical ensembles" and to include processes that have traditionally been seen as within the orbit of the impact of technology on society (Bijker, 1995a).

Beginnings

With these general themes in mind, let us now return to the history of the synthesizer. The types of synthesizers that Moog and Buchla developed in 1964 were radical departures from the preexisting technology such as the

RCA Mark II synthesizer found at the Columbia–Princeton electronic music studio (Pinch & Trocco, 1999). The RCA Mark II was a massive room-sized dedicated piece of equipment that had to be laboriously programmed with punched cards. Electronic music composers needed a studio full of equipment to make their music. Moog and Buchla's synthesizers were much smaller and much easier to control. They used transistors (which had dramatically fallen in price in the early 1960s), and were modular in design, enabling standard modules to be added as required. Each module was voltage controlled and modules were linked by a patch board (that resembled an analog telephone switch board).[1] A voltage could be used to control the functioning of a number of different modules, either signal generators or processors (Moog, 1965). For instance, the pitch of a voltage-controlled oscillator could be varied as its control voltage varied. The output of any unit could be fed as an input into any other module or several at the same time. This means the slowly varying beat of an oscillator could be fed as a control voltage into another oscillator to provide, say, vibrato.

The early units comprised voltage controlled oscillators, amplifiers, filters, and envelope shapers. It was not at all obvious that these new machines should even be called synthesizers as this connoted the huge RCA Mark II synthesizer. Indeed, Moog debated with colleagues whether to use the word synthesizer at all for his device; Buchla initially rejected the label and preferred to refer to his instruments as "music boxes." For a while, synthesizers, regardless of which company made them, were called "Moogs," indicating Moog's early dominance of the market.

It is important to realize just how marginal electronic sounds and the means to produce them were before 1964. Electronic music studios were used either for avant garde music or special effects. Electronic music was often associated with "science fiction," again promulgating its association with things "weird." For instance, in Britain, the popular children's science fiction show, "Dr Who," was introduced with a piece of electronic music made at the BBC's own radiophonic workshop (which used a mixture of electronic devices but no formal synthesizer). Electronic sounds that today dominate our soundscape were a rarity.

In 1964, the path to be created for the synthesizer was not at all obvious. This is evident in Moog's own description of how the instrument came into

---

[1]Voltage control was something discussed in electrical engineering at the time and there was great interest in it largely stemming from the development of the transistor. Because of the exponential relationship between input voltage and output current in a transistor, you can get dramatic increases in current for small voltage changes over the range of frequencies for musical pitch. Most of the parameters in music vary exponentially (e.g., for pitch, the diachronic scale increases in ratios, and amplitude, or loudness, where the scale in decibels is an exponential scale). Moog, who had a background in electrical engineering, saw that voltage control could be applied to musical instruments with likely dramatic results.

being. Moog was working in Trumansburg, in upstate New York, and had been asked by an experimental musician, Herb Deutsch, to try and produce certain yelping sounds electronically. He soon put together a circuit. Here is how Moog describes what happened next:

> Herb, when he saw these things sorta went through the roof. I mean he took this and he went down in the basement where we had a little table set up and he started putting music together. Then it was my turn for my head to blow. I still remember, the door was open, we didn't have air conditioning or anything like that, it was late Spring and people would walk by, you know, if they would hear something, they would stand there, they'd listen and they'd shake their heads. You know they'd listen again—what is this weird shit coming out of the basement? (Moog, interview)

This "weird shit" came from the prototype of the first ever commercial synthesizer. No one, especially in Trumansburg, had heard such sounds before.

Moog, who had earlier made a hobbyist instrument, the Theremin (named after its inventor, Leon Theremin, and used in early science fiction movies, some classical pieces and on "Good Vibrations" by the Beach Boys), started commercial production of his synthesizer in 1967. The machines in those early days were exclusively custom built for academic studios or individually wealthy musicians. (They cost $10,000 each.) In the early days Moog had no idea how successful synthesizers would become:

> This was sorta stuff I did for the hell of it while everybody else tried to make the shop go.... It was fun and interesting and maybe it would lead somewhere ... it was a hoot ... it was something very neat ... not quite thrilling and not quite amazing, a little bit of all those and fun too. (Moog, interview)

Path Creation

Moog and Buchla had very different visions of the synthesizer. I already mentioned that Buchla rejected the label synthesizer, preferring to call his instruments music boxes. Buchla, who was himself a musician, regarded himself as an instrument maker. Although he did for a short while license CBS to produce one of his early synthesizers, he had always rejected mass production:

> I'm an instrument builder.... I don't build machines, I never have built a machine. I build only things that you play. I don't build things that you program, I don't build things that are involved in issues. I am an old fashioned builder of instruments. And it's a niche market. (Buchla, interview material)

Moog, on the other hand, always wanted to mass produce. He had tried building amplifiers before he developed synthesizers. Unlike Buchla, he saw it as one of his main jobs to respond to what musicians wanted: "And we just responded to demand. I didn't stand there and say, you know, 'I'm going to do it the right way, and you can't have it any other way' ..." (Moog, interview)

Moog and Buchla were both facing the fog of uncertainty that accompanies any genuinely new technological innovation. Through this fog they were, each in their separate way, trying to shape the new path. And each pioneer had a different view of what that path should be. For Buchla, the synthesizer should be a unique instrument intentionally designed to enable the performer to make the best use of the new source of electronic sound. He did not want to mass produce and each instrument would be custom built. For Moog the path was one where he would eventually be able to mass produce. Right from the start, Moog maintained a close collaboration with musicians, such as Deutsch, and was prepared to learn from them what they wanted. As we shall see, these different attempts at path creation had radically different consequences for the role to be played by keyboards on synthesizers.

Path Dependence and the User

The early instruments produced by Moog and Buchla were very different in appearance from today's digital keyboard synthesizers (see Fig. 14.1, which is a picture of the popular digital synthesizer, the Yamaha DX7). Figure 14.2 is a picture of the Moog 900 Series and Fig. 14.3 the Buchla 200. As can be seen, the Moog synthesizer resembles an old telephone exchange.

The photograph of the Moog shown in Fig. 14.2 is actually a mock-up used for promotional purposes. Jon Weiss, who is one of the musicians shown in the picture with his hands on the keyboard, did not even use the keyboard in his own electronic compositions. Weiss, like many other electronic musicians, preferred to make each unique sound by adjusting potentiometers and patch cords.

The pose of the two people playing the synthesizer in the background is quite striking. Here is Moog's own commentary on this photograph:

> The keyboards were always there, and whenever someone wanted to take a picture, for some reason or other it looks good if you're playing a keyboard. People understand that then you're making music. You know [without it] you could be tuning in Russia! This pose here [acts out the pose of the left arm extended while the right hand plays a keyboard] graphically ties in the music and the technology. So there are probably a zillion pictures like that. [Moog, interview]

The promotional literature must somehow convey that this new machine can actually be used to play music. The synthesizer at this stage of its development is an odd instrument because, unlike most instruments, it is very hard to knock out a tune on it. Jon Weiss was one of Moog's house musicians

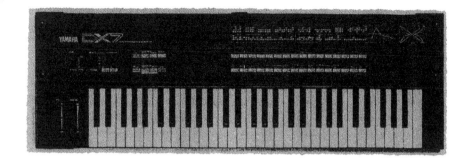

FIG. 14.1.

FIG. 14.2.

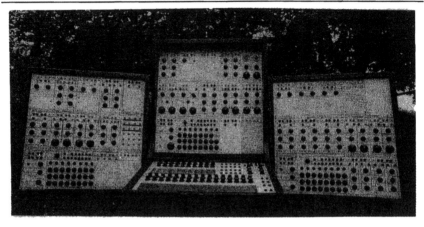

FIG. 14.3.

and sales rep (known as "The Man From Moog" because he was often on the road explaining the intricacies of the Moog to rock musicians) encountered this reaction when he delivered a synthesizer to Mick Jagger for use in the movie *Performance*:

> People don't understand … that the synthesizer itself doesn't produce any sound. When I got the synthesizer to London and I brought it onto the set … I had to go through this with the English workers saying "Agh it's a fabulous sanitizer [sic] and what does it do?" you know "Play us a tune" … Moog heard that so much that in one series of synthesizers he put a little speaker and amplifier in one so that you could actually hear something. People couldn't conceive that this is an instrument but it doesn't do anything. (Weiss, interview)

Another problem the pioneers faced in promoting the early synthesizers was that, at that time, they conceived of the market as consisting mainly of academic studios and electronic music composers. Other sorts of musicians unfamiliar with the arcane world of electronic music studios would have to learn to make music in altogether new ways. When used in the studio, a complicated "patch" could take hours to set up and then the sound had to be recorded on magnetic tape. A single piece of electronic music could require many laborious tape splices.

To understand how a path for a new technology develops, it is important to focus on the use to which a new technology is put. Eastman, the inventor of the first portable camera with roll film, was well aware that developing a technology was co-extensive with finding a new set of uses and users for it (Jenkins, 1975). The previous high-end image of photography was transformed by Eastman into a popular hobby in which anyone could take part. Clearly part of the business of path creation for Eastman was creating a new kind of user to whom the technology could be marketed.

In Moog's marketing of his early synthesizers, the keyboard became a dominant icon because the keyboard is symbolic of music in our culture. Symbolic use must at some point cross over into actual use; purchasers of the early synthesizers who chose the keyboard option began to conceive of the machine in terms of it being a keyboard instrument. This was a new use for the synthesizer.

A second new use of the synthesizer was for live performance a rarity in those early days because of the time needed to set up and change patches. One of Moog' studios musicians, David Borden, had to acquire a whole new set of practices to use the synthesizer for live performance. Borden found his all synthesizer band, Mother Mallard, had to actually rehearse patch changes. These patch changes would take so much time that, during live performances, they would play cartoons between pieces to keep the audience occupied.

Other solutions were adopted by pioneering bands that used the modular Moog for live performance. Moog's engineers and sales reps were often bemused to discover that some rock groups used the synthesizer essentially with only one patch setting, which would be the only sound they would use in their performances. On one occasion a well-known British rock group became distraught when Moog's New York sales representative, Walter Sear, started to take out their patch wires. To Sear's amazement, he found that they feared they would never again be able to make the one sound they used in live performance.

The use of the synthesizer for live performance is important; previously, synthesizers had been exclusively for studio use. With the smaller size of the Moog and Buchla synthesizers and their easier 'real time' control system, live performance was a possibility. Also important at this same time was the emergence of a new breed of musician—the rock musician. This helped spread this new use of the synthesizer. In the late 1960s, with the success of the Beatles' pioneering "Sergeant Pepper's Lonely Hearts Club Band" album, and with the new sound of feedback used by Jimi Hendrix, rock groups were searching for ways to make new and more interesting sounds. The Moog synthesizer, with its array of interesting new sounds, fitted the bill perfectly (rock stars were also one group of users who could actually afford this expensive item of equipment) and soon groups like the Beatles and the Rolling Stones purchased Moog synthesizers. It was one such pioneering rock star, Keith Emerson, of the group Emerson, Lake and Palmer, who most developed the live use of the Moog synthesizer, and Emerson was the first major rock musician to take a modular Moog synthesizer on tour.

Interpretative Flexibility of the Synthesizer

As we trace the early history of the synthesizer, we start to see a path dependence developing. The Moog synthesizer is starting to become a keyboard instrument to be used in live performance. But we should not fall into the deterministic trap of assuming that this particular path was somehow inevitable. The purport of the synthesizer as a keyboard instrument on which to play twelve tone scale music was not shared by everyone. Moog himself was initially ambiguous about the keyboard. He had for years manufactured Theremin kits; the Theremin was played by the operator moving his or her hands in space between two metal rods. Moog provided controllers other than keyboards on his early synthesizers, including one known as the "ribbon controller" (a form of sliding potentiometer). Moog's aim was to provide musicians with what they wanted; thus, as his synthesizer increasingly started to be defined as a keyboard instrument, he was happy to supply keyboards.

Not all musicians wanted to use the synthesizer with the keyboard. For an experimental composer, such as Jon Weiss, it was an instrument on which to explore new sounds and with which to produce non-standard music. Don Buchla saw no need for conventional keyboards on his early machines (see Fig. 14.3). He instead favored a touch-plate device that controlled a range of different parameters. The difference between Buchla's synthesizers and those of Moog was described by Weiss:

> He [Buchla] had a distaste of the keyboard and I think for a legitimate reason. In that he didn't want his machine to be a glorified electric organ. So the only controller that he provided like that was a touch-sensitive pressure pad.... His designs were wild and wonderful. Moog's were conservative, rigorous, and well-controlled ...

> Pinch: *What do you mean by conservative and well-controlled?*

> Weiss: Everything under exact control of one vote per octave, and everything will change exactly the same, and laying everything out in octaves, dividing twelve discrete steps, all that he carried through in the whole design of the machine. The Buchla had—all you had to do was look at some of the names of his modules. Moog on the 900 Series had VCO, VCF, envelope generator, blah, blah. Buchla has "Source of Uncertainty" and sample and hold circuits and bizarre things that he designed ...

> The whole curve of ... [Moog's] company was geared toward a marketable item, while Buchla never had an interest in that. His interest was in making something he could perform on personally and other musicians could perform on. As long he had enough to get by ... it was a whole different mind set. (Weiss, interview)

Buchla was a musician with a background in electronics who was part of the 1960's West Coast music scene (he used to hang out at Grateful Dead concerts playing his synthesizers). He had very little interest in business. What he did have was a definite vision of what electronic music was all about; his instruments were part of bringing that vision to fruition:

> To me it meant simply the source of sound was electronic rather than vibrating strings, membranes and columns of wind. And to me that meant that it was a potentially new source and therefore instruments based on it would be probably new and different. I saw no reason to borrow from a keyboard, which is a device invented to throw hammers at strings, later on operating switches for electronic organs and so on. But I didn't particularly want to borrow the keyboard to control. To me a keyboard is just a way, a nice way, of dealing with harmonic music, polyphonic, harmonic twelve-tone music …. I tried once to put a keyboard on my system, my studio system, and I was able to configure each key to be a totally different sort of response. And I found myself *overwhelmed* by the psychological aspect of looking at this very familiar twelve tone structure and wanting to do music that was very much against what I was conditioned to do … (Buchla, interview)

For Buchla, the synthesizer manifestly was not a keyboard instrument. In terms of social construction, there was interpretative flexibility over the meaning of the synthesizer. The Moog synthesizer itself exhibited this interpretative flexibility. When Jon Weiss used the Moog for composition, he did not use the keyboards. But Weiss seems to have been an exception; other musicians, as we shall see, took to the Moog keyboards with a vengeance.

As was previously mentioned, Moog's vision of his new instrument was to give musicians the tools that they needed. This attitude of responding to musicians led Moog to come up with the first truly portable synthesizer—the Minimoog:

> It began strictly as separate parts … nothing biasing one way of hooking up versus another way of hooking up. And we just responded to demand…. These standard modular synthesizers that we started making, beginning in 1967, began to establish the most commonly used connections. So that began a bias, and things like the Minimoog that were hard-wired, hard built into a particular configuration carried it forward. You know it has been going from there … (Moog, interview)

The Minimoog had a keyboard built into it and used switches rather than patches to change sounds (see Fig. 14.4). Different sounds were "hard wired" into the synthesizer and could be quickly obtained. The overall range of sounds was restricted. By learning from musicians, Moog was able to hard wire in the most commonly used sounds. Also at $1,000 per unit, the Minimoog was much cheaper than the modular Moog. It was also a lot smaller and ideal for taking on tour or to recording studios. In the early 1970s, it became an important instrument for many newly emerging rock groups and has even been described as the first "classic" synthesizer.

The Minimoog is a crucial instrument in the synthesizer's development. It is a machine that starts to constrain the interpretative flexibility found in

FIG. 14.4.

the modular equipment. Its success is evidence that one predominant interpretation of the synthesizer—as a machine to be used with a keyboard, and with an easily available set of preprogrammed sounds—is becoming manifest. Other paths are at the same time starting to vanish. Of course experimental musicians who wanted a broader range of sounds bought other synthesizers. For instance, over the years, Buchla remained faithful to his vision of the keyboardless synthesizer and successfully carved out a niche market for his instruments. There is no doubt, however, that it is the keyboard synthesizers that have become most popular.

Constructing a New Market for the Minimoog

As has been mentioned, in order for a technology path to be developed, a market for that technology has to be developed. It is a mistake to think of a market as somehow miraculously coming into being with a new product or somehow waiting for the right product to come along. Social constructivists view markets as things that have to be actively constructed. The development of the Minimoog market provides a nice illustration of this point.

Until the Minimoog came along, no one from the Moog company had attempted to sell synthesizers through retail music stores. As has been emphasized, they conceived of their market as being almost exclusively made up of academic studios and wealthy individual musicians. The Minimoog changed all that, or rather one pioneer salesman did. Dave Van Koevering, the son of a southern preacher and a demonstrator of novelty musical instruments, formed a partnership with a piano and organ businessman in St Petersberg, Florida, and decided to sell Minimoogs. Van Koevering believed that Minimoogs could be sold in music stores, even though no synthesizer had ever been sold that way before.

> He would take a Minimoog into a music store somewhere in central Florida and show it to the proprietor and salesmen, and almost invariably they'd throw him out. He'd then go to the local Holiday or Ramada Inn and check out the bands. During the break he'd ask the keyboard player aside, show him the Minimoog—the guy would flip—and Dave would say, "I'm going to arrange for you to have this thing." The next morning he'd walk into the store he'd been kicked out of, musician in tow, and say "Here's your first customer, so now you'll have to place an order for two." That's how Minimoogs got to be carried in musical instrument stores. And it began in the absolute most unlikely place, central Florida. (Cochran & Moog, 1993, p. 35)

It can be seen that new, and rather unorthodox, marketing practices helped bring in the new market and create a new form of user. Once Moog saw the success Van Koevering was having, and as word started to spread about the appeal of Minimoog, Moog appointed Van Koevering as his marketing manager. The next year, Moog attended the annual National Association of Music Merchants convention (NAMM) for the first time. By that point the synthesizer had become a musical instrument; it was marketed and sold as one, and musicians were using it to make music.

The Sound of Music

The role of advertising symbolism, different designs, user practices, and marketing I talked about has been outlined. Another important part of the story is the type of sound that becomes associated with the synthesizer. The sound of music is, needless to say, an integral part of the process of the social construction of music technology. People start to associate certain sorts of sound with the synthesizer and this further defines the capabilities of the instrument. For instance, the sounds of the Moog and Buchla synthesizers were quite distinct.

There is one particular record which for most people captures the early sound of the Moog synthesizer. This is "Switched on Bach" by Walter (now Wendy) Carlos. This record, created entirely on the Moog synthesizer, was the recording sensation of 1968. It became the best-selling Bach record ever and is one of the best-selling classical records of all time. It made Moog and his synthesizer famous with appearances on the *Today* TV show, coverage in popular magazines and so on. Here is how Moog described the reaction he received when he first played this record to his fellow engineers:

> In 1968 at the AES [Audio Engineering Society] ... I gave a paper ... this was a couple of weeks before "Switched on Bach" was to be released. And I had probably 100–150 audio-engineers listening. And Carlos let me play a track from "Switched On Bach" as an example and she was in the audience.... I put the tape on and just walked off the stage. And I can remember people's mouths dropping open, and I swear I could see a

couple of those cynical old bastards starting to cry.... At the end she got a standing ovation. Those cynical experienced New Yorkers, they had their minds blown ... [Moog, interview)

The effect on the synthesizer business was equally dramatic:

The conventional wisdom back in the music business ... before "Switched on Bach" came out ... nobody believed that this kind of thing could be used as anything more than a novelty. You couldn't make real music with it. You couldn't be expressive with it. You couldn't make it swing. And then Carlos and a few other people demonstrated they were wrong. You know they just did an end-run around the music business. And *then* in 1969 all hell broke loose. Everybody had to have a, you know every commercial musician had to have a synthesizer. (Moog, interview)

The effect of "Switched on Bach" on sleepy Trumansburg was described by many people who worked at the Moog factory. According to Jon Weiss:

I could see the difference, and there was a world of difference pre SOB and post SOB. Before SOB came out the synthesizer was basically resigned to well-to-do academic institutions, a few private individuals, very few ... it's an expensive machine at 68 dollars it was $10,000 for a synthesizer which was a lot of bucks. And it was pretty much considered lunatic fringe, there's no question about it you know, weird space sounds you know, there wasn't ... there was some rigid thinking about what's music and what isn't music, what's permissible and what's not, and then Carlos came along. Like wham and then suddenly the world thought "Oh Yeah, this is great and its basically like an electronic organ, sorta glorified organ" and so that pigeonholed the whole idea of the synthesizer which was great for the I guess the cause of electronic music, but in a way, it was unfortunate because people didn't realize that that wasn't necessarily what it was about ... (Weiss, interview)

The effect was not always positive in terms of the music that was played.

No question it hit and it hit big, but then from that moment on I saw this influx of the most disgusting copycat efforts, like "Moog Espagana," boy if you could dig up some of those records, some of the most insipid garbage. (Weiss, interview)

The success of "Switched on Bach" helped reinforce the meaning of the synthesizer as a keyboard instrument—it might have been "switched on" but it was still Bach! And now the huge recording industry and associated media were enlisted in the conception of the synthesizer as a keyboard instrument. Although the copycat efforts did not succeed, further success for the Moog synthesizer in the world of popular music came in 1968 when George Harrison purchased one. The Beatles used it on four tracks on "Abbey Road," the best known being "Here Comes the Sun," where the deep throaty warble of the Moog is evident, and "Maxwell's Silver Hammer," where a French horn is emulated. This latter use is significant because it shows that the synthesizer is now starting to acquire a new use among musicians (emulating other instruments).

Later on this became one of the predominant uses of the synthesizer as the type of sounds it produced got ever and ever closer to the "real" thing. Back in the early days, many of the pioneering musicians disparaged these attempts to copy conventional instruments. They were more interested in using the synthesizer to produce new sounds and could see little point in trying to copy something.

> I was interested in the synthesizer because it gave the potential of creating sounds that couldn't be produced by any other means. And create kinds of sounds, kinds of music that you just simply couldn't orchestrate.... When I used the synthesizer I didn't use the keyboard ever.... I was an instrumental player, I was a violinist, so I had no interest in using the synthesizer to create instrumental sounds. Because as far as I'm concerned even when you are using the modern digital generation stuff, the sounds are never as good as the original acoustic sounds, they are so many degrees more complex. I figured what's the point of that—if you wanted something to sound like a French horn, then play a French horn.... Why use this machine to do just that? (Weiss, interview)

The keyboardless Buchla instrument was mainly used by avant garde composers of electronic music. The sound it makes is much more in that genre. A good point of listening comparison is electronic composer, Mort Subotnik's, "Silver Apples of the Moon" played on a Buchla and released at the same time as "Switched on Bach." Subotnik's record, with its assemblage of electronic-type bleeps exploring sound colors was *not* a best seller!

The Moog Goes to Hollywood

Other social groups also played a role in the early history of the synthesizer. Like the Theremin before it, the synthesizer started to be used in movies for both music and special effects. For example, David Borden made some of the scarier sounds for *The Exorcist* on a modular Moog and synthesizer player Don Preston made some of the music and special effects for *Apocalypse Now* on a Moog. Wendy Carlos has gone on to make several pieces of film music including the music for *Clockwork Orange* and *Tron*. A Moog synthesizer was used for the "freaky party" scene music in *Midnight Cowboy*. The synthesizer has dramatically changed the way that film music is made today; often, rather than hiring a full orchestra, a single synthesizer player and a "specialist instrumentalist" are used.

If we were to follow the history further, we would see also the importance of other social groups. In the 1980s, with the era of cheap digital synthesizers manufactured by companies such as Casio, synthesizers become the ubiquitous instrument to be used by anyone. For example, I recently bought a drum synthesizer for my 5-year-old child. It was manufactured by Fisher Price and was to be found in the toy section of my local department store!

The story of the early synthesizer is like that of many technologies: One meaning and set of practices stabilizes and the other meanings and practices slowly vanish or play a smaller role with niche markets. The meaning of synthesizer as a keyboard instrument, a meaning that slowly got embedded in the technology and that was reinforced in the way the instrument was used and the type of music performed on it, which in turn was responded to by changes in the technology (e.g. the Minimoog) in the end wins out although Moog isn't there for the pay day. That only comes in the early 1980s with cheap digital synthesizers, when Japanese companies such as Yamaha, Korg, and Roland start to dominate and every pop star has to have a keyboard synthesizer. Moog is by then out of business as were many of the early companies. By this time, many of the essential components of the path development of the synthesizer as a keyboard instrument had been developed.

Social Construction and Path Dependence

Part of the goal of this chapter is to see how social constructivist work on technology can enrich the notion of path dependence. Path dependence is used by economists in a rather restricted set of cases—those where there is a nonoptimal outcome. It is to be used by economists to explain cases where "history matters;" the assumption being that in most cases, history does *not* matter. Path dependence in this sense can be seen as a form of the sociology of error. That is to say history has to be evoked to explain something, which, according to strict neoclassical theory, ought not to happen.

Constructivist sociologists start off from the opposite assumption; that history always matters. Hence, it is important to seek symmetrical explanations. In other words, we must not reserve historical or sociological explanations for cases that are somehow in "error" or that do not have optimal outcomes.[2] Within the social constructivist approach to technology, the same sorts of social explanations are used for technologies that succeed as for those that fail. The neoclassical economists' use of path dependence seems to be profoundly asymmetrical. History is brought in as an additional explanation for suboptimal outcomes and is not used for most cases of technological innovation.

The social construction of technology approach also differs in methodology. Its methodological rule of thumb is to "follow the actors" (Latour, 1987). In other words, we use qualitative methods to try and recover the sociotechnical frameworks within which the key actors worked. We try and

---

[2]For arguments pointing to the importance of overcoming the sociology of error and developing a fully symmetrical sociology of science see, Bloor (1976).

look at the world through their eyes, and look at the process of innovation as they faced it at the time, including all the uncertainties. It is this methodology that has been used in this paper.

If we are to treat path dependence symmetrically, then every time a particular path appears, it is necessary to ask what, if any, the alternative paths are. Was, for instance, an alternative path possible and, if so, what happened to that path? Dominant historical accounts provided by participants will often not suffice for this task, as most such history is "winner's" history. The alternative path, the failed path, quickly vanishes with little historical trace. By problematizing the notion of a dominant path and examining what the alternatives are, we are able to draw attention to the underlying dynamics, even of dominant stable paths of technological development. The symmetrical approach to path analysis, which draws attention to the alternatives, should apply to cases even when, according to economists, we have optimal outcomes.

Thus, in the case examined in this chapter, I have contrasted Moog's approach with that of Buchla's in order to show what was at stake in terms of the path that the synthesizer might have taken.

For social constructivists, the stability of a technology always seems fragile. What stability there is is threatened by the heterogeneous social world within which such technologies are embedded. If it is possible to identify a stable path, or a "technological trajectory," it is usually because enough of the rest of the world is stable.[3] In the case analyzed in this chapter, there was much stability in the world that accounted for the predominant path the synthesizer took. For example, the dominant musical culture and its conventions and the recording and movie industries all helped constrain the path taken by the synthesizer.

As the variety of chapters in this volume shows, some hard questions are being asked of path dependence. For instance, by introducing the parallel idea of "path creation," Karnoe and Garud (2000) took us away from the rigid deterministic conception of path dependence. Path creation is the idea of "creative forces of agency ... [which] subjects the future to control and redesign" (Mouritsen & Dechow, chap. 13, this volume). I agree with Mouritsen and Dechow that path creation and path dependence are best seen as two sides of the same coin of structure and agency and that they are always related to one other. Although at any one moment it may be possible to identify what particular actors are doing to shape a technological path

---

[3]Of course, standardization is part of the process that leads to stability. Sociologists and historians of science and technology have in recent years devoted much attention to the process of standardization (e.g., Adler, 1998). Also, it is clear that the system feature of technology is another part of the story of how stability arises. For instance, an electric car as a technology cannot reach stability unless there is a network of charging points in existence. For the systems perspective, see Hughes (1983).

(path creation) it should always be kept in mind that the actors are not autonomous and are themselves being shaped by the path (path dependence). Thus, Moog was not only shaping the path the synthesizer was taking by such actions as using keyboards in his advertizing but he was, at the same time, responding to what musicians wanted.

Path creation, is a useful notion because it draws attention to the issue of agency. By looking at path creation and in particular, alternative paths, we can see how actors struggle to maintain a particular path. For example, Hughes notion of a "reverse salient" can be recast within the language of path creation (Hughes, 1983). An actor like Edison facing a reverse salient is, in essence, trying to solve the critical problems around a technology in order to maintain its path. Dealing with reverse salients can be seen as part and parcel of path creation. New paths and potential paths are threatening to open everywhere. Similarly, the notion of "path destruction" (Hirsch & Gillespie, chap. 3, this volume) is useful because it points to the active struggle of technological actors to bring to an end alternative paths that might threaten their own enterprise.

We have not seen any direct manifestation of path destruction in the case of the synthesizer thus far, but it is worth pointing out that with the advent of digital synthesizers, many analogue modular units were ripped from academic studios to be replaced by new portable digital equipment, which the students demanded. Buchla, in particular, found that his own modular equipment was a victim of this path destruction.

Within SCOT it seems that processes of path creation, path dependence, and path destruction are equivalent to closure and stabilization processes that are extended over time. In some ways, these notions (provided we give them the social constructivist slant advocated in this chapter) add to the rather static notion of closure in SCOT. Within SCOT, closure often seems to be a "one hit deal." The closure mechanism acts and then we have stability.[4] It is, I think, more useful to see closure as something that is continually in operation. Following a particular set of path creation and path dependence processes is a way of seeing how closure and stability evolve over time. So, in this case, we have seen how closure started very early on with path creation processes, such as the choice of promotional material to advertize synthesizers.

I have used this case to show how a path dependence can be traced of the synthesizer as a portable keyboard instrument that plays prepackaged sounds. The type of path dependence notion I have used is consistent with the social constructivist framework and indeed adds to that framework. The consequences of the traced path dependence are quite dramatic. Today we are saturated with electronic sound and devices for producing them. We

---

[4]But as Bijker (1995b) notes stabilization may in practice follow more slowly than closure.

hear new sounds in new ways and buy instruments in new ways. But, the revolution in sound since the 1960s was neither accidental nor predetermined; in the end, history always matters.

REFERENCES

Adler, K. (1998). Making things the same. Social Studies of Science, 28, 499–546.

Arthur, B. A. (1988). Self-reinforcing mechanisms in economics. In P. Anderson et al. (Eds.), The economy as an evolving complex system. Reading Ma: Addison-Wesley.

Bijker, W. (1995a). Sociohistorical technology studies. In S. Jasanoff, G. E. Markle, J. C. Petersen, & T. J. Pinch (Eds.), Handbook of Science and Technology Studies (pp. 229–256). Thousand Oaks, CA: Sage.

Bijker, W. (1995b). Bicycles, bulbs and bakelite: Towards a theory of sociotechnical change. Cambridge MA: MIT Press.

Bijker, W., Hughes, T. P, & Pinch, T. J. (Eds.). (1987) The social construction of technological systems: New directions in the sociology and history of technology. Cambridge MA: MIT Press.

Bloor, D. (1976). Knowledge and social imagery. London: Routledge.

Cochran, C. F., & Moog, R. (1993). The rise and fall of Moog music. In M. Vail (Ed.), Vintage synthesizers: Groundbreaking instruments and pioneering designers of electronic music synthesizers, San Francisco, CA: Miller Freeman.

David, P. (1985). Clio and the Economics of QWERTY. Economic History, 75, 227–332.

Garud, R., & Ahlstrom, D. (1997). Researchers' roles in negotiating the institutional fabric of technologies. American Behavioral Scientist, 40, 523–538.

Garud, R., & Rappa, M. A. (1994). A Socio-cognitive model of technology evolution: The case of cochlear implants. Organization Science, 5, 344–362.

Hughes, T. P. (1983). Networks of power. Baltimore, MD: Johns Hopkins University Press.

Jenkins, R. V. (1975). Technology and the market: George Eastman and the origins of mass amateur photography. Technology and Culture, 16, 1–19.

Karnøe P., & Garud, R. (2000). Path creation and dependence in the Danish wind turbine field. In J. Porac & M. Ventrusca (Eds.), The Social construction of industries and markets, NY: Pergamon.

Kline, R. K., & Pinch, T. J. (1996). Users as agents of technological change: The social construction of the automobile in the rural United States. Technology and Culture, 37, 763–779.

Latour, B. (1987). Science in action, Milton Keynes, England: Open University Press.

Mogg, R. (1965). Voltage-controlled electronic music modules. Journal of Audio Engineering Society, 13, 200–206.

Pinch, T. J. (1996). The social construction of technology: A review. In R. Fox (Ed.), Technological change. Methods and themes in the history of technology (pp. 17–35). London: Harwood.

Pinch, T. J., & Bijker, W. (1987). The social construction of facts and artifacts. In W. Bijker, T. P. Hughes, & T. J. Pinch (Eds.), The social construction of technological systems: New directions in the sociology and history of technology (pp. 17–50). Cambridge, MA: MIT Press.

Pinch, T. J., & Trocco, F. (1999). The social construction of the electronic music synthesizer. ICON Journal of the International Committee for the History of Technology, 4, 9–31.

Rosen, P. (1993). The social construction of mountain bikes: Technology and postmodernity in the cycle industry. Social Studies of Science, 23, 479–513.

Author Index

T

Tajfl, H., 338, 350
Takens, F., 174, 175, 209
Takeuchi, H., 217, 242, 339, 350
Tamura, R., 51, 64
Taylor, C., 174, 207
Teece, D. J., 10, 38, 153, 158, 168
Tell, F., 45, 67
Temin, P., 98, 122
Teresi, D., 185, 186, 207
Theiler, J., 190, 206, 209
Theobald, R., 308, 323
Theoret, A., 24, 37
Thietart, R. A., 170, 173, 183, 209
Thirlwall, A. P., 51, 67
Thirtle, C. G., 93, 102, 122
Thomson, R., 330, 350
Tilton, J., 140, 148
Tong, H., 171, 209
Toninelli, P. A., 105, 118
Townes, C. H., 186, 208
Trajtenberg, M., 187, 209
Trickett, A., 58, 67
Trist, E. L., 339, 351
Trocco, F., 382, 397
Truffer, B. P., 277, 297
Tryggestad, K., 18, 37
Tsiddon, D., 51, 64
Tsoukas, H., 6, 38
Tucker, D. J., 245, 262, 266
Turner, J. C., 338, 350, 351
Tushman, M. L., 55, 56, 68, 78, 85, 90, 159, 168, 183, 209, 216, 227, 242, 245, 266, 328

U

U.S. Congress, 187, 209
Usher, A. P., 14, 17, 18, 38, 85, 90, 104, 122
Utterback, J. M., 55, 63, 160, 166, 183, 209, 213, 216, 227, 241, 242
Uzzi, B., 83, 90

V

van den Ende, J., 276, 297
van der Loo, F., 275, 288, 296
Van de Ven, A. H., 6, 23, 27, 36, 38, 81, 85, 90, 111, 118, 122, 173, 183, 184, 185, 190, 200, 205, 207, 246, 266, 333, 348, 351
van Est, R., 278, 279, 280, 281, 282, 297
Van Looy, B., 346, 347, 349, 350
Vastano, J. A., 204, 209

Venkataraman, S., 23, 38
Ventresca, M., 8, 38
Vernon, R., 92, 122
Verspagen, B., 46, 65
Vinten, G., 174, 209
Voltaire, F., 127, 148
von Burg, U., 134, 148
von Hippel, E., 157, 168
von Tunzelmann, N., 46, 65
Vygotsky, L., 347, 351

W

Wagner, S., 187, 208
Waldrop, M. M., 173, 174, 178, 183, 209
Walker, G., 132, 147, 148
Walker, R., 82, 90, 128, 130, 148
Wallace, D., 293, 297
Walsh, V., 93, 122
Wangensteen, O. H., 159, 165, 168
Wangensteen, S. D., 159, 165, 168
Wan, H. Y., Jr., 95, 98, 122
Ward's Auto World, 226, 235
Warren, J. F., 306, 323
Waterman, R. H., 18, 23, 24, 38
Watzlawick, P., 320, 325
Weber, M., 154, 168, 287, 297
Weick, K. E., 2, 7, 8, 11, 12, 14, 24, 27, 38, 82, 90, 223, 231, 242, 307, 320, 325, 340, 341, 343, 349, 351
Weingart, P., 330, 350
Weiss, J., 132, 133, 148
Welch, J., 356, 377
Wenger, E., 339, 340, 350
Wernerfelt, B., 354, 377
Westley, F., 340, 351
West, L. J., 44, 68
Wheelwright, S. C., 329, 351
White, A. O., 46, 68
White, E. B., 28, 38
White, G. R., 305, 325
White, H. C., 216, 218, 221, 223, 242
Whitley, R. D., 5, 38
Whitston, C., 112, 120
Wicken, O., 275, 288, 296
Wiederholt, B., 3, 38
Wilcoxen, P. J., 103, 120
Williamson, O. E., 153, 168
Wilson, J., 142, 148
Winter, S. G., 5, 37, 49, 67, 104, 106, 107, 112, 121, 122, 123, 129, 148, 152, 153, 159, 167, 168, 196, 207, 245, 266, 269, 270, 296, 328, 354, 376
Witt, U., 104, 123
Wolf, A., 204, 209
Woolgar, S., 339, 349
Wright, G., 103, 115, 117, 123

Subject Index

411

For Product Safety Concerns and Information please contact our EU
representative GPSR@taylorandfrancis.com
Taylor & Francis Verlag GmbH, Kaufingerstraße 24, 80331 München, Germany

www.ingramcontent.com/pod-product-compliance
Ingram Content Group UK Ltd.
Pitfield, Milton Keynes, MK11 3LW, UK
UKHW021605240425
457818UK00018B/396